Springer Series in Surface Sciences

Volume 62

This series covers the whole spectrum of surface sciences, including structure and dynamics of clean and adsorbate-covered surfaces, thin films, basic surface effects, analytical methods and also the physics and chemistry of interfaces. Written by leading researchers in the field, the books are intended primarily for researchers in academia and industry and for graduate students.

More information about this series at http://www.springer.com/series/409

Farzad Nasirpouri

Electrodeposition of Nanostructured Materials

Springer

Farzad Nasirpouri
Faculty of Materials Engineering
Sahand University of Technology
Tabriz
Iran

ISSN 0931-5195 ISSN 2198-4743 (electronic)
Springer Series in Surface Sciences
ISBN 978-3-319-44919-7 ISBN 978-3-319-44920-3 (eBook)
DOI 10.1007/978-3-319-44920-3

Library of Congress Control Number: 2016950382

Printed on acid-free paper

This Springer imprint is published by Springer Nature
The registered company is Springer International Publishing AG
The registered company address is: Gewerbestrasse 11, 6330 Cham, Switzerland

Preface

Nanoscience and nanotechnology have continued to grow at a rapid pace since their emergence a few decades ago. Nanostructured materials ranging from zero-dimensional, one-dimensional and two-dimensional to bulk nanomaterials are central to nanotechnology. It is crucial these days to find more cost-effective and versatile methods of fabrication and synthesis to compete with existing technologies. There is a tremendous amount of interest in chemical or electrochemical approaches for the discovery and development of novel nanostructured materials and applications. Electrochemical deposition or electrodeposition has been widely used for years in the plating industry for anti-corrosion and decorative applications in metals and alloys. It is a cost-effective, flexible and capable method for producing a variety of materials. Given these characteristics, the basic understanding of electrodeposition has been considerably enhanced over the last few decades, and it has been successfully downscaled for use in the deposition of a wide range of nano-sized materials with outstanding functional and mechanical properties.

The importance of electrodeposition in nanoscience and nanotechnology, and in particular in the development of nanostructured materials, has been widely acknowledged in recent decades, and further progress is certainly inevitable. Interest in the technology and the exploitation of its potential by both the academic and industrial sectors is anticipated in the coming years. Therefore, I felt it was essential to collect the most important advances and findings in the field of electrodeposition of nanostructured materials in the form of a monograph. Having more than 15 years of research experience on the electrodeposition of coatings and nanostructures, I came to the idea of a book covering the electrodeposition of nanostructured materials from fundamental principles to the most recent progress. This book is prepared to disseminate the major factors and principles of electrodeposition towards the fabrication of nanostructured materials as a unique reference. I outlined the pathways of this strategy as mentioned.

The book starts with the fundamental aspects of nanostructured materials, and develops the principles of electrochemistry, followed by the basics of electrodeposition. It then discusses a number of the most interesting electrodeposited nanostructured materials. Specifically, Chap. 1 gives an overview of the principles of materials science and nanostructured materials. A general discussion of the basic structures of materials from atoms, atomic binding, physics of solid state materials or condensed matter physics, the band theory of solids and crystallography, classification of nanostructured materials, the dimensionality and quantum size effect in nanostructures, and the electronic properties of three types of nanostructures is expanded in detail. Chapter 2 deals with the principles of electrochemistry in materials science. The electrochemical aspects of electrodeposition are central to process control and development. Three concepts are covered: equilibrium electrochemistry, dynamic electrochemistry and instrumental electrochemistry. In Chap. 3, I explain the fundamentals of electrodeposition, including the principles of electrolysis, electrodeposition cells, nucleation and growth of electrodeposits, basics of overpotential (OPD) and underpotential (UPD) deposition, methods of characterizing the initial stages of electrodeposition, and electrodeposition through the application of external stimuli such as magnetic or ultrasound forces. Chapter 4 deals with electrodeposited two-dimensional (2D) and three-dimensional (3D) meso- and nanostructures. UPD is associated with 2D and OPD is related to 3D nucleation and growth. Free-standing metal or alloy mesocrystals and nanostructures on a substrate and multiple 3D nuclei and consequent coalescence for film are described. Template electrodeposition for the fabrication of nanowire arrays is explained in Chap. 5. The most common templates will be exploited, and electrochemical deposition mechanisms by means of direct current, pulse current and alternating current will be discussed. Recent progress in electrodeposited magnetic nanowires including metal and alloy, multilayered, core–shell and diameter-modulated nanowires are presented. Chapter 6 deals with electrocrystallization at the nanoscale. In this chapter, I focus on pulse electrodeposition; principles and important particulars will be discussed, including a comparison of Ni nanocrystalline films electrodeposited by direct and pulse techniques. Metal matrix–particulate nanocomposites are the topic of Chap. 7. Co-electrodeposition of metal matrix with particulates involves the electrochemical fundamental where mass transport and kinetics play important roles. Various theoretical models and the most important factors of co-electrodeposition will be discussed and compared in an example system: nickel–carbon nanotube nanocomposite coatings. Lastly, Chap. 8 concerns particular methods and techniques of electrodeposition of nanostructures, including scanning tunneling microscopy (STM)-assisted electrodeposition, lithographically patterned nanowire electrodeposition, electrodeposition of mesoporous films from lyotropic liquid crystals and electrodeposition of nanostructures by galvanic displacement. This book will be useful for different groups of readers, from academics or technologists, researchers and graduate students, to engineers and professionals, enabling the correlation of the principles of electrodeposition with their use in developing new strategies for the fabrication of numerous cutting-edge nanostructures. This book can also be used as a textbook for undergraduate and graduate students in related disciplines.

I have attempted to discuss the most promising electrodeposited nanostructured materials, with an emphasis on those for which we have direct research experience. Therefore, so many people have helped to make this book possible, and I am sincerely indebted to them. I am grateful to my mentors and colleagues in the past and present, including Prof. Walther Schwarzacher, Prof. Laurie Peter, Prof. Simon Bending, Prof. Alain Nogaret, Prof. Mohammad Ghorbani, Prof. Azam Irajizad, Dr. Alexander Samardak, Dr. Alexey Ognev and his colleagues for their advice and fruitful collaboration on many topics of this book. I am also very thankful to all members of my research group at Sahand University of Technology for their cooperation, their efforts, constant research work and encouragement, in particular Mr. Hamed Cheshideh, Mr. Seyed-Majid Peighambari, Mr. Ali Fardi-Ilkhichy, Mr. Amin Pourandarjani, Ms. Sanam Abedini, Mr. Seyed-Mehdi Janjan, Mr. Farhad Daneshvar-Fattah, Mr. Mohammad-Reza Sanaeian, Ms. Masoumeh Nadi, Ms Soheila Beheshti, Mr. Mehdi Hadizadeh, Ms. Katayun Alipour, Mr. Navid Alinedjaidian, Mr Alireza Sarshar-Noshar, Ms. Saedeh Barzegar, Mr. Hossein Firouzi, Mr. Enayatollah Panahi, Ms. Aysan Hadighe-Rezvan, Mr. Sajjad Nasiri-ahmadabad, Ms. Rana Mahmoodi, Ms. Mina Abdollahzadeh and so many others that I may have forgotten. In addition, the financial support and research grants from the Research Affairs of Sahand University of Technology, Iranian Ministry of Science, Research and Technology, the Iran Nanotechnology Initiative Council and others are acknowledged.

I would very much welcome any comments and discussions from peers in this field for further development and improvement of this book.

Tabriz, Iran Farzad Nasirpouri

Contents

Chapter 1
An Overview of Nanostructured Materials

Abstract In this chapter, an overview of the principles of materials science and nanostructured materials is presented. The chapter begins with a general discussion to explain the basic structures of materials from atoms, atomic binding, physics of solid state materials or condensed matter physics, the band theory of solids and crystallography. This is followed by a classification of materials at the nanometer scale, where dimensionality and quantum size effect play important roles. The electronic properties and applications of three types of nanostructures are explained and compared.

1.1 Introduction to Materials Science

In principle, *materials science* is defined as the science of solid materials or condensed matter which describes the relationship between the structure, properties, processing and performance of materials. An understanding of the relationship creates novel science by nature of developing new materials for high-technology applications and better way of life. From a structural point of view, all materials can generally be divided into two classes: *crystalline* and *non-crystalline*. Materials are traditionally classified as metals, semiconductors, ceramics and polymers. However, considered as crystalline solids, the physics of materials may be well described by solid state physics.

As a result of developments in this field, a new class of materials, known as nanomaterials or nanostructured materials, has recently emerged. The development of this new group of materials has been inspired by very rapidly evolving science and technology at the nanometer scale. In general, a *nanomaterial* is defined by its dimensions and size, where at least one dimension must be in the range of 0–100 nm. This critical size range is associated with many interesting phenomena which in principle obey nanophysics. The physical behavior of materials at nanoscale is explained by quantum mechanics as the center of the field of nanomaterials and nanotechnology. Thus, we start this book with an introduction to this extremely fascinating field of science and technology.

F. Nasirpouri, *Electrodeposition of Nanostructured Materials*,
Springer Series in Surface Sciences 62, DOI 10.1007/978-3-319-44920-3_1

1.2 Physics of Solid State Materials

1.2.1 Atoms and the Periodic Table of Elements

The structure of atoms consists of a nucleus surrounded by electrons. For the simplest atom, hydrogen, the electrical potential energy between the negative charge of the electron moving in a circle with radius *r* around the nucleus and the positive charge of the nucleus is defined as:

$$U = -\frac{1}{4\pi\varepsilon_o}\cdot\frac{e^2}{r} \tag{1.1}$$

where $\varepsilon_o = 8.854187817 \times 10^{-12}$ F/m is the permittivity of free space or of the classical vacuum. This classical relation leads to the collapse of atoms, because the potential energy existing between the nucleus and electrons becomes null at an infinite distance. Thus, the movement of the classical electron becomes spiral instead of circular towards the nucleus, releasing the excess energy in the form of electromagnetic radiation. Therefore, no stable atom would exist under the laws of classical physics.

The structure of atoms requires a quantum mechanical interpretation for its description. Bohr [1] discovered that the structure of a hydrogen atom needed the quantization of angular momentum and energy levels to become stable. Bohr's semi-classical model of the atom considers its planetary nature, with an electron circling the nucleus, which was postulated with two conditions. The first involves the quantization of the angular momentum of the electron of mass *m* circling the nucleus in an orbit of radius *r* and speed *v*:

$$mvr = n\hbar = n\frac{h}{r} \tag{1.2}$$

Here, *h* is the Planck constant, $\hbar$ is the angular Planck constant and *n* is the arbitrary integer quantum number defined as *n* = 1, 2, 3, …

Second, electrons possess definite and distinct energy values. Total energy is the sum of kinetic and potential energy:

$$E = K + U \tag{1.3}$$

Considering (1.1)–(1.3), one obtains:

$$E = -\frac{1}{8\pi\varepsilon_o}\cdot\frac{e^2}{r} \tag{1.4}$$

or

$$E_n = -\frac{me^4}{8\varepsilon_o^2 h^2} \cdot \frac{1}{n^2} \tag{1.5}$$

At the ground state, $E_0 = -13.6$ eV, and the subsequent energy levels would attain a certain value depending on the quantum number n. Despite its success in explaining the properties of hydrogenic electrons bound to donor impurity ions in semiconductors, or in analyzing the optical spectra of semiconductors exposed to radiation, the planetary nature of Bohr's model was unable to precisely explain the states of electrons and their location and dispersion in an atom.

Quantum mechanics proved the particle–wave nature of light and matter, giving rise to an understanding of the behavior of atomic-scale particles by a wave function, $\psi_{(r,t)}$. Schrödinger [2] described the location of a particle having a wave function, $\psi_{(r,t)}$ in a specific situation by postulating the particle–wave nature of a traveling wave in one dimension with a length of L:

$$\psi_{(x,t)} = A.e^{\mathrm{ikx}-i\omega t} \tag{1.6}$$

where $k = \frac{2\pi}{\lambda}$ is the wave number for wavelength λ, and ω is the angular frequency. Schrödinger proposed a statement for the energy of the particle (1.3) by the application of the correct particle–wave equation, and obtained a time-independent equation in one direction, x:

$$\frac{\hbar^2}{2m}\frac{\mathrm{d}^2\psi_{(x)}}{\mathrm{d}x^2} + (E - U)\psi_{(x)} = 0 \tag{1.7}$$

The Schrödinger equation can be applied to derive the electronic quantum states in hydrogen or any one electron atom having a Coulomb potential energy (1.1). It is written in the case of spherical polar coordinates which are more suitable for explaining the electron motions in an atom [3] as:

$$\frac{-\hbar^2}{2m}\frac{1}{r^2}\frac{\partial}{\partial r}\left(\frac{r^2\partial\psi}{\partial r}\right) - \frac{\hbar^2}{2mr^2}\left[\frac{1}{\sin\theta}\frac{\partial}{\partial\theta}\left(\sin\theta\frac{\partial\psi}{\partial\theta}\right) + \frac{1}{\sin^2\theta}\frac{\partial^2\psi}{\partial\varphi^2}\right] + U(r)\psi = E\psi \tag{1.8}$$

In the spherical symmetry of the atom, the solution is achieved by separating the variables so that the wave function is represented by the product:

$$\psi = R(r)P(\theta)F(\phi) \tag{1.9}$$

The separation leads to three equations for the three spatial variables, the solution of which gives rise to three quantum numbers: $R(r)$ for the principal quantum number n, $P(\theta)$ for the orbital quantum number l, and $F(\phi)$ for the

1 H 1.0079 $1s^1$								Atomic number Symbol Atomic mass Electron configuration									2 He 4.00 $1s^2$
3 Li 6.941 $2s^1$	4 Be 9.0121 $2s^2$											5 B 10.81 $2p^1$	6 C 12.011 $2p^2$	7 N 14.0 $2p^3$	8 O 15.999 $2p^4$	9 F 15.99 $2p^5$	10 Ne 20.17 $2p^6$
11 Na 29.939 $3s^1$	12 Mg 24.305 $3s^2$											13 Al 26.9 $3p^1$	14 Si 28.08 $3p^2$	15 P 30.973 $3p^3$	16 S 32.06 $3p^4$	17 Cl 35.45 $3p^5$	18 Ar 39.94 $3p^6$
19 K 39.098 $4s^1$	20 Ca 40.078 $4s^2$	21 Sc 44.955 $3d^1 4s^2$	22 Ti 47.88 $3d^2 4s^2$	23 V 50.941 $3d^3 4s^2$	24 Cr 51.995 $3d^5 4s^1$	25 Mn 54.938 $3d^5 4s^2$	26 Fe 55.847 $3d^6 4s^2$	27 Co 58.93 $3d^7 4s^2$	28 Ni 58.69 $3d^8 4s^2$	29 Cu 63.54 $3d^{10} 4s^1$	30 Zn 65.39 $3d^{10} 4s^2$	31 Ga 69.723 $4p^1$	32 Ge 72.61 $4p^2$	33 As 74.921 $4p^3$	34 Se 78.95 $4p^4$	35 Br 79.9 $4p^5$	36 Kr 83.80 $4p^6$
37 Rb 85.467 $5s^1$	38 Sr 87.62 $5s^2$	39 Y 88.905 $4d^1 5s^2$	40 Zr 91.224 $4d^2 5s^2$	41 Nb 92.906 $4d^4 5s^1$	42 Mo 95.94 $4d^5 5s^1$	43 Tc (98) $4d^5 5s^2$	44 Ru 101.07 $4d^7 5s^1$	45 Rh 102.9 $4d^8 5s^1$	46 Pd 106.42 $4d^{10}$	47 Ag 107.86 $4d^{10} 5s^1$	48 Cd 112.41 $4d^{10} 5s^2$	49 In 114.82 $5p^1$	50 Sn 118.71 $5p^2$	51 Sb 121.75 $5p^3$	52 Te 127.60 $5p^4$	53 I 126.90 $5p^5$	54 Xe 131.30 $5p^6$
55 Cs 132.90 $6s^1$	56 Ba 137.32 $6s^2$	57-71 La	72 Hf 178.49 $5d^2 6s^2$	73 Ta 180.94 $5d^3 6s^2$	74 W 183.8 $5d^4 6s^2$	75 Re 186.20 $5d^5 6s^2$	76 Os 190.2 $5d^6 6s^2$	77 Ir 192.2 $5d^7 6s^2$	78 Pt 195.08 $5d^9 6s^1$	79 Au 196.96 $5d^{10} 6s^1$	80 Hg 200.59 $5d^{10} 6s^2$	81 Tl 204.38 $6p^1$	82 Pb 207.2 $6p^2$	83 Bi 208.98 $6p^3$	84 Po (200) $6p^4$	85 At (210) $6p^5$	86 Rn (222) $6p^6$
87 Fr (223) $7s^1$	88 Ra (226) $7s^2$	89-103 Ac															

57 La 138.90 $5d^1 6s^2$	58 Ce 140.11 $4f^1 5d^1 6s^2$	59 Pr 140.9 $4f^3 6s^2$	60 Nd 144.24 $4f^4 6s^2$	61 Pm (145) $4f^5 6s^2$	62 Sm 150.36 $4f^6 6s^2$	63 Eu 151.96 $4f^7 6s^2$	64 Gd 157.25 $4f^7 5d^1 6s^2$	65 Tb 158.92 $4f^9 6s^2$	66 Dy 162.5 $4f^{10} 6s^2$	67 Ho 164.93 $4f^{11} 6s^2$	68 Er 167.26 $4f^{12} 6s^2$	69 Tm 168.93 $4f^{13} 6s^2$	70 Yb 173.04 $4f^{14} 6s^2$	71 Lu 174.96 $4f^{14} 5d^1 6s^2$
89 Ac (227) $6d^1 7s^2$	90 Th 232.03 $6d^2 7s^2$	91 Pa 231.03 $5f^2 6d^1 7s^2$	92 U 238.02 $5f^3 6d^1 7s^2$	93 Np (237) $5f^4 6d^1 7s^2$	94 Pu (244) $5f^6 7s^2$	95 Am (243) $5f^7 7s^2$	96 Cm (247) $5f^7 6d^1 7s^2$	97 Bk (247) $5f^9 7s^2$	98 Cf (251) $5f^{10} 7s^2$	99 Es (252) $5f^{11} 7s^2$	100 Fm (257) $5f^{12} 7s^2$	101 Md (258) $5f^{13} 7s^2$	102 No (259) $5f^{14} 7s^2$	103 Lr (260) $5f^{14} 7s^2$

Fig. 1.1 Periodic table of elements illustrating the electronic structure and quantum numbers of atoms

magnetic quantum number ml. For each electron there is also a spin quantum number with projection of $\pm 1/2$ as a result of relativistic corrections to the Schrödinger equation. The quantum numbers associated with electron arrangements and labels of atomic orbitals are shown in Fig. 1.1, and can be found in greater detail with the shape and equation of wave functions in [3, 4].

Following these rules, it is predicted that the number of distinct quantum states for a given n is $2n^2$. According to Pauli's exclusion principle, only one electron is allowed in each distinct quantum state. Thus $2n^2$ is the total number of electrons which can be accommodated in the nth electron shell of an atom. Occupation rules exist for quantum states with electrons, where filling of orbitals with electrons in one-electron atoms takes place by dramatically increasing the quantum numbers. For poly-electronic atoms, the interactions between them must be considered by occupation rules such as Hund's rules [4].

This is the basis for the chemical table of the elements, the so-called Mandeleyev table (Fig. 1.1), where the wave function ψ_{n,l,m_l,m_s} provides quantum states, and Pauli's exclusion principle states that only one electron can be accommodated in each quantum state. There exists a specific location for each individual element owing to its atomic structure. The table has rows and columns called periods and groups, respectively, each of which has specific characteristics. All of the elements located in a period have the same number of atomic orbitals. Each group consists of elements with the same number of electrons in the outer orbital valence electrons. There are exceptions to the order in the case of transition elements. Transition elements add electrons to the penultimate orbital. The two rows separate from the table are for lanthanides (or rare earth metals or inner transition elements) and actinides, which are radioactive and are not often found in nature. The electronic states and orbital arrangements of each atom are shown in Fig. 1.1.

1.2.2 *Atomic Bonds and Condensed Matter*

Condensed matter or solids are formed by the accumulation of atoms. The properties of bulk matter depend upon the electronic structure of the atom incorporated into the matter. In the periodic table of elements, each individual atom in the period is compared with its neighbor atoms having smaller or larger atomic numbers or with other atoms in its group. But what forms a solid or condensed matter? In other words, what is the main reason for the cohesion of the atoms to be condensed in either liquid or solid form?

In principle, the attractive electrostatic energy between the electrons and protons of atoms plays a vital role in their cohesion. However, other energy factors such as magnetics, exchange interaction, van der Waals' forces and covalent force impose other effects. The formation of a solid by bonding of the atoms reduces the energy level. Figure 1.2 depicts a generally adopted energy plot of possible interactions occurring between two binding atoms. Total energy results from the sum of the

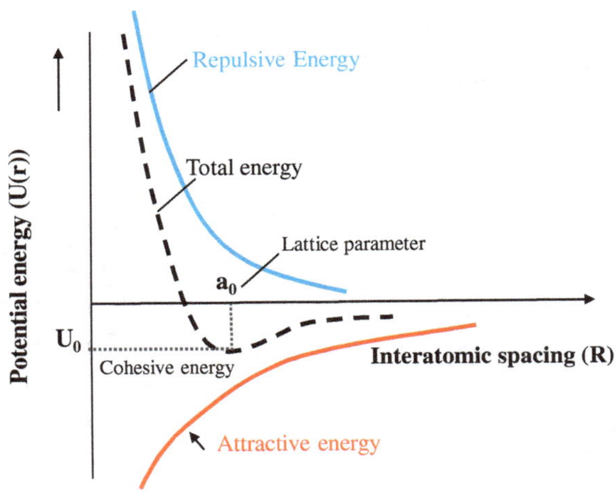

Fig. 1.2 Variation in interacting energies during binding of atoms for all classes of solid materials. a_0 is the lattice parameter and U_0 is the cohesive energy

attractive energy and repulsive energy terms. The binding of atoms takes place at the minimum energy at the point where cohesive energy and interatomic distance are obtained. Despite the different mathematics and relationships for various classes of solid materials, this plot is common to all types.

Generally, cohesive energy is an energy term used to describe the strength of a given solid material. Provided that such energy is applied, the matter will be divided into its components in their free electronic ground states. For instance, an ionic solid is separated into the ions forming the ionic crystal. Figure 1.3 shows the cohesive energy, melting point and elastic modulus for all possible solid crystals on a periodic table of elements. Here, different materials can be seen to exhibit various magnitudes of cohesive energies as well as other intrinsic physical properties which substantially depend upon the bonding types of their atoms or components.

Here we aim to briefly explain general atomic bonds in solids and their principles and diversity in materials science. There are generally four types of atomic

0.002 H						Bulk Modulus (10^{15}N.m^{-2}) Symbol Cohesive Energy (Kcal/mol) Melting Point (K)											0.00 He
0.116 Li 37.7 453.7	1.003 Be 76.5 1562											1.78 B 134 2365	4.43 C 170	0.012 N 113.4 63.15	O 60.03 54.36	F 19.37 53.48	0.010 Ne 0.46 24.56
0.068 Na 25.67 371.0	0.354 Mg 34.7 922											0.722 Al 78.1 933.5	0.988 Si 106.7 1687	0.304 P 79.16 W 317	0.178 S 65.75 388.4	Cl 32.2 172.2	0.013 Ar 1.85 83.81
0.032 K 21.54 336.3	0.152 Ca 42.5 1113	0.435 Sc 89.9 1814	1.051 Ti 111.8 1946	1.619 V 122.4 2202	1.901 Cr 94.5 2133	0.596 Mn 67.4 1520	1.683 Fe 98.7 1811	1.914 Co 101.3 1770	1.86 Ni 102.4 1728	1.37 Cu 80.4 1358	0.598 Zn 31.04 692.7	0.569 Ga 64.8 302.9	0.772 Ge 88.8 1211	0.394 As 68.2 1089	0.091 Se 56.7 494	Br 28.18 265.9	0.018 Kr 2.68 115.8
0.031 Rb 19.64 312.6	0.116 Sr 39.7 1042	0.366 Y 100.8 1801	0.833 Zr 144.2 2128	1.702 Nb 174.5 2750	2.725 Mo 157.2 2895	(2.97) Tc 158 2477	3.208 Ru 155.4 2527	2.704 Rh 132.5 2236	1.808 Pd 89.8 1827	1.007 Ag 68.0 1235	0.467 Cd 26.73 594.3	0.411 In 58.1 429.8	1.11 Sn 72.4 505.1	0.383 Sb 63.4 903.9	0.230 Te 50.34 722.7	I 25.62 386.7	Xe 3.80 161.4
0.20 Cs 18.54 301.6	0.103 Ba 43.7 1002	0.243 La 103.1 1194	1.09 Hf 148.4 2504	2.00 Ta 186.9 3293	3.232 W 205.2 3695	3.72 Re 185.2 3459	(4.18) Os 188.4 3306	3.55 Ir 160.1 2720	2.783 Pt 134.7 2045	1.732 Au 87.96 1338	0.382 Hg 15.5 234.3	0.359 Tl 43.4 577	0.430 Pb 46.78 600.7	0.315 Bi 50.2 544.6	(0.26) Po 34.5 527	At	Rn 4.66
(0.020) Fr	(0.132) Ra 38.2 973	(0.25) Ac 98 1324															

0.239 Ce 99.7 1072	0.306 Pr 85.3 1205	0.327 Nd 78.5 1290	(0.35) Pm	0.294 Sm 49.3 1346	0.147 Eu 42.8 1091	0.383 Gd 95.5 1587	0.399 Tb 93.4 1632	0.384 Dy 70.2 1684	0.397 Ho 72.3 1745	0.411 Er 75.8 1797	0.397 Tm 55.8 1820	0.133 Yb 37.1 1098	0.411 Lu 102.2 1938
0.543 Th 142.9	(0.76) Pa	0.987 U 128	(0.68) Np 109	0.54 Pu 83.0	Am 63	Cm 92.1	Bk	Cf	Es	Fm	Md	No	Lr

Fig. 1.3 Periodic table of elements for relevant crystals consisting of the cohesive energy, melting point and Young's elastic modulus. Plotted from data given in [5]. For research purposes refer to the original references

bonds in solid materials: van der Waals–London bond, covalent bond, ionic bond and metallic bond. Each type of atomic binding creates a specific class of solids.

1.2.2.1 Inert Gas (van der Waals) Solids

Inert gas elements form the simplest type of solids. These solids are transparent and electrically insulating, with a low melting point and weak atomic binding. A large number of solids fall within this category, including inert gases, polymers and organic molecules. The ionization energy of such solids is quite high [5]. The electronic structure of free atoms taking part in this type of solid is spherically symmetric due to the completely filled outermost electron shells. In solid form, the electronic distribution around each atom does not change significantly from that of the free atom, as the cohesive energy of these solids is approximately 1% of the ionization energy of free atoms. This atomic binding is caused by the *van der Waals–London* interaction, which occurs due to dipole–dipole charge interactions between atoms. Each atom preserves dynamic oscillating charge dipoles as a quantum effect, forming simple harmonic oscillators. This leads to the formation of intact dipole–dipole interactions which cause atoms to be attracted to each other, developing attractive energy. On the other hand, as atoms become closer, their electron distribution overlaps. The *Pauli* exclusion principle asserts that multiple-electron occupation of quantum states is not possible. Thus, the overlapped electrons must travel to higher energy levels. This forces atoms to separate, generating repulsive energy. In short, the attractive van der Waals energy offsets the repulsive energy where the atomic bond is formed. A general equation describing the variation among energies in this form of atomic binding can be written as:

$$U_{(R)} = -\frac{A}{R^6} + \frac{B}{R^{12}} \tag{1.10}$$

in which A, B are constants defined according to the atom electronic properties, R is atom distance, and $U_{(R)}$ is the total energy.

1.2.2.2 Ionic Solids

Atoms next to the inert gases on the periodic table of elements are easily ionized by accepting or releasing an electron to obtain completely filled outermost shells like their inert neighbors. The ionic bonds are generally formed between alkaline metals and halogens, where the electronegativity and electropositivity from each individual element are combined to form the solids. For alkaline metals, the outermost shell electron weakly connected to the nucleus is easily donated, and constitutes the positive ionic element. Halogens located in the group after the inert gas elements need only a single electron to completely fill the outermost shell, where they become the negative ionic element. Some examples of these solids are alkali halides (sodium chloride, lithium fluoride, etc.), oxides and sulfides, as well as some complex salts of inorganic chemistry. The binding of the metal and halogen atoms happens by electron exchange between them, forming ionic elements as the main constituents of the ionic solids. The ionic binding energy is called *Madelung* energy, defined by summing the long-term electrostatic interactions between negative and positive ion charges in a solid material: $\pm\frac{q^2}{R}$.

In addition, there exists repulsive energy due to the Pauli exclusion quantum effect in the form of:

$$\lambda.\exp\left(-\frac{R}{\rho}\right),$$

where λ and ρ are empirical parameters related to the amplitude and range of the repulsive quantum interaction. The total lattice energy of an ionic solid comprising 2 N ions (or N molecules) at their equilibrium atomic distance (R_0) is obtained:

$$U_{\text{total}} = -\frac{N\alpha q^2}{R_0}\left(1 - \frac{\rho}{R_0}\right) \tag{1.11}$$

α is called the Madelung constant and ρ is found to be about 0.1 R_0 [5].

1.2.2.3 Covalent Solids

Ionic and van der Waals bonds cannot be the reason for the formation of substances such as O_2, N_2, diamond and Si, or of III–V compounds such as GaAs. Another type of atomic bond, called a covalent bond, is responsible for the formation of these solids. Because covalent bonds are strong, solids are quite hard materials. These solids are used in the development of semiconductor materials, which will be introduced later.

To interpret covalent atomic binding, one simply needs to consider the molecular hydrogen bond. The probability of electron sharing, called *transition frequency*, increases markedly as the hydrogen free atoms come closer. For example, the transition frequency of an electron of one hydrogen atom in another atom is 10^{12} per year when the atomic separation is 50 Å. If the hydrogen atoms come much closer, with interatomic distance of 2 Å, the transition frequency increases considerably, up to 10^{14} s^{-1}. Under such circumstance, it is not clear which electron belongs to which hydrogen atom, making the bi-atom system indiscrete; this is called electron sharing. The shared electrons will change the electron probability function $|\psi^2|$. The binding energy for covalent solids was first expressed by Heitler and London [6] (Fig. 1.4).

Fig. 1.4 Variation in total energy for **a** antiparallel and **b** parallel electron spin states in a bi-atom hydrogen system

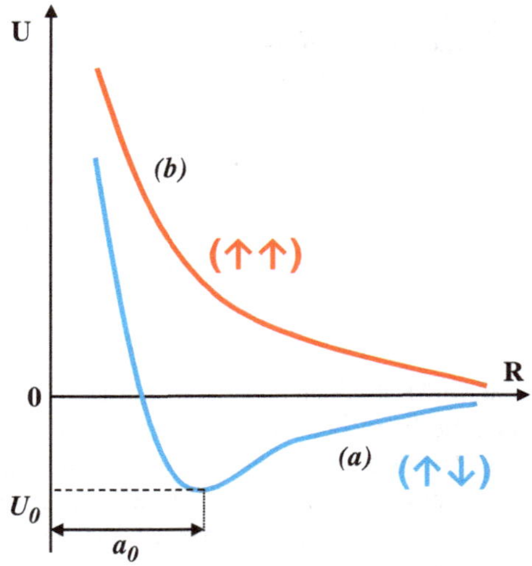

The total energy (U) of the hydrogen bi-atomic system can be written in two different states:

(a) When the spins of the two electrons are in bonding or antiparallel states:

$$U_{ap} = 2E_0 + \frac{K + A}{1 + S^2} \tag{1.12}$$

(b) When the spins of the two electrons are in anti-bonding or parallel states:

$$U_p = 2E_0 + \frac{K - A}{1 - S^2} \tag{1.13}$$

Here, E_0 is the energy of a single free hydrogen atom, K is the electrostatic coulomb interaction, A is the exchange coupling interaction and S is the electron overlapping integral. Considering the values of these factors, the covalent bond forms when an antiparallel states exist where $U_{ap} < 2E_0$. Figure 1.3 shows the variation in total energy as a function of interatomic distance for the two states introduced above. The minimum energy is seen on the curves dedicated to the antiparallel states.

1.2.2.4 Metallic Solids

Atoms of the top periods in the periodic table of elements are bonded together via metallic bonds forming all metals and alloys. The valence electrons located in the

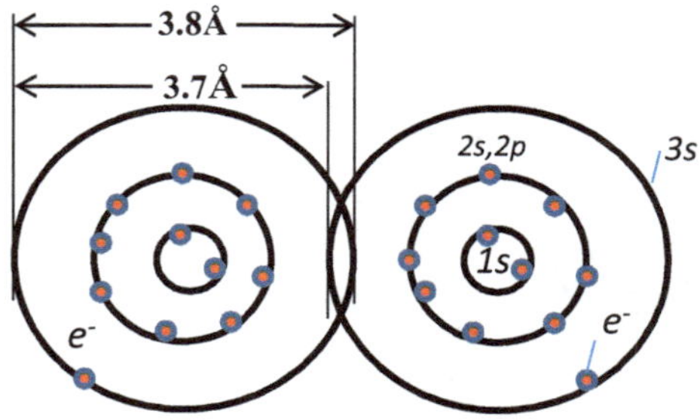

Fig. 1.5 Scheme of electronic arrangement of a sodium atom forming a metallic bond

outermost shell of the metal atoms are weakly bonded to the nucleus. When the metallic bond occurs, the outer orbitals of the atoms are overlapped, as shown in Fig. 1.5, and the valence electrons form an electron gas that permeates the entire solid lattice. Thus a combination of negatively charged electron gas as delocalized electrons and positively charged ions as localized electrons plus nuclei exists inside the entire metals. The delocalized electrons which form the electron gas are called conduction electrons. These electrons determine the characteristic properties of metals such as high electrical and thermal conductivity and high optical reflectivity.

1.2.2.5 Crystalline and Amorphous Solids

An ideal crystal is constructed by the infinite repetition of identical structural units in space. In the simplest crystals such as in nickel, cobalt iron, copper and the alkali metals, the structural unit is a single atom. However, the smallest structural unit may comprise many atoms or molecules.

The structure of all crystals can be described in terms of a lattice consisting of a group of atoms attached to every lattice point.

The group of atoms is called the basis; when repeated in space it forms the crystal structure. The crystalline materials are realized in one or more crystalline forms known as *Bravais* lattices, as shown in Fig. 1.6.

A three-dimensional crystal can be defined by three translational vectors a, b and c. The location of each point is given by arbitrary digit numbers of h, k, and l for x, y, z axes:

$$r = h \cdot \boldsymbol{a} + k \cdot \boldsymbol{b} + l \cdot \boldsymbol{c} \tag{1.14}$$

Every lattice developed by translating one of these vectors constitutes a Bravais lattice. The smallest parallelepiped formed is called the *unit cell*. All unit cells in the crystal have the same shape and volume. All vertices are similarly occupied by atoms of one or more elements constituting lattice sites. As shown in Fig. 1.6, each unit cell may be defined by six factors, a, b, c, α (angle between b and c), β (angle between a and c), and γ (angle between a and b). There are three Bravais cubic lattices: simple cubic (sc), base-centered cubic (bcc) and face-centered cubic (fcc). In the *sc* lattice, the

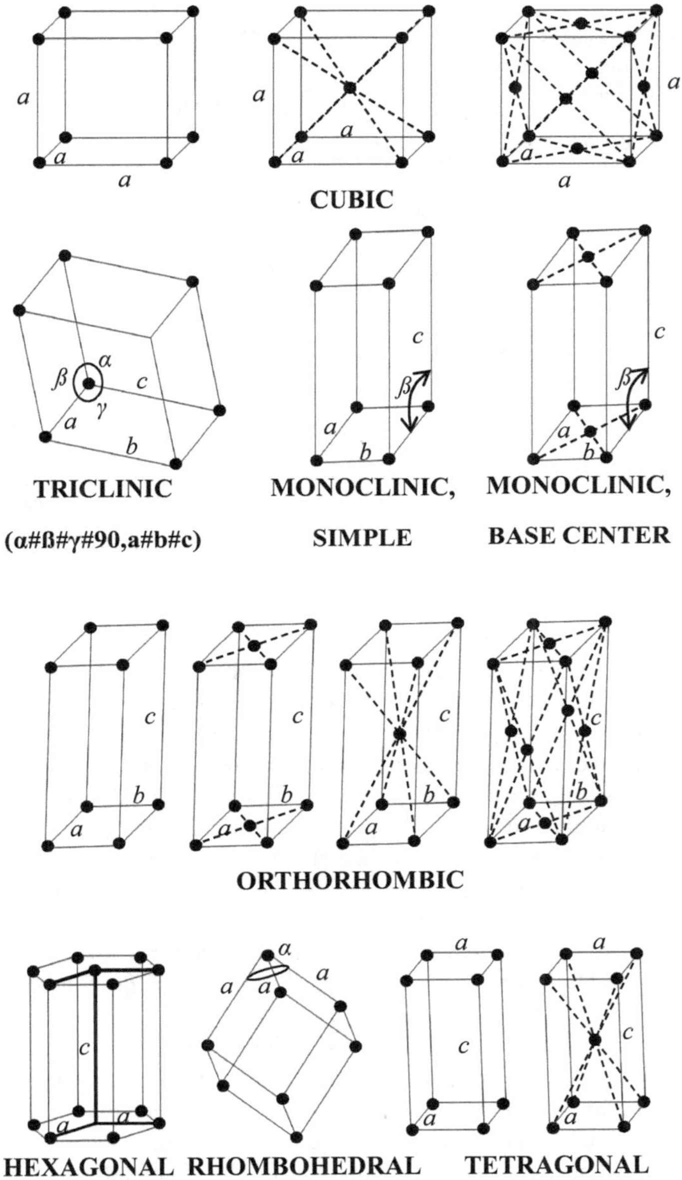

Fig. 1.6 The Bravais space lattices for all possible crystalline materials

atoms are located in the eight apices of the cubic unit cell. In the bcc lattice, in addition to the eight atoms placed at the apices, there is one atom in the center of the unit cell. An fcc lattice comprises eight atoms at the apices and eight atoms in the centers of the faces, as indicated in Fig. 1.6.

Fig. 1.7 Atomic arrangements in close-packed lattices in **a** two-dimensional (2D) and **b** three-dimensional (3D) systems. In 3D systems, *ABAB*… and *ABCABC*… repetitions of atoms make different close-packed lattices

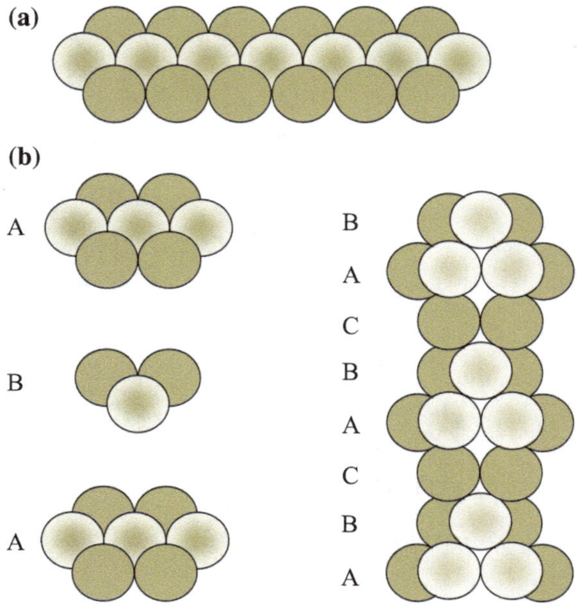

Close-packed lattices, an efficient packing of spherical atoms, can take place in two or three dimensions. Figure 1.7 shows the possible close-packed lattices in these dimensions. For two-dimensional packing systems, there are two primitive cells with three and six vertices. For three-dimensional close-packed lattices, there are spaces or sites between the atoms to be filled by atoms with different arrangements. The three-dimensional close-packed arrangement can be obtained by the placement of the atomic layers, namely A and B, in two ways, generating hexagonal close-packed (hcp) and face-centered cube (fcc) structures. Hexagonal close-packed structures are formed by the repetition of A and B atomic layers in the form ABAB…, whereas fcc close-packed lattices are generated by the repetition of ABCABC… atomic layers. The coordination number, or the number of nearest atoms, for both hcp and fcc structures is 12.

In reality, crystals exist in different crystalline systems. Most metals are crystallized in hcp, fcc and bcc lattice systems. Semiconductors form diamond structures with complex atomic arrangements. A space lattice of diamond is fcc with tetragonal bonding associated with the face-centered atoms. Each atom has four neighbor and 12 next-nearest neighbor atoms. The diamond structure may be viewed as two fcc lattices displaced from each other by one quarter of the length of the body diagonal. The structure of zinc sulfide results from the combination of two fcc lattices, each of which has Zn and S atoms.

The crystal structures may be studied through the diffraction of photons, neutrons and electrons. The diffraction results from the superposition of the waves scattered elastically by the individual atoms of a crystal.

W.L. Bragg described the principles of the diffraction of beams from a crystal. The Bragg law can be explained simply: Suppose that the incident waves are reflected from parallel planes of atoms in the crystal, with each plane reflecting only a very small fraction of the radiation. The reflection occurs with the same incidence and reflection angles from the atomic base plane. Provided that the reflections from parallel planes of atoms interfere constructively, the diffracted beams are exploited. The Bragg law is stated as:

$$2d \cdot \sin(\theta) = n\lambda \tag{1.15}$$

where d is the spacing between parallel atomic planes, θ is the incident angle and λ is the beam wavelength. This is valid for $\lambda < 2d$.

A perfect crystal structure, repeating a particular geometric pattern of atoms without interruption or error, is quite unusual in reality unless it is grown under careful growth conditions—for example single crystal bulk materials grown by directional solidification. This means that crystals in reality certainly consist of structural defects. These structural defects are classified into three categories: zero-dimensional defects, such as vacancies and interstitial and substitutional impurity atoms; one-dimensional defects, such as dislocations; and two-dimensional defects such as grain boundaries. The schematics of these defects are shown in Fig. 1.8.

Grain boundaries are surface or area defects that constitute the interface between two single-crystal grains of different crystallographic orientation. Atomic bonding in particular grains terminates at the grain boundaries. Due to the higher number of broken or dangling bonds of such atoms on grain boundaries, they are necessarily more energetic than those within the grain interior. Thus the grain boundaries become heterogeneous regions for atomic reactions and processes, and favor their acceleration or decay as appropriate. For example, electronic transport in metals is weakened due to increased scattering at grain boundaries, which also serve as charge recombination centers in semiconductors [7].

Dislocations are one-dimensional or line defects that arise from a particular crystallographic rearrangement in the lattice. The two basic types of dislocations are the edge and the screw. The edge dislocation results from wedging in an extra row of atoms; screw dislocations require cutting followed by shearing of the perfect crystal lattice. The geometry of a crystal containing a dislocation is such that when a simple closed traversal is attempted about the crystal axis in the surrounding lattice, a closure failure occurs. The displacement from the starting position is obtained by a lattice vector, the so-called Burgers vector. Dislocations are important because of their role in mechanical properties such as plastic deformation and work hardening.

The last type of defect considered is the non-dimensional or zero-dimensional defect. Vacancies or atomic impurities are considered zero-dimensional or point defects. Vacancies are simply point defects that arise when lattice sites are

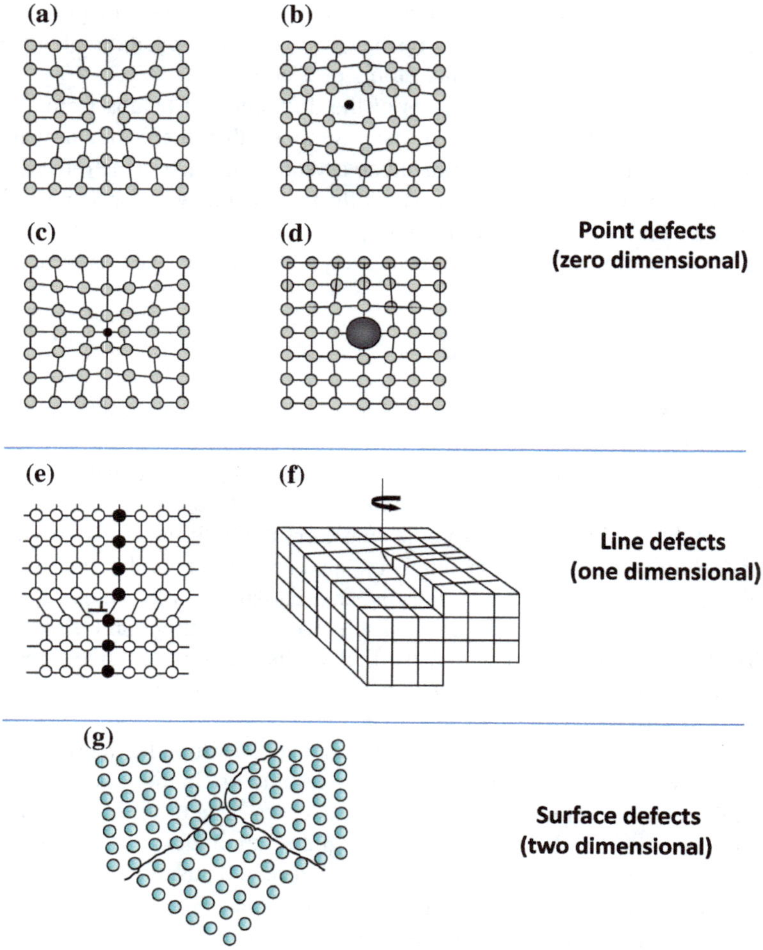

Fig. 1.8 Schematic pictures of different categories of structural atomic defects in solids. *Zero-dimensional* **a** vacancy, **b** interstitial atom, **c** small substitutional atom, **d** large substitutional atom. *One-dimensional* **e** edge dislocation, **f** screw dislocation. *Two-dimensional* **g** grain boundary

unoccupied by atoms. They form because the energy required to remove atoms from their sites and locate them on the surface is not remarkably high. This low energy, coupled with the increase in the statistical entropy of mixing vacancies among lattice sites, gives rise to a thermodynamic probability that an appreciable number of vacancies will exist, at least at elevated temperature. Vacancies are different from dislocations, which are not thermodynamic defects. Because dislocation lines are oriented along specific crystallographic directions, their statistical entropy is low. Coupled with high formation energy due to the many atoms involved, thermodynamics would predict a dislocation content of less than one per

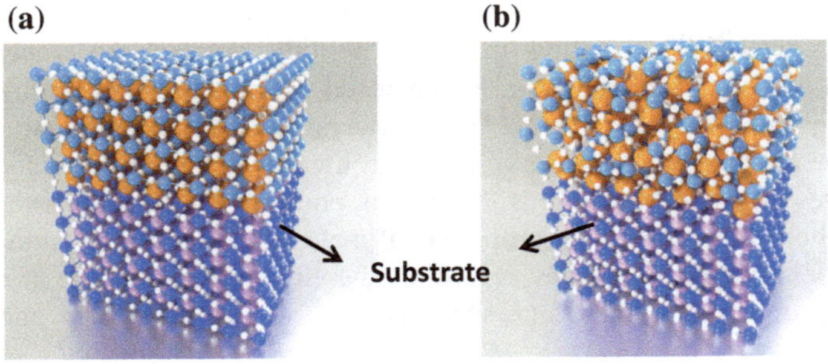

Fig. 1.9 Schematic representation of **a** crystalline long-range order and **b** amorphous short-range order in a three-component ABO_3 perovskite (A = La, Pr, Nd (*orange color*), B = Al, Ga (*blue color*), O = oxygen (*white color*)). This picture shows the interface of a polar perovskite oxide on an oxide substrate. Adapted from [8] with the permission of Nature Publishing

crystal. Thus, although it is possible to create a solid devoid of dislocations, it is impossible to eliminate the vacancies.

The vacancies play an important role in all processes related to solid state diffusion, including recrystallization, grain growth, sintering and phase transformations. In semiconductors, vacancies are electrically neutral as well as charged and can be associated with dopant atoms. This leads to a variety of normal and anomalous diffusional doping effects [7].

Other types of non-dimensional defects occur due to the addition of foreign atoms into the lattice. Interstitial defects are produced when an atom is placed into the crystal at a site that is normally not a lattice point. Substitutional imperfections are produced when an atom is removed from a regular lattice point and replaced with a different atom, usually of a different size, which expands or shrinks the lattice around the imperfection.

In contrast to crystalline solids with long-range order, there is another group of condensed matter, amorphous solids, in which the predictable long-range atomic order breaks down—for example glass, inorganic oxides and polymers. The atoms of these materials follow their random positions after solidification from the melt, even at low rates. A few metals exhibit this property, including certain alloys composed of transition metal (iron, nickel) and metalloid (phosphors, boron) combinations through extremely rapid quenching of melts (e.g. 10^6 C/s) [7]. Figure 1.9 compares the atomic arrangement of a typical crystal and amorphous solid.

1.2.3 The Band Theory of Solids

Before we proceed to an introduction of nanostructures, it is necessary to understand the classification of materials from a solid state physics point of view. So far we have explained the atomic arrangement of bonding types of solid materials. Thus, it is well understood that there is a periodic order in the arrangement of atoms. In addition, a crystal solid is composed of the sum of electrons and ions interacting with each other for all types of crystals over the entire structure. For metals, the entire crystal consists of an electron gas which permeates alongside the atomic sites or ions. This figure is the starting point for understanding the physical interpretation of the electronic properties of materials. This can be further used to answer the principle questions regarding the differences between the physical properties of various materials such as electrical conductivity and thermal coefficients.

1.2.3.1 Free Electron Gas Model

In this model, the conduction electrons are freely distributed in a metal or a semiconductor, and the surface of the solid is the only confining effect against the movement of electrons. All possible interactions between the participating particles in the electron gas/ion system are neglected here. We regard this system as a three-dimensional box containing the delocalized free conduction electrons. For metals, the energy barrier holding the electrons is called the work function, which is on the order of only a few electron volts. Considering the wave–matter characteristics, one can apply the Schrödinger equation to estimate the wave function ($\psi_{(rorx,y,z)} = \exp(ik \cdot r)$ of electrons trapped in a three-dimensional bulk with size length L. The states of the three-dimensional (3D) trap are:

$$\psi_{(x,y,z)} = \left(\frac{2}{\mathrm{L}}\right)^{3/2} \sin\left(\frac{n_x \pi x}{\mathrm{L}}\right) \sin\left(\frac{n_y \pi y}{\mathrm{L}}\right) \sin\left(\frac{n_z \pi z}{\mathrm{L}}\right) \tag{1.16}$$

With the kinetic energy for free electrons:

$$E_k = \frac{\hbar^2}{2\,\mathrm{m}} \cdot \left(k_x^2 \cdot k_y^2 \cdot k_z^2\right) \tag{1.17}$$

For multi-electron systems with a large number of electrons, the quantum number will become very large, and the energy of the successively filled states will also be large. The energy is given by:

$$E_n = \frac{h^2}{8\,\mathrm{mL}^2} \cdot \left(n_x^2 \cdot n_y^2 \cdot n_z^2\right) \tag{1.18}$$

This equation resembles a sphere on a space lattice of wavenumber k, with quantum integers n_x, n_y, and n_z on the three axes where the outer surface is defined as the Fermi sphere on $k_F = \frac{n_F \pi}{L} \cdot n_F$ is the quantum number of the highest energy level filled. The number $N_{(E)}$ of states out to k_F can be obtained:

$$N_{(E)} = 2 \cdot \frac{\frac{4\pi}{3} k_F^3}{(2\pi/\mathrm{L})^3} = \frac{L^3}{3\pi^2} k_F^3 \qquad (1.19)$$

(The factor of 2 comes from the fact that N electrons of spin quantum number $m_s = -1/2$ and N electrons of spin quantum number $m_s = -1/2$ are allowed for Ni sites, the so-called degeneracy factor). Thus:

$$E_F = \frac{\hbar^2}{2\,\mathrm{m}} \cdot \left(\frac{3\pi^2 N}{L^3}\right) \qquad (1.20)$$

The distribution of energy levels can be readily defined by differential $\mathrm{d}N_{(E)}/\mathrm{d}E$ defined as the density of electron states per unit volume at energy E; $\rho_{(E)}$:

$$N_{(E)} = \frac{L^3}{3\pi^2} \cdot \left(\frac{2mE}{\hbar^2}\right)^{3/2} \qquad (1.21)$$

Thus:

$$\rho_{(E)} = \frac{N_{(E)}}{E} = \frac{L^3}{3\pi^2} \cdot \left(\frac{2m}{\hbar^2}\right)^{3/2} \cdot E^{1/2} \qquad (1.22)$$

Figure 1.10 illustrates the density of electron states for bulk three-dimensional materials. The *Fermi–Dirac distribution* ($f_{F-D(E)}$) function explains the occupation rules for fermions (i.e. electrons) as expressed by:

$$f_{F-D(E)} = \left(1 + \exp\left(\frac{E - \mu}{K_B T}\right)\right)^{-1} \qquad (1.23)$$

in which μ is the chemical potential of metal, which is identical to the Fermi energy level. This function is sketched in Fig. 1.10 for zero and non-zero temperatures, and the occupied energy levels are shown for different temperatures [5].

1.2.3.2 Nearly Free Electron Gas Model and Period Structures

The free electron gas model is unable to describe the differences in electronic behaviors between materials—for example, the difference between metal and semiconducting materials or the origin of the band gap. In order to understand these featured differences, we must consider the motion of electrons in solids influenced by the crystal potential or energy. This will change the distribution of these electrons, thus introducing the ***band theory*** of solids.

The free electron model defines the energy values allowed to distribute essentially continuously from zero to infinity. In contrast, with the nearly free electron

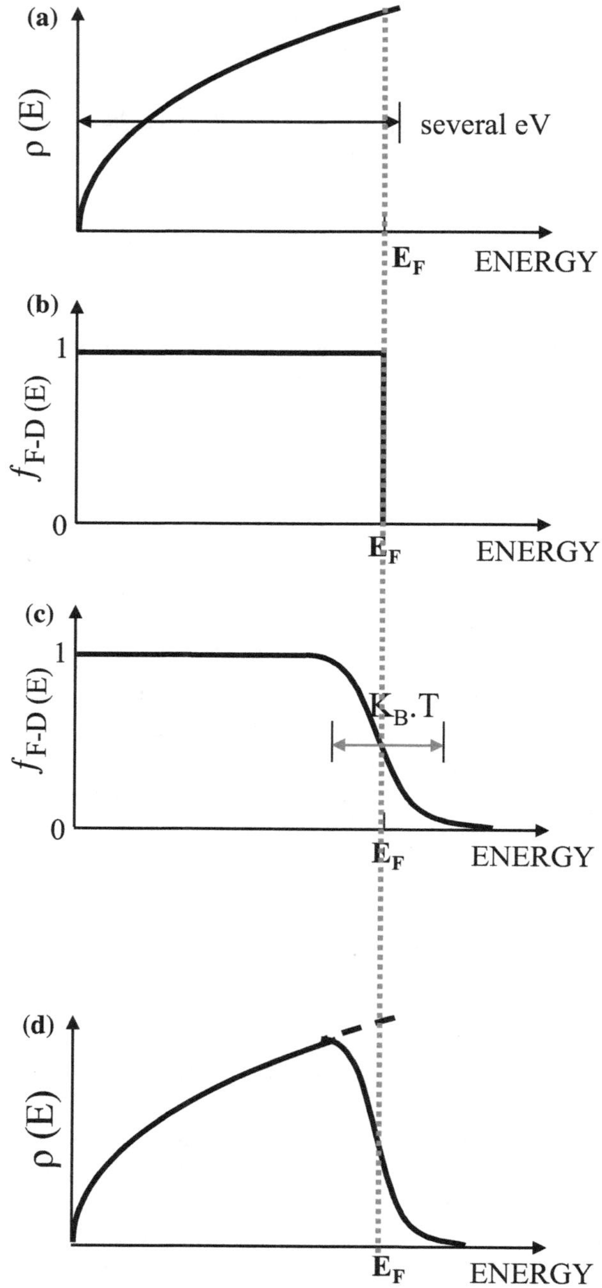

Fig. 1.10 **a** Density of electron states ($\rho_{(E)}$), **b** occupation ($f_{F-D(E)}$) at zero temperature, **c** occupation ($f_{F-D(E)}$) at non-zero T temperature (K_BT is the thermal energy; K_B is the Boltzmann constant), and **d** density of occupied (*dashed*) electron states for a bulk *three-dimensional solid*

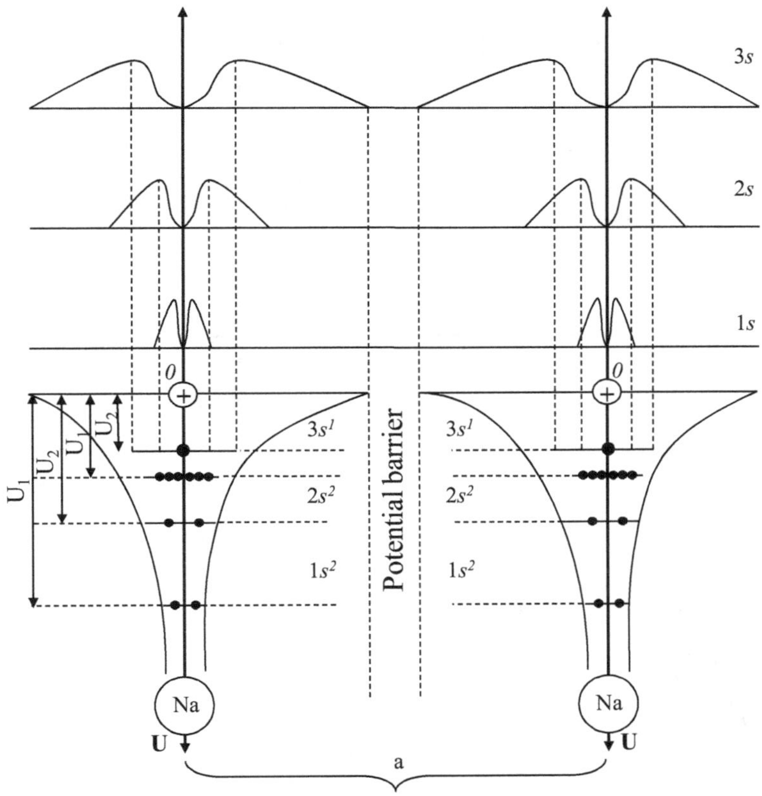

Fig. 1.11 The variation in electron distribution probability for sodium free atoms with infinite separation

model, the electron bands are defined such that the free electron gas is perturbed only weakly by the periodic potential of the ion cores. This concept is schematically explained for sodium atoms in ground free states and in binding in a typical sodium metal crystal shown in Fig. 1.11. The separation of Na atoms is large enough that each Na atom exhibits the electron arrangement of a single free atom, consisting of $1s^2, 2s^2, 2p^2 \cdot 3s^1$. If the energy barrier between the two neighbor atoms is infinite, then no electron can transfer from one atom to another. When the atoms attract each other, the electron distribution changes so that a metal bond forms between them. This is associated with the formation of conduction electrons (e.g. $3s^1$ for Na), so-called electron gas, because the energy of these electrons is higher than the energy barrier formed between the ions. Furthermore, the energy barriers between the two atoms produce a periodic energy potential, as indicated in Fig. 1.12 [6].

This schematically explains the concept of the periodic energy potential in a crystal lattice. As is evident, the infinite energy barriers in the single atoms are

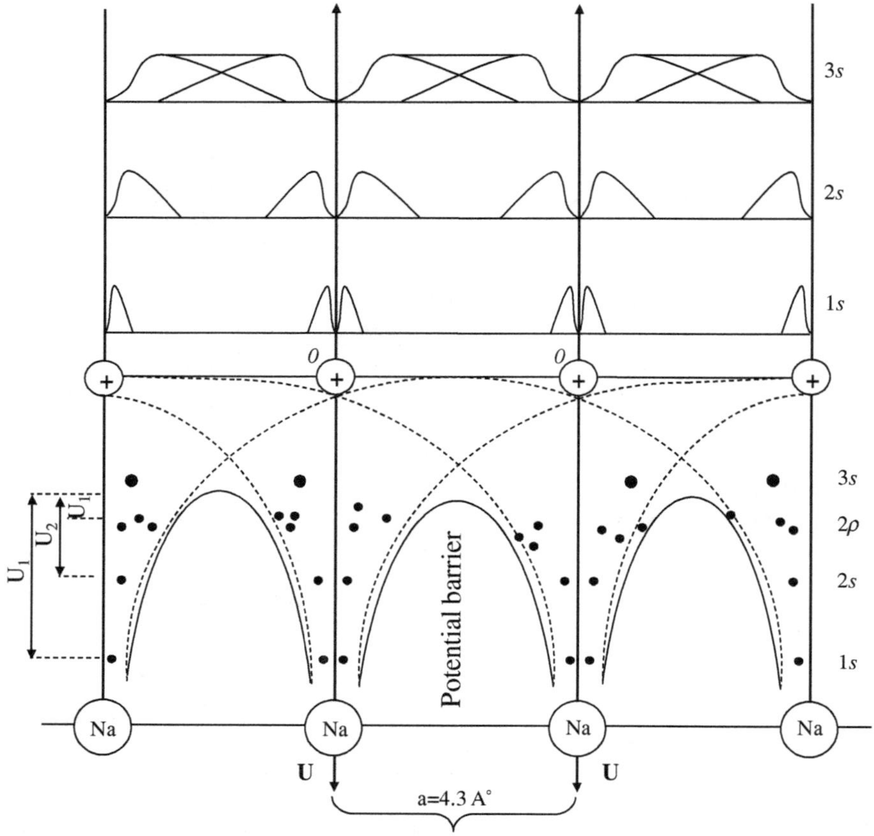

Fig. 1.12 The variation in electron distribution of sodium atoms in a metallic crystal indicating of the periodic potential barrier

changed to periodic finite energy barrier whose height is shorter that the energy level of the conduction electrons, i.e. 3*s* orbitals in Na atom crystal.

The wave function of the nearly free electrons under the effect of the periodic potential energy may be described by the Bloch theorem as:

$$\psi_k(r) = u_k(r) \cdot \exp(i\boldsymbol{k} \cdot \boldsymbol{r}) \tag{1.24}$$

where $u_k(r)$ is the period of the crystal lattice with $u_k(r) = u_k(r+T)$. Here, T is a translation vector of the lattice. The solution to the Schrödinger equation on the wave function (1.24) is that the free electron behavior is still valid for certain energy ranges, namely the ***allowed band***, while in other ranges of energy, namely forbidden energy bands or so-called ***energy gaps***, no electron states are allowed. These allowed and forbidden bands will appear on the occupied electron density of states.

1.2.3.3 Electron Bands and Gaps: The Origin of Band Gap in Solids

The formation of energy bands may be well understood by assuming the energy levels of atoms before and after binding. In each isolated atom, the electron energy levels are discrete, as shown on the right-hand side of Fig. 1.13. As the atoms approach one another, the individual energy levels split as a consequence of Pauli's exclusion principle as shown on the left-hand side of Fig. 1.13. Level splitting and broadening occur first for the valence or outer electrons, since their electron clouds are the first to overlap. During bonding of atoms, electrons populate these lower energy levels, reducing the overall energy of the solid. Upon further dimensional shrinkage, the magnitude of the overlap rises, causing a strong repulsive force between atoms. At the equilibrium, where the solid has actually formed, some of the levels have broadened into bands of energy levels. The bands span different ranges of energy, depending on the atoms and specific electron levels involved. Sometimes, as in metals, bands of high energy overlap. Insulators and semiconductors have energy gaps of varying widths between bands in which the electron states are not allowed [6, 7].

Here we explain the reasons for the energy level splitting, band structure evolution and implications with regard to properties as the most fundamental and difficult questions in solid state physics. We briefly return to the subject of periodic potential energy and *Bloch* function in solids.

In a crystal with a periodic arrangement of atoms, Bragg scattering of electron waves take place. The periodic potential energy for N atoms with interatomic separation a is determined by the modulation of an energy barrier with definite height located in the middle of an atomic separation, a. When a traveling wave, $\exp(i\boldsymbol{k}\cdot\boldsymbol{r})$, is scattered by the atoms, a coherent reflected wave $\exp(-i\boldsymbol{k}\cdot\boldsymbol{r})$ may be

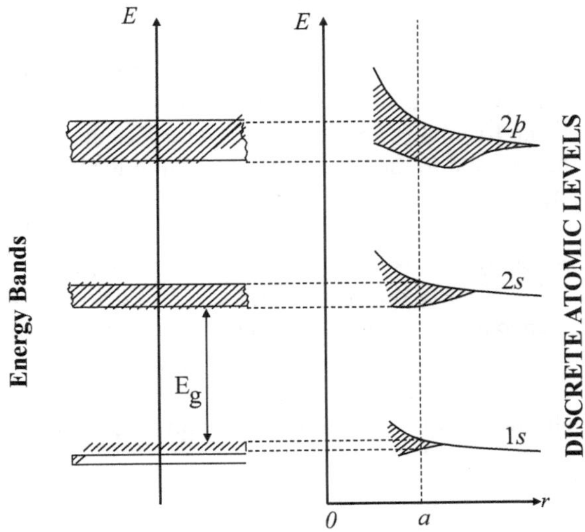

Fig. 1.13 Schematic picture of the formation of band structure from discrete atomic level in a given crystal after binding

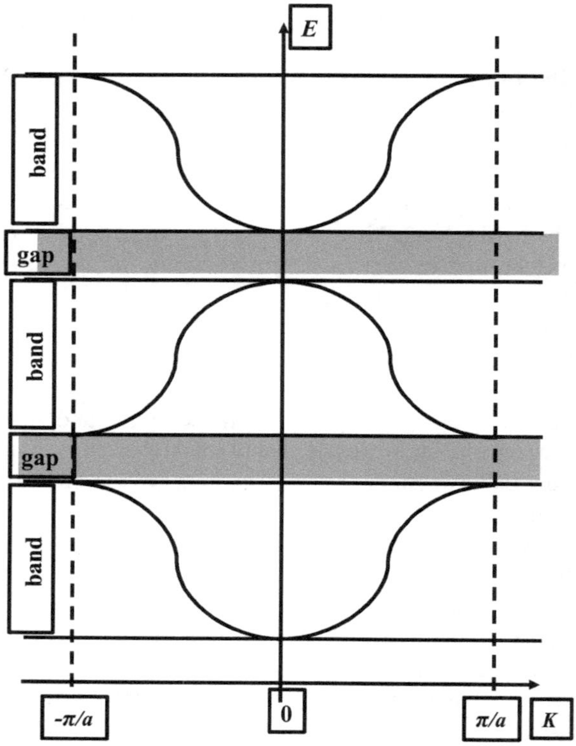

Fig. 1.14 Schematic representation of electron bands and gaps for the first three energy bands according to the *Kronig-Penny* model

generated, leading to a standing wave. The condition of the Bragg law, the coherent scattering, is very clear (the path difference between a wave back-scattered at $x = 0$ and one back-scattered at $x = a$ must be an integer number of electron wavelengths). Thus, one can write: $k = \frac{n\pi}{a}$. Therefore, any linear combination between incident and reflected waves present at integer n of $\frac{\pi}{a}$ on the wavenumber, k, space lattice would be an answer to the equation.

The Kronig–Penney model [9] describes the formation of energy bands in solids. The solution to the Schrödinger equation for this wave function under the periodic potential is illustrated in Fig. 1.14. The coherent scattering wave function of electrons generates the bands restricted in k to values less than π/a. Energy gaps happening at $k = n\pi/a$ are the result of Bragg scattering at $k = \pm n\pi/a$ [5].

For N sites, the theory predicts 2 N delocalized electron states based on the degeneracy factor of 2 for electrons. The Pauli exclusion principle works here for filling of the bands. For example, the hydrogen atom states can have only one electron. The electrons, of course, occupy from the lowest energy states preferentially until all the electrons are accommodated in the quantum states.

1.2.3.4 Metals, Semiconductors and Insulators

Based on the occupation of energy bands by electrons, we can categorize solids into two major groups. The first group consists of solids that exhibit a band structure consisting of a semi-filled band located on top of a completely filled band. Alkaline and rare earth metals are among such materials where substantially incomplete bands or hybridized semi-filled bands exist. The second group comprises solids characterized by an empty band located on top of a fully filled band. Diamond, germanium and silicon are simple examples of this group. The highest energy band is called the conduction band, and the fully filled bottom one is called the valence band. The energy gap between these bands is called the band gap, E_g. Figure 1.15 compares the band structure of these two groups schematically.

The former resembles the band structure of conducting materials, namely metals. In metals, the motion of electrons is freely taking place regardless of their dependence on the crystal lattice, as there are empty energy levels in the semi-filled band to move from one to another under external electrical fields. This develops high conductivity magnitudes for metals. This is not the case for the second group of solids, semiconductors and insulators. At absolute zero temperature and ground state (without any external physical or chemical stimulation), the band structure of these solids suggests that no electron mobility can take place due to the lack of any empty states. However, the stimulated electron conductivity is developed under certain conditions, which categorize this group of solids into two sub-groups, namely semiconductors and insulators. In general, solids having a band gap exceeding 3 eV, such as diamond, are called insulators, whereas those with a band gap of 1–3 eV, such as Si, Ge, and GaAs, are called semiconductors [6].

In semiconducting materials, an external energy such as thermal or optical energy must be exerted to push electrons from filled to empty levels. The ways in which semiconducting materials are created are shown in Fig. 1.16. The ordinary

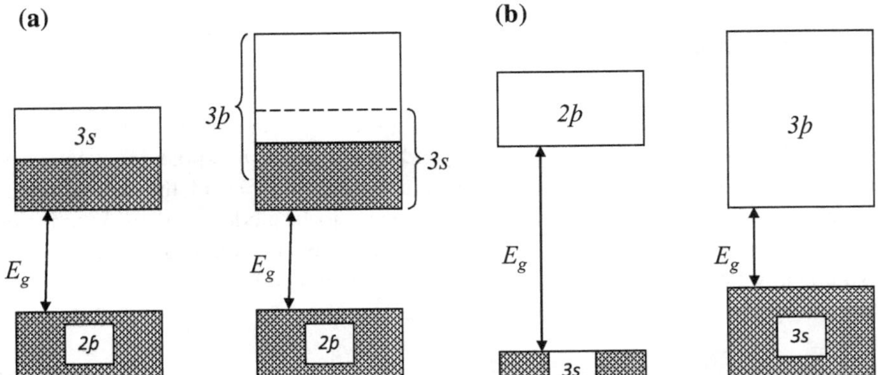

Fig. 1.15 Schematic picture of band structure of all solids classified into two groups: **a** semi-filled top band above a completely filled band, **b** empty top band on fully filled bottom band

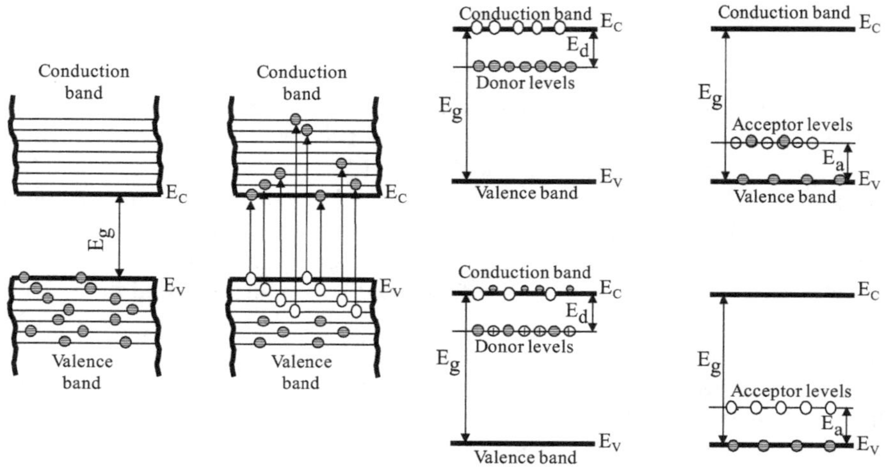

Fig. 1.16 Schematic presentation of band structures of **a** intrinsic semiconductor with electron-hole pair, and **b** impurity semiconductors with donor and acceptor levels

semiconductor materials, namely intrinsic semiconductors, are realized by applying any external physical energy such as irradiation of photons or heating. This generates electron–hole pairs, where electrons jump over the band gap and transfer to the conduction band, and empty states—so-called holes—remain in the valence band. This will cause electrical conductivity for intrinsic semiconductors like Si and GaAs. This process is schematically shown in Fig. 1.16a, in which thermal energy is the stimulating force to increase the electron energy. However, the doping of impurities such as boron or phosphorous into group *V* elements is another method of producing impurity semiconductors.

1.3 Thermodynamics of Materials

Thermodynamics is able to predict the feasibility of chemical reactions or phase transformations in materials, such as the oxidation of a metal exposed to weather or the transport of atoms in a typical phase transformation as in the formation of eutectoid phase. However, all of the events need to be understood by kinetics as well, because some processes may not occur at all even though they are thermodynamically favored. The basics of thermodynamic feasibility of a phenomenon are defined by the free energy Gibbs function G expressed by:

$$G = H - TS \tag{1.25}$$

where H is the enthalpy, S the entropy, and T the absolute temperature. A change in the Gibbs free energy occurring during a process is given by:

$$\Delta G = \Delta H - T\Delta S \tag{1.26}$$

where ΔH and ΔS are the corresponding enthalpy and entropy changes. One important result of the second law of thermodynamics is that spontaneous reactions are predicted to occur only when $\Delta G < 0, T$ and $P =$ constant. This means that the energy of the system becomes a more negative value G. Provided that $\Delta G = 0$ there is no driving force for any change, implying that the equilibrium state persists. In contrast, if $\Delta G > 0$, no reaction takes place. This is an important criterion by which scientists may determine or analyze many problems in materials science-related phenomena such as phase transformation and phase diagrams, nucleation and growth of crystals [7].

1.3.1 Nucleation and Growth of Solids

The formation of solids can take place by nucleation under favorable thermodynamic conditions. Two types of nucleation are feasible: homogenous and heterogeneous (or non-homogeneous). For the homogenous nucleation of materials, a supersaturation of growth species is necessary. The reduction in temperature of an equilibrium mixture such as a saturated solution of mineral salts, the formation of metal quantum dots in glass matrix by annealing at moderate temperatures, and the solidification of metal dendrites from melt are good examples. For heterogeneous nucleation, a surface serves as a substrate on which solid materials start to nucleate and grow. Fabrication of thin films and coatings are the most common examples of this mechanism.

On the mathematical interpretation of thermodynamic expressions, there is a vast amount of literature, which can be found in [10–14]. Here, we concentrate on the simplest nucleation system. We consider the case of thermodynamic nucleation of a spherical solid phase under homogenous nucleation conditions and a semi-spherical solid phase under heterogeneous nucleation conditions, as shown in Fig. 1.17a, b. In such a process, the transformation from the source phase (i.e. gas, solution or solid) to the result phase, i.e. solid condensate, must be associated with a reduction in the chemical free energy by volume contraction ΔG_v which is expressed by:

$$\Delta G_v = -\frac{K_B T}{\omega} \ln\left(\frac{C_0}{C}\right) \tag{1.27}$$

where C is the concentration of growth species in supersaturation, and C_0 is the concentration of growth species at equilibrium (namely the solubility), K_B is the Boltzmann constant, T is temperature, and ω is the atomic volume. The formation of a solid surface, on the other hand, is associated with an increase in the free

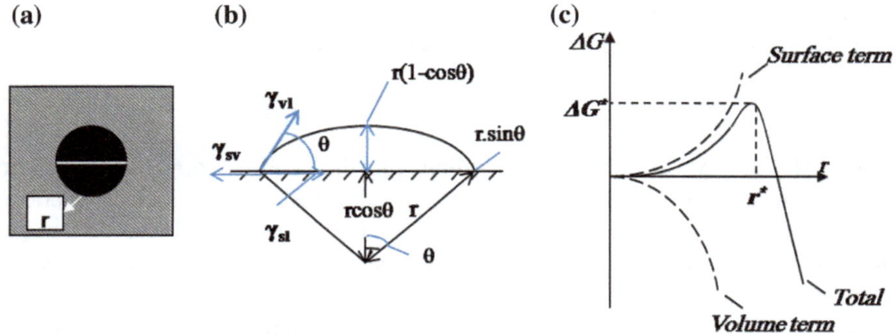

Fig. 1.17 Schematic picture of a nucleus formed under **a** homogenous nucleation of a spherical solid nucleus (*black*) in a supersaturated vapor, solution or solid media (*gray*), **b** heterogeneous nucleation in the case of semi-sphere on a substrate, and **c** variation in Gibbs free energy during nucleation

energy defined by the surface energy density γ. In the presence of the substrate, the surface energy of each incorporating interface will be applied to the surface energy term.

For homogenous nucleation, the change in total Gibbs free energy is written as:

$$\Delta G = \frac{4}{3}\pi r^3 \Delta G_v + 4\pi r^2 \gamma \tag{1.28}$$

At the critical point, one writes: $\frac{d\Delta G}{r} = 0$. Thus, the critical radius r^* and the critical Gibbs energy so-called energy barrier ΔG^* can be obtained:

$$r^* = \frac{2\pi\gamma}{\Delta G_v}, \Delta G^* = \frac{16\pi\gamma^3}{3 \cdot (\Delta G_v)^2} \tag{1.29}$$

For heterogeneous nucleation, we find:

$$\Delta G = a_3 r^3 \Delta G_v + a_1 r^2 \gamma_{vf} + a_2 r^2 \gamma_{fs} - a_2 r^2 \gamma_{sv} \tag{1.30}$$

γ_{vf}, γ_{fs} and γ_{sv} are the surface or interface energy of the vapor–nucleus (f), nucleus–substrate and substrate–vapor, as indicated in Fig. 1.17b. The geometric constants are calculated as:

$$\begin{aligned} a_1 &= 2\pi(1 - \cos\theta) \\ a_2 &= \pi \sin^2\theta \\ a_3 &= 3\pi\left(2 - 3\cos\theta + \cos^2\theta\right) \end{aligned}$$

in which θ is called the wetting or contact angle and is defined by Young's equation:

$$\gamma_{sv} = \gamma_{fs} + \gamma_{vf} \cos\theta \tag{1.31}$$

At the critical point, the nucleus size and the barrier energy are:

$$\begin{aligned} r^* &= \frac{-2\left(a_1\gamma_{vf} + a_2\gamma_{fs} - a_2\gamma_{sv}\right)}{2a_3\Delta G_v} \\ \Delta G^* &= \frac{4\left(a_1\gamma_{vf} + a_2\gamma_{fs} - a_2\gamma_{sv}\right)^3}{27a_3^2\Delta G_v} \end{aligned} \tag{1.32}$$

After substitution of the geometric constant, these will be written as a function of the contact angle:

$$\begin{aligned} r^* &= \frac{2\pi\gamma_{vf}}{\Delta G_v}\left\{\frac{\sin^2\theta\cos\theta + 2\cos\theta - 2}{2 - 3\cos\theta + \cos^3\theta}\right\} \\ \Delta G^* &= \frac{16\pi\gamma_{vf}}{3\Delta G_v^2}\left\{\frac{2 - 3\cos\theta + \cos^3\theta}{4}\right\} \end{aligned} \tag{1.33}$$

If we consider (1.29) and (1.33), we can rewrite them as:

$$\begin{aligned} r^*_{\text{Het.}} &= r^*_{\text{Hom.}} f(\theta) \\ \Delta G^*_{\text{Het.}} &= \Delta G^*_{\text{Hom.}} f'(\theta) \end{aligned} \tag{1.34}$$

$f'(\theta)$ is called the wetting factor. Depending on the contact angle, the mechanism of nucleation varies. If $0 < \theta < 180°$, Young's equation predicts the formation of the nucleus on the surface, and the nucleation mechanism obeys the heterogeneous equations showing that the energy barrier is smaller than that of the homogenous nucleation. When $\theta = 180°$, the nucleus does not wet the substrate at all, the wetting factor equals 1, and the critical energy barrier becomes the same as that of homogeneous nucleation. When = 0° the wetting factor becomes equal to zero, and there is no energy barrier for the formation of new phase. One example of such a case is when the deposited material is the same as that of the so-called epitaxial growth on the substrate [13, 14].

Thermodynamic interpretation of nucleation must be accomplished by introducing the kinetic information, namely the rate of nucleation, which is defined as the number of stable nuclei per unit volume in unit time. The rate of nucleation varies as a function of three major factors: (i) the thermodynamic fluctuation parameter of Gibbs energy $P^{\Delta G^*}$, (ii) the concentration of growth species C_0, and (iii) the successful jump frequency of growth species f as expressed by:

$$\text{Nucleation Rate} \propto P^{\Delta G^*} C_0 \cdot f \tag{1.35}$$

The growth of nuclei takes place in a multi-step process. The major steps in the growth of nuclei can be divided into two controlling phenomena: (a) the growth species are supplied under a particular driving force such as diffusion (for electrically charged particles like ions, other forces exist such as migration of ions under an external electric field), and (b) accommodation of adatoms on the growth sites on the surface of the solid, which may lead to crystallization of the solid (e.g. in electrodeposition this is called electrocrystallization). The former, namely the diffusion-controlled step, may involve several stages, including the preparation of growth species from precursors in bulk medium, transport from the precursor bulk medium to the growing surface of the solid, and adsorption and surface diffusion of atoms or other particles such as adions in electrochemistry. The latter, namely the surface-limited step, consists of irreversible incorporation of atoms into a solid atomic network, desorption of adatoms and side reactions leading to the generation of by-products.

Crystallization of solids is an interesting and easy-to-understand phenomena. This section answers many questions about substantially why a crystal forms, or why it exhibits a particular crystal habit, and so on. There is a vast and long-standing body of literature on this topic (e.g. see [13–15]). Kossel et al. (KSV) [17] proposed a classic step-wise growth model, which determines a growth mechanism based on the surface defects on different crystal planes. For instance, on a flat surface as shown in Fig. 1.18a, there exist many surface defects or steps, which in turn act as growth sites. The number and types of the defect growth sites differ for different facets of a given crystal. Here it is schematically shown for a simple cube crystal at the (100), (110) and (111) crystalline planes (see Fig. 1.18a). The simplest crystalline plane for a simple cube is (100), making it an ideal example for explaining crystal growth. Even for such a simple crystalline plane or facet, however, there exist many growth sites, as shown in Fig. 1.18b, including terrace,

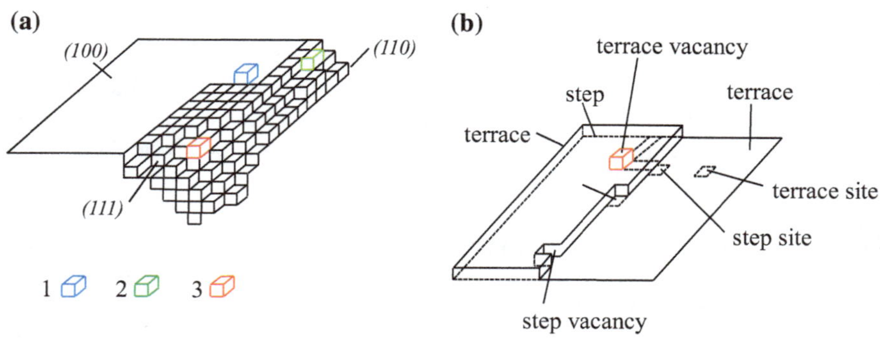

Fig. 1.18 **a** Growth sites on different facets of a simple cube crystal. Replotted from [15] with the permission of Springer. **b** Categories and names of growth atomic sites on (100) surface of a cube. Replotted from [16] with the permission of RSC Publishing

vacancy, ledge, ledge-kink, kink, ledge, step-kink and step-ledge. Each site has an energy defined by the number of broken atomic bonds. These growth sites will accept adatoms traveling on the surface when the growth proceeds to the second major step (step b explained above). Adatoms with appropriate energy will incorporate into the appropriate growth sites: the higher the number of appropriate growth sites, the higher the growth rate. Due to the limits of the KSV model, i.e. the regeneration of growth sites, Burton, Cabrera and Frank (BCF) [18] proposed a model in which the screw dislocations on the growing solid surface are responsible for the continuous generation of growth sites. A picture of screw dislocation was shown in Fig. 1.7, which demonstrates the types of defects on a surface.

The reason for differences in growth rates of a given crystal on different crystalline planes is the differences in accommodation capacity of different facets. The periodic bond chain (PBC) theory developed by Hartman and Perdok [19] explains how different facets of a given crystal have different surface energies. This was explained by the difference in the number of unsatisfied or broken bonds on different atoms. This model categorizes all crystal habits into three different groups of facets based on their broken bonds. As the number of broken bonds increases, the growth rate on such facets increases [13–19].

1.4 Kinetics of Materials

Thermodynamics may describe states of matter only in equilibrium. Understanding a system, even if only a rough estimation, is not possible without knowledge of the kinetics involved. Thermodynamics provides no information about the mechanism required to maintain equilibrium. Kinetics, on the other hand, can be used to describe the intricate balance quantitatively. Kinetic theory contemplates a system at a microscopic level for non-equilibrium states, in contrast with statistical thermodynamics for equilibrium states. In materials science, therefore, thermodynamics predicts the feasibility of chemical reactions or phase transformations, but is unable to answer uncertainties about their rate of progress and mechanisms. Kinetic theory is thus able to provide an analytic view regarding the sequence of all the processes and evolutions in a step-by-step manner. For kinetics studies, two types of motions are considered. The first concerns the conduction of the chemical reactions which lead to changes in the properties of materials. The second type is related to the movement or transport of particles or matter.

Kinetics for chemical reactions determines a constant concentration ratio at equilibrium, just as thermodynamics does. Such agreement is required of any kinetic theory. In the limit of equilibrium, the kinetic equations must collapse to relations of the thermodynamic form; otherwise, the kinetic picture cannot be accurate. Let us consider a reaction under the dynamic equilibrium between two substances A and B which are linked by simple unimolecular elementary reaction:

$$A \underset{K_r}{\overset{K_f}{\rightleftarrows}} B \tag{1.36}$$

Both elementary reactions are active at all times, and the rate of reaction is:

$$v_i = k_j \cdot C_i \tag{1.37}$$

where i is indicative of substances A and B, and k_j is the rate constant for each reaction direction; j denotes the forward (f) and reverse reaction (r) directions. The rate of the forward process is v_f, whereas the rate of the reverse reaction is v_r in molarity per second (mol s^{-1}).The rate constant k_j has dimensions of s^{-1}. The net rate of the process of conversion of A to B is:

$$v_{net} = k_r \cdot C_B - k_f \cdot C_A \tag{1.38}$$

At equilibrium one finds: $v_{\text{net}} = 0$. The rate constant is therefore defined as:

$$K = \frac{k_f}{k_r} = \frac{C_B}{C_A} \tag{1.39}$$

For the kinetics of chemical processes, we understand the effect of temperature on the rate constant of reactions. The Arrhenius equation applies:

$$k = A \cdot e^{-\frac{E_a}{RT}} \tag{1.40}$$

Here, E_a is called activation energy, which is the barrier energy present between the initial and final states for an individual reaction, as shown in Fig. 1.19. It can be the change in the Gibbs free energy or enthalpy associated with an evolution or transformation. A is the amplitude of the fluctuations of particles, generally known as the frequency factor, and R is the universal gas constant (8.3144598 J mol^{-1} K^{-1}).

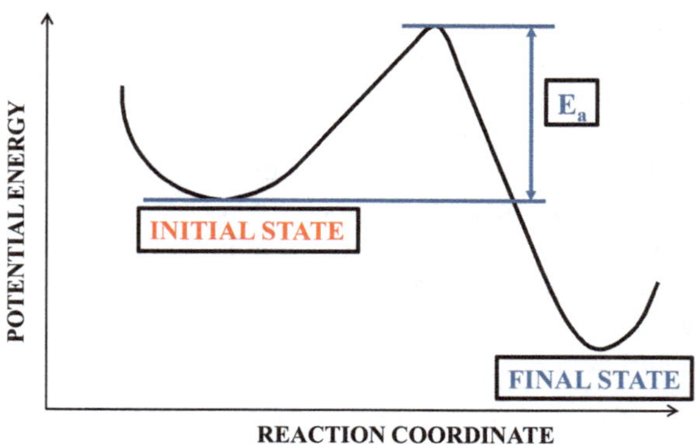

Fig. 1.19 The barrier energy or activation energy for a given chemical reaction converting a substance from initial state to final state

Kinetics is also capable of describing the evolution of mass flow throughout the system, including both the approach to equilibrium and the dynamic condition. Whenever a material system is not in thermodynamic equilibrium, driving forces naturally arise to push it towards equilibrium. Such a situation can occur, for example, when the free energy of a microscopic system varies from point to point because of compositional inhomogeneity. The resulting atomic concentration gradients generate time-dependent mass transport effects that reduce free energy variations in the system. Examples of such processes include ion transport in electrolytes under *migration* and *diffusion* mechanisms, and phase transformations. One example of kinetics control on processes in solids is mass transport by diffusion, which may be defined as the transport of an atomic or molecular species within a given matrix under the influence of a concentration gradient. Fick established the phenomenological connection between concentration gradient and the resulting diffusional transport through the equation:

$$J = -D\frac{\partial C}{\partial x} \tag{1.41}$$

where J is the flux, and D in cm^2/s is the diffusion coefficient under the defined concentration gradient. D is dependent on temperature, according to the Maxwell–Boltzmann relation:

$$D = D_o \cdot e^{-\frac{E_D}{RT}} \tag{1.42}$$

E_D is the activation energy for diffusion.

For time-dependent diffusion processes, Fick's second law gives the diffusion equation as:

$$\frac{\partial C}{\partial t} = D\frac{\partial^2 C}{\partial x^2} \tag{1.43}$$

The diffusion equation is a second-order differential equation with respect to space and a first-order differential equation with respect to time [7, 20, 21].

1.5 Nanostructures and Bulk Nanostructured Materials

From here, we will focus on a new class of materials, namely nanostructures. In general, a nanostructure is a material with at least one dimension in the nano size regime. The definition of nano size regime is controversial, as dimensions typically range from sub-nanometer to several hundreds of nanometers. A continuous transition of the properties and physics of materials takes place when moving from macro to nano size. Bulk materials at the macro size level obey the classical mechanics of physics or related physical phenomena such as electromagnetism or electrostatics.

Even micron-sized materials behave largely similarly to bulk materials. However, there is an intermediate group of materials, mesostructures, in which groups of hundreds of atoms exhibit the same properties as those observed in their bulk materials. Further miniaturization of materials has led to the emergence of a new generation of materials, called nanomaterials or nanostructured materials, which exhibit physical properties distinctly different from those of other groups. Several remarkable specific properties have been exploited through the realization of nanomaterials.

1.5.1 Dimensionality in Nanomaterials

In order to simplify the classification of nanomaterials, we will divide them into major groups where at least one dimension of nanomaterial lies in the nanometer range. We start with a three-dimensional (3D) bulk solid constructed on three dimensions x, y and z at macroscale. Size confinement is an essential part of our definition of nanostructures (i.e. generally accepted as 1–100 nm for nanoscale size regime, depending upon the physical characteristics in question). Some examples of the critical size lengths of different phenomena are indicated in Fig. 1.19. The length scales introduced for a particular physical concept vary independently according to their specific phenomena. Several physical properties and applications with their critical length at nanoscale are illustrated.

For electron transport in particular, which will later be utilized to determine the electron density of states in nanostructures, the quantum mechanical length scale applies. In general quantum or wave mechanics, a wavelength is ascribed to a particle with kinetic energy E. This is called the de Broglie wavelength ($\lambda_{\text{de-Borglie}}$):

$$\lambda_{\text{de-Borglie}} = \frac{h}{(2mE)^{\frac{1}{2}}} \tag{1.44}$$

Let us return to the rules of electron occupation of bands. At the Fermi level, the quantum mechanical wave function of an electron with energy E_f has a wavelength λ_f. The electrons at the Fermi level have the highest energy and can gain small amounts of energy in an accelerating electric field or other wave-scattering process to transfer to any empty state at higher orbitals. As such, the electrons at the Fermi level play an essential role in transport properties. Thus, λ_{mfp} is defined as the length scale at which a diffracting obstacle will reveal the wave-like quantum mechanical nature of the particle; therefore, the mean free path would be the critical scattering length for electrons, which varies under different conditions. The scattering length scale depends upon temperature, electron density, impurity concentration, kinetic energy and external forces such as magnetic fields. Let us define the most common critical length for electron transport to solve the Schrödinger equation in order to find the wave functions of electrons confined in the nanostructures. We define the electron mean free path as the average distance covered by electrons between two scattering events.

We now explain ***dimensionality*** effects, or size confinement, for nanostructures. Downscaling of materials can be conducted on the three dimensions one by one. Three groups, including two-dimensional (2D) achieved by confinement along the z-axis, one-dimensional (1D) by confinement along both the y and z axes, and zero-dimensional (0D) by confinement along all three x, y and z axes, will result. The direct consequence of size confinement for materials is the generation of "nanostructures". However, there is another group designated as nanomaterials, that of bulk nanostructured materials. This fourth group, namely three-dimensional (3D) or bulk nanostructured materials, has exterior dimensions in macro size, but its interior constituents comprise nanostructures of the first, second and third groups. Figure 1.20 shows the schematic definition of dimensionality of nanostructures starting from bulk 3D solids and leading to 0D, 1D and 2D nanostructures.

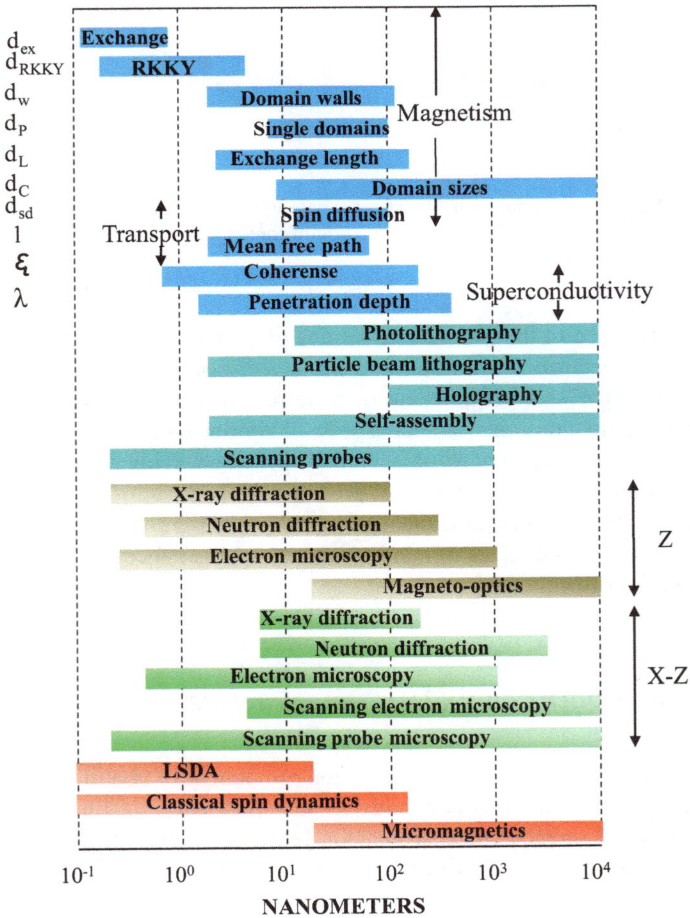

Fig. 1.20 Diagram representing the critical length scales of different physical properties and applications of materials. Replotted from [22] with the permission of Elsevier

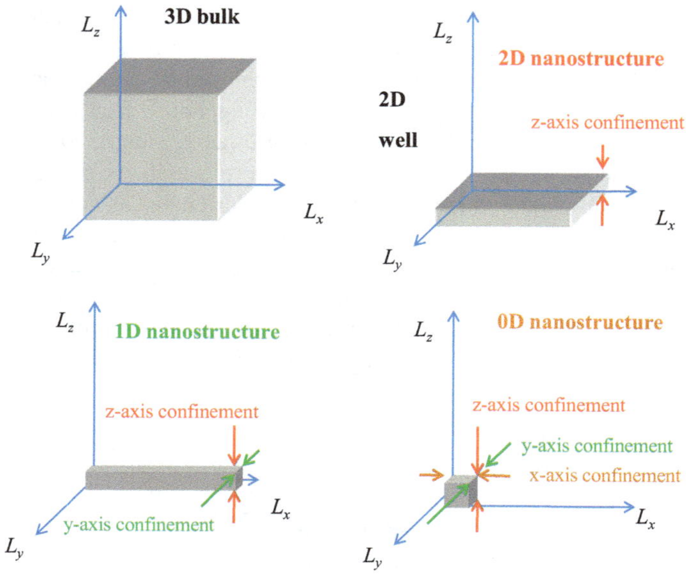

Fig. 1.21 Schematics of dimensionality in nanomaterials. Size confinement in each dimension promotes the dimensionality effect in the formation of three types of nanostructures (2D, 1D and 0D)

We now proceed to the physical properties of nanostructures and compare them with regard to electron distribution and the density of states for 0D, 1D and 2D nanostructures. The size confinement of materials at nanoscale can be understood by the principle quantum mechanical aspects of confined electrons discussed earlier in this book, using the picture of a particle in a box. This simplified model of elementary quantum mechanics can predict the density of electron states in nanostructures having precise dimensions in the quantum nanoscale size regime. Figure 1.21 shows schematics of nanostructures and the effect of dimensionality on the electron density of states.

1.5.2 *Two-Dimensional (2D) Nanostructures*

Two-dimensional nanostructures are a major component of coatings and thin films where nano size confinement is conducted. Coatings and thin films have been an important field of science and research for years, and developments in the deposition of thin films have met with considerable success. Film growth methods can generally be divided into two groups: vapor phase deposition and liquid-based growth. The former includes, for example, evaporation, molecular beam epitaxy (MBE), sputtering, chemical vapor deposition (CVD) and atomic layer deposition

(ALD). Examples of the latter are electrochemical deposition or electrodeposition, chemical solution deposition (CSD), Langmuir–Blodgett films and self-assembled monolayers (SAMs).

In all deposition processes, the formation of thin films involves heterogeneous nucleation and growth associated with a wide variety of processes, including heterogeneous chemical reactions, evaporation, and adsorption and desorption on growth surfaces. The early stages of deposition are accompanied by distribution of small but highly mobile atomic clusters or islands that form on the substrate surface. The change in the volume of Gibbs free energy defines the initial size of the nuclei. However, consequent growth determines whether the solid forms a crystalline lattice, and how shape and morphology change from one system to another. In the first stage, the effect of interface energies introduced by Young's (1.31) determines the morphology and structure of films grown by deposition. The next stage occurs when the growth continues until the nuclei start to coalesce, forming a continuous film on the substrate surface. This process is typically able to produce films from a few hundred angstroms up to microns in thickness.

Figure 1.22 illustrates the three principle modes of nucleation that occur during film growth, which obey the following basic mechanisms:

(1) Island or Volmer–Weber growth,
(2) Layer or Frank–van der Merwe growth, and
(3) Island-layer or Stranski–Krastanov growth.

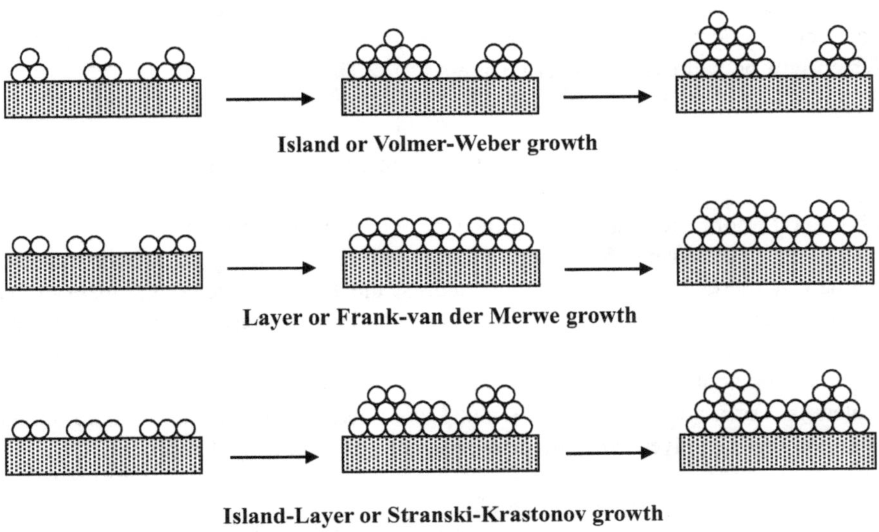

Fig. 1.22 Schematic representation of three principle nucleation modes for thin film growth

According to Young's equation, the mechanism of nucleation may be determined. When the growth species are likely to bond to each other instead of to the surface atoms, island growth takes place. For island growth, the wetting angle must be larger than zero, and therefore we write: $\gamma_{sv} < \gamma_{fs} + \gamma_{vf}$. If the deposit does not wet the substrate at all or the wetting angle equals 180°, the nucleation mechanism will become independent of the surface, and homogeneous nucleation will occur. In contrast, a layer growth mode is feasible where the deposit wets the substrate completely and the contact angle equals zero; thus Young's equation becomes: $\gamma_{sv} = \gamma_{fs} + \gamma_{vf}$. The layer growth mode becomes very important in the deposition of single-crystal films through either homoepitaxy or heteroepitaxy. Epitaxy refers to extended single-crystal film formation on top of a crystalline substrate. Two types of epitaxy can be distinguished, and each has important scientific and technological implications. Homoepitaxy refers to cases in which the film and substrate are the same material. Heteroepitaxy refers to films and substrates composed of different materials, and it is of course the more common phenomenon.

The third mode of nucleation is the layer-plus-island or Stranski–Krastanov (SK) growth mechanism, which is a consequence of an intermediate combination of the aforementioned modes. In this case, after one or more monolayers have been formed, subsequent layer growth becomes unfavorable, and islands form. The transition from two- to three-dimensional growth is not completely understood, but any factor that disturbs the monotonic reduction in binding energy characteristics of layer growth may be the cause. For example, film–substrate lattice mismatch causes strain energy to accumulate in the growing film. When released, the high energy at the deposit–intermediate layer interface may trigger island formation. This growth mode is fairly common and has been observed in metal–metal and metal–semiconductor systems. It is particularly important for the electrodeposition of films, which will be addressed in later chapters [7, 13, 14].

1.5.2.1 Quantum Wells or Nanofilms: Dimensionality Effect

Figure 1.23 compares the electron density of states for 3D bulk, 2D, 1D and 0D nanostructures. Starting from the density of electronic states for a bulk material defined by the (1.15)–(1.17), we now address the effect of size confinement in 2D nanostructures, so-called quantum wells or nanofilms. It must be stressed here that the 2D nanostructures fabricated by epitaxial growth can be considered in this calculation model. We start with (1.22) for the density of electron states for 3D bulk materials. For a quantum well of width l_z confined in z direction, the Schrödinger equation can be solved, and the answer which must be considered is:

$$\psi_n(x,y,z) = \left(\frac{2}{l_z}\right)^{\frac{1}{2}} \sin\left(\frac{n_z \pi z}{l_z}\right) \cdot e^{(ik_x.x)} \cdot e^{(ik_y.y)} \tag{1.45}$$

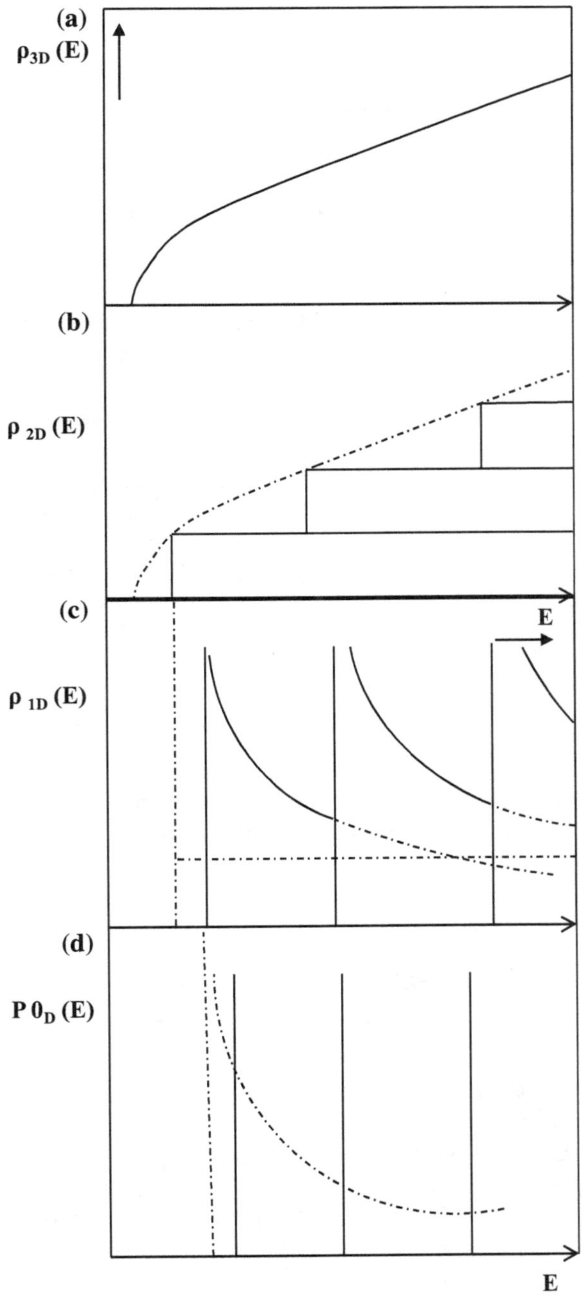

Fig. 1.23 Diagrams of the electron density of states for bulk solids and 2D, 1D and 0D nanostructures. From [23] with the permission of Springer

The kinetic energy of electrons is written as:

$$E_v(k) = \varepsilon_v + \frac{\hbar^2 k^2}{2m}, k^2 = k_x^2 + k_y^2 \tag{1.46}$$

With discrete energies, ε_v is written due to size confinement in the z direction as

$$\varepsilon_v = \frac{\hbar^2}{2m}\left(\frac{v\pi}{l_z}\right)^2, v = 1, 2, 3, \ldots \tag{1.47}$$

This quantum size effect works only if the electron mean free path is larger than the film thickness l_z. Otherwise, the quantum states will be broadened. ε_v is defined as the bottom of a 2D subband with respect to dispersion in k_x and k_y directions. According to the 2D quantum confinement applied here, we need to consider the total surface area of a 2D k^2 circle to obtain the number of quantum states which can be further filled with electrons. By adding the spin degeneracy and dividing by $l_x l_y$, we reach the number of quantum states in an individual 2D subband:

$$n_{2D}(E) = 2 \cdot \frac{l_x l_y}{4\pi^2}\left(\pi k^2\right) = \frac{k^2(E)}{2\pi} = \frac{m}{\pi\hbar^2} \cdot E \tag{1.48}$$

Thus the density of electron states in 2D subbands is written as:

$$\rho_{2D}(E) = \frac{m}{\pi\hbar^2} \cdot \sum\nolimits_v \Theta(E - E_v) \tag{1.49}$$

$\Theta(x)$ is a heavy side step function which determines several steps starting from each ε_v. Each time the energy reaches a new subband, the density of states jumps by $\frac{m}{\pi\hbar^2}$. The density of states for 2D subbands is shown in Fig. 1.23b [23].

1.5.3 *One-Dimensional Nanostructures (Quantum Wires or Tubes)*

One-dimensional nanostructures are categorized by a wide variety of names: whiskers, fibers or fibrils, nanowires and nanorods, nanotubes and nanocables. Among these, nanowires or quantum wires are frequently used, because size confinement exists in two dimensions, as shown in Fig. 1.23c. Nanowires are synthesized or fabricated via many techniques based on chemical or physical routes, depending on the nature of the materials and their application. Major routes include vapor–solid and vapor–liquid–solid (VLS), mechanical recrystallization, and template-based synthesis using electrodeposition and electrophoretic deposition, colloidal dispersion, melt or solution filling, conversion by chemical reaction, electrospinning and lithography. The application of electrodeposition in the fabrication of nanowires will be discussed in later chapters.

Spontaneous growth, template-based synthesis and electrospinning are considered bottom-up approaches, whereas lithography is a top-down technique. Spontaneous growth commonly results in the formation of single-crystal nanowires or nanorods along a preferential crystal growth direction, depending on the crystal structure and surface properties of the nanowire materials. Template-based synthesis produces mostly polycrystalline or even amorphous products [13, 14].

1.5.3.1 Quantum Wires: Dimensionality Effect

Here we need to apply the size confinement in two directions, y and z. This leads to the generation of a wave spectrum with two quantum numbers, each of which is specified for each confined direction. The Schrödinger equation gives this wave function for an electron wave confined in two directions y and z:

$$\psi_n(x, y, z) = \left(\frac{2}{l_z}\right) \sin\left(\frac{n_z \pi z}{l_z}\right) \cdot \sin\left(\frac{n_y \pi y}{l_y}\right) \cdot e^{(ik_x . x)} \quad (1.50)$$

and the kinetic energy of electrons is written as:

$$E_{\nu,\mu}(k) = \varepsilon_{\nu,\mu} + \frac{\hbar^2 k^2}{2m}, k^2 = k_x^2 \quad (1.51)$$

With discrete energies, ε_ν is written due to size confinement in the z-direction as:

$$\varepsilon_{\nu,\mu} = \frac{\hbar^2}{2m}\left(\frac{\nu\pi}{l_z}\right)^2 + \left(\frac{\mu\pi}{l_y}\right)^2, \nu, \mu = 1, 2, 3, \ldots \quad (1.52)$$

This energy disperses in only one direction, x, whereas it is confined in other directions. Therefore, subbands with dispersion energy in one direction are defined, and the number of quantum states for quantum wires is written as:

$$n_{1D}(E) = \frac{2}{\pi} \cdot (K(E) = \frac{2}{\pi} \cdot \left(\frac{2mE}{\hbar^2}\right)^{\frac{1}{2}} \quad (1.53)$$

Thus the density of electron states in 1D subbands is written:

$$\rho_{1D}(E) = \frac{1}{\pi}\left(\frac{2m}{\hbar^2}\right)^{\frac{1}{2}} \cdot \sum\nolimits_{\mu,\nu} (E - E_{\mu,\nu}) \cdot \Theta(E - E_{\mu,\nu}) \quad (1.54)$$

The subbands of the electron states will follow the plot shown in Fig. 1.23c, and $\varepsilon_{\nu,\mu}$ is the quantum step for each subband [23].

1.5.4 Zero-Dimensional Nanostructures

Categorized into nanoparticles and quantum dots, 0D nanostructures are one of the main groups of nanomaterials. Several techniques, including both top-down and bottom-up approaches, have been developed and applied for the synthesis of nanoparticles. Milling or attrition, repeated quenching and lithography are top-down approaches. Bottom-up approaches are the most popular methods for the synthesis of nanoparticles, and many techniques have been developed. These are basically divided into two major groups: thermodynamics-based and kinetics-based methods. The former includes the synthesis of nanoparticles via homogeneous and heterogeneous nucleation from liquid or vapor. The latter is based on the confinement of chemical reactions, nucleation and growth processes in a small space. Aerosol, micelle and other template-based methods are the most famous techniques.

The most important properties that nanoparticles must possess include small size in the range of nanometers, identical particle size (or monosized particles), narrow size distribution, identical shape or morphology, identical chemical composition and crystal structure, and individually dispersed or monodispersed particles [13, 14].

1.5.4.1 Quantum Dots or Nanoparticles: Dimensionality

Lastly, size confinement or dimensionality can be performed for 0D nanostructures, referred to here as quantum dots. In this circumstance, confinement exists in all three dimensions, and we have quantum dots with a completely discrete spectrum:

$$\varepsilon_{\nu,\mu,\lambda} = \frac{\hbar^2}{2m}\left[\left(\frac{\lambda\pi}{l_x}\right)^2 + \left(\frac{\mu\pi}{l_y}\right)^2 + \left(\frac{\nu\pi}{l_z}\right)^2\right], \nu, \mu, \lambda = 1, 2, 3, \ldots \quad (1.55)$$

The density of electron states in 0D subbands is written as:

$$\rho_{0D}(E) = 2 \cdot \sum\nolimits_{\nu,\mu,\lambda} \delta(E - E_{\mu,\nu}) \quad (1.56)$$

$\delta(x)$ is zero for $x \neq 0$ and is infinity for $x = 0$. The diagram for the density of electron states is shown in Fig. 1.23d. The distribution of electrons is completely discrete, recalling the electron arrangement in a single atom [23].

1.5.5 Bulk Nanostructured Materials

Bulk nanostructured materials are defined as bulk solids constructed with nanoscale or partially nano-scale building blocks. Let us first illustrate schematically in

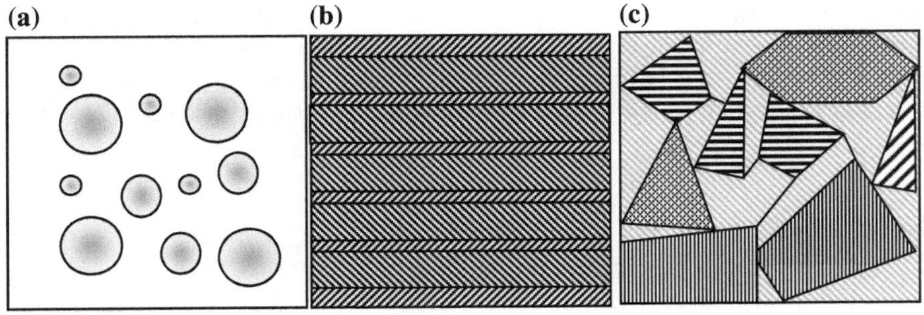

Fig. 1.24 Examples of bulk nanostructured materials: **a** nanocomposite, **b** superlattice, **c** nanocrystalline materials

Fig. 1.24 a number of major microstructures of bulk nanostructured materials. The left panel shows a composite material consisting of a metal or non-metal matrix filled with nanostructures. The nanostructures used as filling constituents can be any of the types introduced in the above sections. The panel in the middle of Fig. 1.24 shows a bulk nanostructured material with a superlattice microstructure consisting of repetitive nanolayers of different materials. The right panel shows nanocrystalline bulk materials. Other types of bulk nanostructured materials may exist that are not included in this figure.

Nanocomposite materials have become a major component of nanomaterials, where a matrix filled with a large variety of systems—including one-dimensional, two-dimensional, three-dimensional and amorphous nanostructures—is used to improve a particular property of the material. The matrix can be metal or non-metal, such as organic or ceramic materials. The properties of nanocomposites depend not only on the properties of their individual parents, but also on their morphology and interfacial characteristics.

Another type of bulk nanostructured material is nanocomposites exhibiting a superlattice—sometimes called multilayer or lamellar structures. In metal or oxide form, the superlattice forms by the repetition of the two metal or ceramic constituents. For non-metals like organic materials, lamellar composite intercalated and exfoliated structures can be obtained. If the polymer chains alternate with the inorganic layers in a fixed compositional ratio with a well-defined number of layers, an intercalated composite forms. However, if the number of polymer chains between the layers is almost continuously variable and the layers stand >100 Å apart, the structure is called an exfoliated composite.

Nanocrystalline materials constitute a major group of bulk nanostructured materials. The unique properties of nanocrystalline materials are derived from their large number of grain boundaries compared to their coarse-grained polycrystalline counterparts. In nanocrystalline solids, a large fraction of atoms (up to 49%) are boundary atoms. Thus the interface structure plays an important role in determining the physical and mechanical properties of nanocrystalline materials. The field of nanocrystalline (or nanostructured or nanophase) materials is a major area of activity in modern materials science. Some featured examples are enhanced mechanical properties of nanostructured materials for a variety of potential structural applications, and ferromagnetic materials with nanoscale microstructures for potential application as soft magnetic and permanent magnet materials [24, 25].

References

1. N. Bohr, Phil. Mag. **26**, 1 (1913)
2. E. Schrödinger, Ann. Phys. **79**, 361 (1926)
3. E. Wolf, *Nanophysics and Nanotechnology* (Wiley-VCH, Weinheim, 2004)
4. L. MiesslerGary, J. FischerPaul, A. TarrDonald, *Inorganic Chemistry*, 5th edn. (Pearson Pubs, USA, 2013)
5. C. Kittel, *Introduction to Solid State Physics*, 6th edn. (Wiley, New York, 1986)
6. G.I. Epifanov, *Solid state physics (translated from the Russian by Mark Samokhvalov)* (Publisher, Moscow, Mir 1979)
7. M. Ohring, *The Materials Science of Thin Films* (Academic Press, San Diego, California, 1992)
8. Li Changjian et al, Scientific Reports **5**, 13314 (2015)
9. R.L. de Kronig, W.G. Penny, Proc. Roy. Soc. Ser. A **130**, 499 (1931)
10. R.E. Reed-Hill, *Physical Metallurgy Principles* (Van Nostrand Publications, 1972)
11. R. Abbaschian, R.E. Reed-Hill, (Cengage Learning publications, 2008)
12. D.A. Porter, K.E. Easterling, M. Sherif, (CRC Press, 2009)
13. G. Cao, *Nanostructures and Nanomaterials Synthesis, Properties, and Applications*, 1st edn. (Imperial College Press, London, 2004)
14. G. Cao, Y. Wang, *Nanostructures and Nanomaterials Synthesis, Properties, and Applications*, 2nd edn. (World Scientific Publication, Singapore, 2011)
15. C.H.L. Goodman, *Crystal Growth: Theory and Techniques*, vol. 1 (Springer, 1974)
16. L. Bahrig, S.G. Hickey, A. Eychmüllera, Cryst. Eng. Comm. **16**, 9408–9424 (2014)
17. M. Volmer, *Die Kinetik der Phasenbildung* (Steinkopff, Dresden, 1939)
18. W. Burton, N. Cabrera, F.C. Frank, Phil. Trans. Roy. Soc. **243**, 299 (1951)
19. P. Hartman, W.G. Perdok, Acta Crystal. **8**, 49 (1955)
20. Allen J. Bard, Larry R. Faulkner, *Electrochemical methods: fundamentals and applications*, 2nd edn. (John Wiley & Sons, USA, 2001)
21. P.W. Atkins, *Physical Chemistry*, 5th edn. (Oxford University Press, Oxford, 1994)
22. M.R. Fitzsimmonsa, S.D. Bader, J.A. Borchers, J. Magn. Magn. Mater. **271**, 103–146 (2004)
23. U. Rossler, *Quantum Transport in Ultrasmall Devices, Editeded,* ed. by Ferry, D.K. et al., Springer US. Reprint permission copyright, 2016 with the permission of Springer
24. C. Kosh, *Nanostructure Science and Technology, R&D Status and Trends in Nanoparticles, Nanostructured Materials, and Nanodevices, Chapter 6* (Springer, Netherlands, 1999)
25. S.C. Tjong, H. Chen, Mater. Sci. Eng. R **45**, 1–88 (2004)

Chapter 2
An Overview to Electrochemistry

Abstract This chapter deals with the principles of electrochemistry in materials science. The electrochemical aspects of electrodeposition are central to control and development of this process. Three concepts are covered: equilibrium electrochemistry, dynamic electrochemistry and instrumental electrochemistry. Equilibrium electrochemistry is based on the thermodynamics of electrolyte and electrochemical cells working at equilibrium and no kinetical information is provided. Dynamic electrochemistry is based on kinetics in which the electrode/electrolyte interfacial phenomena, electrode processes, and rate of reactions are described. In instrumental electrochemistry, the electrochemical cells and techniques are classified.

2.1 Introduction to Electrochemistry in Materials Science

Electrochemistry of materials is a wide field for study which encompasses the electrochemical processes for fabrication, characterization, functionalization and modification of materials in any possible aspects. Electrochemistry generally deals with the relationship between the chemical and electrical phenomena. The energy of certain chemical reactions involving ionic species is readily converted to electromotive forces such as batteries or fuel cell. In contrast the application of electromotive force in a solution consisting of ions or in molten salts favors chemical reactions such as in electrorefining, electroforming or electrodeposition.

Electrochemistry is concerned with either physical or chemical processes when a solid electron conductor known as electrode is brought to contact with an ionic conductor known as electrolyte the set of which forms a half-cell. Provided that two half cells connect each other through the similar phases, an electrochemical cell will from in which an electrical charge involving the transfer of electrons and ions is passed.. This comprises a large number of individual and interfacial characteristics including, the solid state physics, the thermodynamics and kinetics of electrolyte, charge and electron transfer processes, and the interfacial electrode processes. We shall, however, restrict the discussion in this chapter to contacts with solids either metals or semiconductors which are or can become electronic conductors.

F. Nasirpouri, *Electrodeposition of Nanostructured Materials*,
Springer Series in Surface Sciences 62, DOI 10.1007/978-3-319-44920-3_2

We will emphasize here to develop the principles of electrodeposition from fundamental point of view to give understandable approaches for advanced concepts. Therefore, after an overview on electrochemistry, the basics of electrodeposition, the thermodynamics and kinetics of electrodeposition, the electrochemical as well as in situ techniques for evaluation of the process, and the effect of other factors will be discussed in this chapter.

2.2 Principles of Electrochemistry

Electrochemistry may be mainly categorized into two sections: equilibrium electrochemistry where it is concerned with the inherent properties of ions, formation of electrolytes, and their thermodynamics as well as the electrochemical cells at equilibrium and dynamic electrochemistry where the interface of electrode/electrolyte and charge transfer processes come in action for any electrochemical processes.

2.3 Equilibrium Electrochemistry

2.3.1 Ions: Formation, Thermodynamics and Interactions

Let's begin this section with the elementary topic on the formation of ions and electrolytes by introducing a subject titled the formation of mobile ions in solution as a base to electrochemistry. Depending on the nature of solvent and solute, we can distinct two ways by which mobile ions may form in a solution to create the ionic conducting phases. The first one applies to ion formation in a solvent where the solute is a neutral molecule for instance acetic acid, CH_3COOH. One proton of neutral acetic acid is transferred to neutral water molecule that lead to the formation of CH_3COO^- in polarised water molecules having a proton excess ion. A characteristic of solutions formed in this way is that usually the ionic concentration is a rather small fraction (e.g., about 0.1% of the solute molecules are ionised to given ions). The second which is the focus of our argument in electrodeposition is the method of forming mobile ions in solution by the dissolution of a solid lattice of ions. A general example of such lattice is the often-cited sodium chloride. Some attempt to show what happens in this type of ion formation is reproduced in Fig. 2.1a. The ions separated from each their arrangement in the crystal are surrounded by the solvent molecules (i.e. here water). This would require an energetically favorable state which occurs via a considerable energy change during interaction between the lattice ions and the solvent molecules.

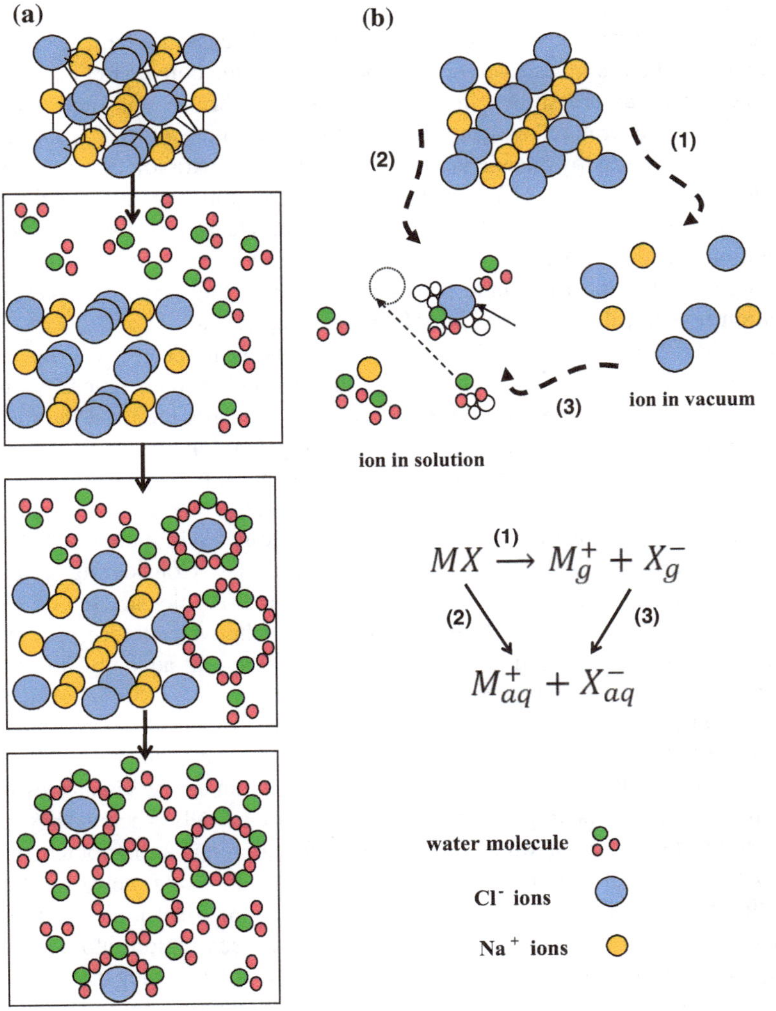

Fig. 2.1 Schematic representation of formation of ions from an ionic salt. **a** Step-by-step procedure of formation of chlorine and sodium ions in water (from *top* to *bottom*). **b** A flowchart of pathways of the formation of aqueous ions from an ionic salt

We now attempt to answer two major questions raised when a typical MX solid ionic salt is dissolved in water as: (i) what is the interaction between ions and solvent molecules, namely ion-solvent interaction and (ii) what is the interaction between the ions dissolved in water namely ion-ion interactions? It is worth noting that the electrolyte is electrically neutral where the anions and cations both exist. There could be two types of ionic species in electrolyte including free ions and pair ions including *free* ions and *pair* ions. These make the principles of the interactions between ions and solvents or each other as will be discussed later.

To answer the first question, we draw pathways of the conversion of a solid salt to aqueous ions as are shown in Fig. 2.1b. At equilibrium state and a constant pressure and temperature, we can define enthalpy change for each pathway. For pathway (1) the enthalpy change is associated with the separation of ions to infinite, so called cohesive energy, U. For pathway (2), dissolution takes place and thus, the dissolution enthalpy change ΔH_{sol} equals to the heat released or absorbed during the dissolution process of MX in a plenty of water where excess amount of water is present (the definition of infinite separation of ions). Pathway (3) represents the hydration process, the enthalpy change of which is ΔH_{hyd}.

The microscopic picture of ion-solvent interaction has been given by Bockris and Reddy in detail long time ago [1]. The interaction is accompanied by dividing the interface of ion-solvent into three distinct regions. Let's focus on the hydration of ions in water. In the region at the vicinity of a typical ion, or the primary region, the ionic field influence determines the water orientation. It is easier to view the ion as a point charge and the solvent molecules as the electric dipoles to make argument about the principal source of ion-solvent interactions. It is mentioned that ions significantly align with dipoles of the water molecules. Considering a spherical symmetry for the electric field around each ion, water dipoles are forced to point towards the central ion without being connected to the water lattice. These water molecules provide a sheath around each ion making it aqueous ion and are immobilized in respect with their ion.

Thus, the primary region next to the ions consists of the water molecules oriented out of the water structure and immobilized by the ionic field. The sheath moves as long as the ions surrounded move. In the immediate vicinity of the ion water molecules may be trapped and oriented in the ionic field whereas they are still connected to the lattice of water. This secondary region has a distributed structure of water. The ion still has some orienting influence on the water network: it tries to align the water dipoles parallel to the spherically symmetrical ionic field, and the water network tries to convert the water in the in-between region into a tetrahedral arrangement. Hence, depending on their distance from the ion, they orient out of the water network to varying degrees. In this intermediate region, the water structure is said to be partly broken down. Thus, they do not obey the translational motion of the ions. For the farthest distances from ions, the electrostatic interaction owing to the ionic fields becomes attenuated to virtually zero and therefore, the structure of water consisting of tetrahedrally bonded networks of molecules exists, namely *bulk water* exists. The interaction between ion and solvent defines an important concept namely *solvation*. These structural changes in the primary and secondary regions are generally referred to as solvation (or as hydration when, as is usual, water is the solvent).

Hess's law of constant heat summation states that the standard enthalpy change of a reaction is the sum of the standard enthalpies of contributing individual reactions whether the reaction is made in one step or in several steps. In our case:

$$\Delta H_{hyd} = \Delta H_{sol} - U \tag{2.1}$$

ΔH_{sol} is experimentally determined. U is also measurable using either calorimetric or theoretical methods. The enthalpies of dissolution and hydration achieved by this method correspond to the formation of entire ionic electrolyte or a typical chemical reaction:

$$\mathrm{MX} \leftrightarrow \mathrm{M}^+ + \mathrm{X}^- \tag{2.2}$$

However, in equilibrium electrochemistry, the thermodynamics may also predict the formation energy of individual ions M^+ or X^- Sin an electrolyte by some assumptions. The enthalpy, entropy and free Gibbs energy of formation of individual ions can be obtained for cations and anions formed in an electrolyte. Let's simply consider the enthalpy of formation of an ion in electrolyte. For a typical dissolution reaction, (2.2), the standard enthalpy is written:

$$\Delta H_f(\mathrm{MX}) = \Delta \mathrm{H_f}(\mathrm{M}^+) + \Delta \mathrm{H_f}(\mathrm{X}^-) \tag{2.3}$$

The enthalpy of an overall reaction is meaningful, but the enthalpies of the individual formation of $\Delta H_f(M^+)$ and $\Delta H_f(X^-)$ are not. The problem is solved by defining an ion to have enthalpy of formation equal to zero, the hydrogen ion:

$$\Delta H_f(H^+) = 0 \text{ at all tempreatures} \tag{2.4}$$

Then in the reaction 2.1, one contributes an auxiliary reaction which constitutes hydrogen with zero enthalpy. Thus, the relative enthalpy of each individual ion can be obtained. By analogy, the entropy and free Gibbs energy of formation of all individual ions can be readily calculated.

The activities of ions *(a)* can be also discussed here for ionic electrolytes. According to the chemical potential of a solute one can write:

$$\mu = \mu^0 + RT\, ln\,(a) \tag{2.5}$$

For M_pX_q electrolyte, the Gibbs function is:

$$G = p\mu_+ + q\mu_- = G^0 + RT \cdot ln\, \gamma_\pm \tag{2.6}$$

$\gamma_\pm$ is the mean activity coefficient for each ion which represents the non-ideality of electrolyte solution and is defined: $= \left(\gamma_+^p \cdot \gamma_-^q\right)^{\frac{1}{p+q}}$.

The long range and strength of the columbic interaction between ions is likely the primary reason for any deviation from ideality in ionic solutions. This is the basis of the *Debye-Hückel* theory [2] of ionic solutions. This theory gives qualitative and quantitative descriptions of the ion-solvent interaction for free ions. The theory pictures that a central ion is electrically balanced by ions of opposite charge in a distance called ionic atmosphere. This argument is supported by this assumption that the energy and therefore the chemical potential of any given central ion are lowered as a consequence of its columbic interaction with its ionic atmosphere. This energy reduction is assumed the difference between the Gibbs functions of real (G) and ideal (G^0) electrolyte which can be identified as: $T.\ ln\,\gamma_{\pm}$.

The theory predicts the activity coefficient at very low concentrations as written:

$$\log(\gamma_{\pm}) = -|z_+ z_-|AI^{\frac{1}{2}} \tag{2.7}$$

in which $A = 0.509\ (\text{mol kg}^{-1})^{1/2}$ for an aqueous solution at 25 °C and I is the ionic strength defined as: $I = \frac{1}{2}\sum_i z_i^2 \cdot m_i$. In this relation, z_i and m_i are the charge number and molality of ion, i.

Relation (2.7) so called the limited Debye-Hückel equation is only able to predict the activity coefficient of very dilute ionic solutions with molalities less than 0.01–0.001 mol/kg^{-1}. The departures from the theoretical model above these molalities are large. For more concentrated ionic solutions, instead, the activity coefficient may be estimated form *extended Debye-Hückel law*:

$$\log(\gamma_{\pm}) = -\frac{|z_+ z_-|AI^{\frac{1}{2}}}{1 + BI^{\frac{1}{2}}} \tag{2.8}$$

where B us another constant as a measure for ion distribution [3].

Following the classic Debye-Hückel theory of dilute electrolytes explained above, Bjerrum [4, 5] suggested a concept of ion pairing as a means of extending its applicability to higher concentrations. This regards the answer to the second question on the ion-ion interaction. We must consider the formation of pair ions in ionic solutions. Bjerrum argued that it is only short-range Columbic interactions that lead to ion-pair formation and, further, when a pair of oppositely charged ions are situated at a certain distance apart, it is more appropriate to consider them free ions. This distance, q, is:

$$q = -\frac{(z_+ e)^2 (z_- .e)^2}{2\varepsilon kT} \tag{2.9}$$

Bjerrum concluded therefore that ion-pair formation occurs when an ion of a particular type of charge (e.g., a negative ion) enters a sphere of radius q drawn around a reference ion of the opposite charge (e.g., a positive ion). However, it is the ion size parameter that defines the distance of closest approach of a pair of ions. The Bjerrum hypothesis can therefore be stated as follows: If a < q, then ion-pair formation can occur; if a > q, the ions remain free [1].

2.3.2 Electrochemical Cells

The basic part of apparatus in electrochemistry is an electrochemical cell. The cell consists of two electrodes (or a conductor of electrons) dipping into an electrolyte (a conductor of ions). This simple combined electrode/electrode system is called half-cell. In an electrochemical system in which an electrochemical reaction takes place, the electrons produced or depleted in an electronically conducting phase, namely electrode, need to be electrically neutralized at the interface. This involves a region of the solvent having an ion as the bearer of empty electronic states which can exchange the exiting electrons. This is electrochemical reduction and the corresponding half cell is called cathode. Conversely, the filled electronic states of these ions are the origin of the electrons that enter the metal in the reverse electron-transfer reactions involved in electrochemical oxidation reactions, called anode. An electrochemical cell that generates electricity as a consequence of a spontaneous reaction is called *galvanic* cell. Provided that a non-spontaneous reaction is driven by supplying current from an external source, the electrochemical cell comprising this reaction then is called *electrolytic* cell. Obviously, for electrodepositing materials, an electrolytic cell is required.

In general as the reaction proceeds, the electrons will be released in the anodic half-cell:

$$Red_1 \rightarrow Ox_1 + ne^-$$

and travel through the external circuit and will be consumed in the cathodic half-cell:

$$Ox_2 + ne^- \rightarrow Red_2$$

For each electrochemical event, a *redox* couple must cooperate to enable the electrochemical process to precede, i.e. Red_1 / Ox_1. The cell reaction is finally attained by the sum of the two half-reactions. In literature, in order to facilitate the concept, all half reactions either anodic or cathodic are written in the reduction reaction. In this case, the overall cell reaction is achieved from cathodic half-reaction subtracted by the anodic half-reaction:

$$Red_1 + Ox_2 \rightarrow Red_2 + Ox_1 \tag{2.10}$$

The cell potential is the difference between anodic and cathodic electrode potentials. Therefore, one can deduce the relationship between the cell potential (E) and the change of free Gibbs energy of the cell reaction:

$$\Delta G_r = -nFE \tag{2.11}$$

F is the Farady's constant (96,485 KC mol^{-1}), and n is the number of electrons. The *Nernst equation* is written for (2.10) as:

$$E = E^0 - \frac{RT}{nF} \cdot \ln(Q) \tag{2.12}$$

Q is the quota of the cell reaction defined as: $\frac{[Red_1].[Ox_2]}{[Red_2].[Ox_1]}$. E^0 is the standard cell potential. At equilibrium the cell produces the maximum electrical work and the quota equals the equilibrium constant of the cell reaction.

2.4 Ion Transport

Let's talk a little about the transportation of ions in electrolytes during occurring of an electrochemical reaction. Three types of transportation mechanisms exist for the movement of ions in electrolytes: *Migration, Diffusion and convection.*

2.4.1 *Migration*

The principle mechanism of ion transport is migration defined as the movement of positive or negative ions under the influence of an applied potential between electrodes in an electrolyte [6]. A cation moves towards the negative electrode, cathode, and an anion moves towards the positive electrode, or the anode. The motion of ions in electrolyte may be readily determined by measuring the electrical resistivity or conductivity. The conductivity of a solution, K, as a function of its molar concentration, c, namely molar conductivity Λ_m is a well developed measure and is defined:

$$\Lambda_m = \frac{K}{c} \tag{2.13}$$

The molar conductivity is generally expressed in S cm^2 mol^{-1}. Provided that K is exactly proportional to electrolyte concentration, it will be independent of concentration. However, this does not happen in practice likely owing to the interactions of ions with solvent and with each other as discussed earlier. One consequence of such interactions is that the number of ions in an electrolyte is not necessarily identical to the concentration of the electrolyte. A good example for this might be weak and strong electrolytes. In strong electrolytes, the molar conductivity decreases only slightly as its concentration is increased, whereas the molar conductivity of weak electrolytes drops sharply from normal values at approximately zero concentration to low values as their concentration increases.

For strong electrolytes, the Kohlrausch's law defines the relationship between molar conductivity and concentration:

$$\Lambda_m = \Lambda_m^o + \chi c^{1/2} \tag{2.14}$$

$\Lambda_{\mathrm{m}}^{\mathrm{o}}$ is the limiting molar conductivity at the boundary of zero concentrations and χ is a coefficient depending on the stoichiometry of electrolytes. $\Lambda_{\mathrm{m}}^{\mathrm{o}}$ can be a measure of conductivity of electrolytes on the basis of ionic molar conductivities of individual ions (λ_+ for cations and λ_- for anions) in electrolyte:

$$\Lambda_m^o = \nu_+\lambda_+ + \nu_-\lambda_- \tag{2.15}$$

v_+ and v_- are the number of cations and anions per formula unit of electrolytes.

The microscopic picture of variation of conductivities of electrolytes as a function their concentration is discussed here. Recall the Debye-Hückel theory (2.7 and 2.8). Here there is the same behavior for the variation of activity coefficients versus concentration owing to the formation of the ionic atmosphere around each ion. For ion transportation under the effect of an electrical field, the picture of ionic atmosphere must be reconsidered as shown in Fig. 2.2a. The shape of the ionic atmosphere in the absence of electrical field is spherical. Upon the application of an electrical field, all ions drift towards the opposite direction of the applied field. In the initial stages of transportation, the ions forming the ionic atmosphere do not immediately follow the movement of the central ion, and therefore, shape of the atmosphere is distorted. For the example picture shown in Fig. 2.2a, the atmosphere remains behind the moving central ion which retards the movement of central ion. This reduction occurred in the mobility of ion is called the *relaxation effect*. The ionic atmosphere also influences the motion of ions in another way. This effect is called the *electrophoretic effect* which reduces the mobility of ions because of the drag force caused by the electrolyte. This will be addressed later. The quantitative explanation of these effects is quite complicated, but an attempt to give a relationship has been made by *Debye-Hückel-Onsager* theory [3].

For weak electrolytes, the degree of ionization α of the electrolyte influences the molar conductivity. For dilute weak electrolytes, one can write:

$$\Lambda_m = \alpha\Lambda_m^o \tag{2.16}$$

and Ostwald's dilution law suggests the relationship between molar conductivity and concentration of a weak electrolyte by knowing the ionization equilibrium constant, K_{eq}:

$$\frac{1}{\Lambda_m} = \frac{1}{\Lambda_{\mathrm{m}}^{\mathrm{o}}} + \frac{\mathrm{c}\Lambda_m}{K_{eq} \cdot \Lambda_{\mathrm{m}}^{\mathrm{o2}}} \tag{2.17}$$

We now attempt to describe the difference between conductivities of various ions and electrolytes using ion mobility. To define this concept, first we need to consider the migration of ions in an electrolyte between two electrodes with a potential difference. Under the effect of the electrical field with strength E, ions

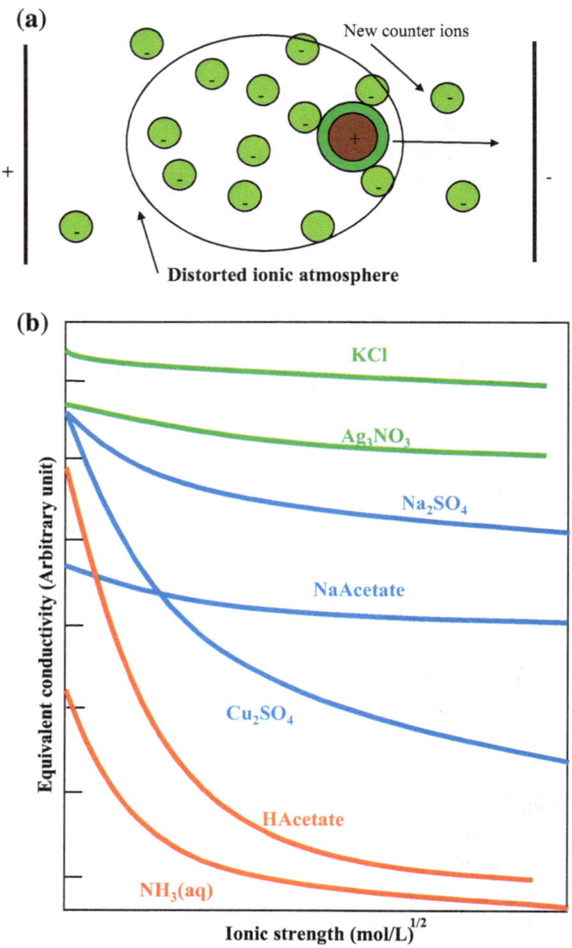

Fig. 2.2 **a** The change of shape of spherical ionic atmosphere in the presence of applied electrical field. **b** The dependence of molar conductivities v square root of the ionic strength for strong (*green*), intermediate (*blue*) and weak (*red*) electrolytes

moves towards the oppositely charged electrode. Conversely, the movement is affected by a frictional force the coefficient of which may be defined by the Stokes law. Provided that these two opposite forces are balanced at a point, the ions move in an electrolyte with viscosity, η, under a constant speed, namely *drift speed, s,* defined in general for an ion with hydrodynamic radius of a and electrical charge of n electrons:

$$s = \frac{neE}{6\pi\eta a} \tag{2.18}$$

The drift speed in turn is proportional to the strength of the electrical applied field:

$$s = uE \tag{2.19}$$

where u is called mobility of the ion and has dimension $m^2\ s^{-1}\ V^{-1}$. Mobility is useful to interpret the magnitude of conductivities of electrolytes and ions. The conductivity, K, is written:

$$K = nuvcF \tag{2.20}$$

and, therefore the molar conductivity of ions is:

$$\lambda = nuF \tag{2.21}$$

In the limit of zero concentrations, the molar conductivity of electrolyte can be obtained:

$$\Lambda_m^o = n_+ v_+ u_+ + n_- v_- u_- \tag{2.22}$$

We now reach a point where it is essential to distinguish the portion of current that is carried by cations and anions in an electrolyte. The transport number, t, is defined as the fraction of total current carried by each type of ions. At the zero limit of concentration, the limited transport numbers, t^o, is defined as:

$$v\lambda = t^o \Lambda_m^o \tag{2.23}$$

which can be used to determine the individual ionic conductivities, and also ion motilities.

2.4.2 *Diffusion of Ions*

In Chap. 1, we discussed the diffusion in materials and introduced Fick's laws which must be here combined with other characteristics of ions like the mobility to understand the conductivity of electrolytes. The *Einstein relation* defines the relationship between diffusion coefficient and the ionic mobility:

$$D = \frac{uRT}{nF} \tag{2.24}$$

The Einstein relation can be further extended to develop the other relationships between diffusion coefficient and the molar conductivity, and frictional force due to the drag force of electrolyte.

The *Nernst-Einstein* equation (2.25) determines the relationship between diffusion coefficient of ions with their molar conductivities:

$$\lambda = nuF = \frac{n^2DF^2}{RT}$$

and the limiting molar conductivity can be then written:

$$\Lambda_m^o = \frac{F^2}{RT}(n_+^2 v_+ D_+ + n_-^2 v_- D_-) \quad (2.25)$$

This is quite useful to apply this equation to find ionic diffusion coefficients from conductivity measurements.

Furthermore, the *Stokes-Einstein* equation can be derived to define the relationship between the diffusion coefficient and the viscosity of solution:

$$D = \frac{K_B T}{6\pi\eta a} \quad (2.26)$$

Diffusion equation mentioned earlier in Chap. 1 may also include convection defined as the movement of particles due to the motion of a streaming fluid flowing at a velocity, *V*. The generalized diffusion equation to determining the change of concentration caused by both diffusion and convection is:

$$\frac{\partial C}{\partial t} = D\frac{\partial^2 C}{\partial x^2} - V\frac{\partial c}{\partial x} \quad (2.27)$$

2.5 Dynamic Electrochemistry: Processes at Electrodes

So far we have discussed the properties of ions, electrolytes and electrochemical cells at the equilibrium where there was no current passing through the electrochemical cell in all the sections. Provided that the electrochemical cell is completed by combing two half-cells, a current starts to pass through the cell. This is accomplished by processes occurring at the electrode/electrolyte interface, called electrode processes. The electrode processes are central to dynamic electrochemistry, where mass transport and electron transfer reactions control the rate of any electrochemical reactions. A measure of this rate is the current density as defined by the electrical current per surface area, or the charge flux. When the cell works at equilibrium, the net current is zero and the cell potential is called *zero-current* cell potential. When the cell potential is changed by means of applying potential, then the half-cell anodic or cathodic half-reactions occur. The change of potential from zero-current potential is called *overpotential*. In this section, we emphasize on a working electrochemical cell and derive the equations for the relationship between potential (or overpotential) and current density.

2.5.1 The Electrode Double Layer

A potential difference exists at the interface of any pair of conducting phases. We simply consider solid electrode/solution electrolyte here to elucidate the interfacial phenomena. Figure 2.3a shows a simple metal/electrolyte interface with oppositely

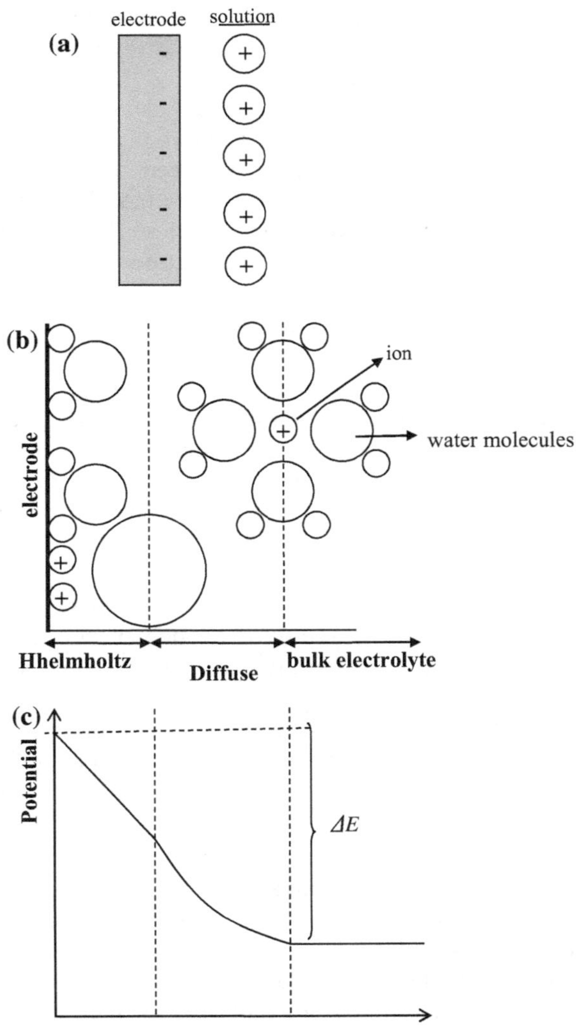

Fig. 2.3 **a** A simple metal/electrolyte interface, **b** A schematic of molecular or atomic level structure of the double layer on metal/electrolyte interface, **c** Variation of potential across the double layer (Stern model)

charged ions attracted on the metal surface. The potential difference of the electrode/electrolyte interface is generated by an electrical *double layer* through which the charge difference between the two phases is balanced.

We described earlier in Chap. 1 the electronic structure of a metal, where metal ions (atom nucleolus and localized electrons) at crystal lattice sites are surrounded by the conduction electron gas. The band structure of a metal ends at a semi-filled band up to the Fermi energy level. When a metal come to contact a solution, transfer of electrons to and from this energy level is reasonable for conducting any electrode processes which will be discussed later. For now, the electrostatic columbic interaction of surface and surrounding ions in electrolyte accumulates the ions on vicinity of electrode surface. This is in the first instant to compensate the surface charge (e.g. negative charges at a metal surface) forming an interfacial region in planar form parallel to electrode surface. This is a general interfacial phenomenon when an electrode is placed into an electrolyte. However, it becomes more important in electrode kinetics.

A detailed picture of the double layer forming on a metal electrode surface dipping into an electrolyte is shown in Fig. 2.3b. The structure of the double layer is speculated to have details introduced by the most common models namely the *Helmholtz* model, the *Gouy-Chapman* model and the *Stern* model.

The *Helmholtz* model of the double layer implies that a compact layer consisting of ions adhere to the surface of electrode via two different approaches. First, *the specific adsorption* of ions occurs through van der Waals and chemical forces on electrode surface. Most ions may be specifically adsorbed, thereby releasing a large fraction of their hydration sheath. This region is located at the vicinity of electrode, the boundary of which is called inner Helmholtz plane (IHP). Next to the IHP, the solvated (or hydrated in aqueous solutions) ions accumulate on the closest possible distance due to the electrostatic interaction. This happens for electroneutrality until the excess electrical charge residing on the electrode is balanced by an equal number of oppositely charged ions. The separation distance is only the water molecules of the hydration spheres encapsulating the ions. This makes a separate region which has a boundary with the rest of the electrolyte as shown by a line (or plane) parallel to the metal surface called outer Hemholtz layer (OHP). The Helmholtz model ignores the disrupting effect of the thermal energy on motion of ions and other species in the system which might damage rigid planar structure of the planes. Furthermore, the size of hydrated ions locating in the OHP may not allow the accumulation of a sufficient number of ions with opposite sign of charge to neutralize the surface electrical charge. Figure 2.3c indicates the variation of potential at the electrode/electrolyte interface. On the basis of the Helmholtz model, the potential variation across the double layer is linear.

In contrast, *Gouy and Chapman* proposed a model based on a diffuse electrical double layer by taking into account the effect of thermal energy. The distribution of ions around the electrode surface was considered like *a giant ion* with one-dimensional distribution almost similar to three dimensional ionic atmosphere proposed by the Debye-Hückel theory. The ions with opposite sign of electrical charge are adhered to the surface with high concentration which subsequently decays as the distance from the electrode surface increases. In this approach the potential variation follow a non-linear function (Fig. 2.3c). The model is not the best approach to elucidate the double layer structure due to lack of understanding of ion distribution at very vicinity of electrode surface. It has been shown that this model has some deficiencies such as neglecting of ion-ion interactions, the assumption of point-charge ions and considering a constant value for the dielectric constant in the region between the electrode and the bulk of the solution [1].

Thanks to advances in the electrochemical experimental studies, it has been clarified that in reality a combination of the Helmholtz and Gouy-Chapman models is closer to actual double layer structure. Stern suggested [7] that the Helmholtz layer of ions is located in contact with the electrode and next to it, the diffuse layer of the Gouy-Chapman theory exist until the variation of potential fall into the bulk electrolyte. Figure 2.3c depicts the variation of potential (ϕ) across the double layer regions consisting of linear and non-linear regions. This model considers the effect of specific adsorption which was further discussed in details by Graham [8].

2.5.2 *Zeta Potential* (ζ)

Considering that the diffuse layer continues towards the bulk electrolyte, the non-linearity of this approach is:

$$\phi = \frac{\sigma}{\varepsilon_o \varepsilon_k} \cdot e^{\chi(a-x)} \tag{2.28}$$

where x is the distance from electrode surface, and a is the distance of closest accommodation of ions from surface, or the position of the OHP and σ is the charge density at the electrode surface defined as the charge per unit area at $x = 0$. χ is the constant defined in similar manner as what proposed by Debye-Huckel theory (i.e. $1/\chi$ was the radius of ion atmosphere). Here we consider it as the thickness of the diffuse layer, δ.

At the position of OHP, x approaches a, then: $a - x \rightarrow 0$. Under these conditions, we write:

$$\phi = \phi_0 = \frac{\sigma}{\varepsilon_o \varepsilon_k} \tag{2.29}$$

Provided that the potential difference across the interface is uniform, the OHP is called the Stern plane. Under stagnant conditions, when the electrode/electrolyte interfacial region is completely kinetically immobilized, ϕ_0 can be identified as zeta potential, ζ. However, we must consider this is not the case for all dynamic electrochemical situations. In particular, for colloidal dispersion of particles in electrolyte, the interface of solid/electrolyte will be readily transferred by any means. If motion happens the position of the Stern plane will change. So a slipping or shear plane distinguishes the interface at which the diffuse and Helmholtz layers interfere.

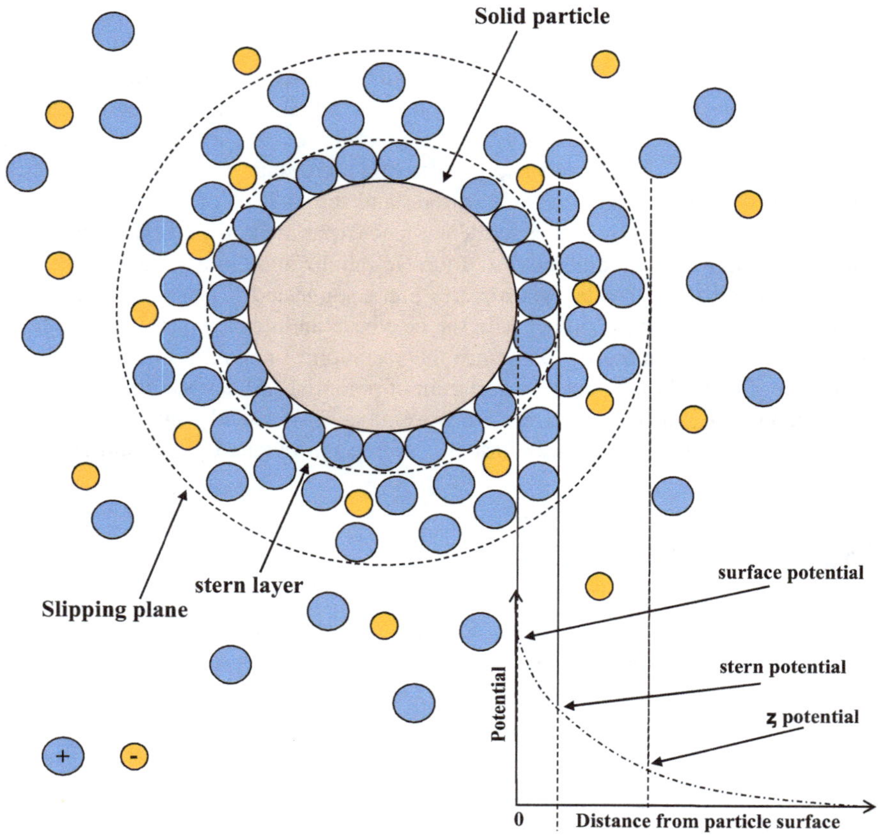

Fig. 2.4 A comparison of the position of slipping plane and OHP for determination of the Zeta potential on a typical *solid surface* dipping in an electrolyte

The slipping plane separate any species like ions or molecules intimately adhered to the solid surface and other species moving relative to them. Figure 2.4 demonstrates the position of the slipping plane relative to the OHP in a double layer of metal particles dipped in an electrolyte. The position of this plane indicates the zeta potential relative to the bulk fluid exactly defining the potential difference between electrode and the stationary layer of fluid around it.

Under the simple electrode/electrolyte conditions, the double layer consisting of the Helmholtz and the diffuse layer possesses a capacity. Considering, C_D as the capacity of diffuse layer, we write:

$$C_D = \frac{\sigma}{\zeta} = \varepsilon_o \varepsilon_k \tag{2.30}$$

And therefore,

$$\zeta = \frac{\sigma\delta}{\varepsilon_o \varepsilon_k} \tag{2.31}$$

For total capacity, *C*, of the double layer which makes the total potential drop across the solid/electrolyte interface, we obtain that there are two components in the sum including diffuse layer capacitance and the Helmholtz layer capacitance:

$$\frac{1}{C} = \frac{1}{C_D} + \frac{1}{C_H} \tag{2.32}$$

In a given electrode/electrolyte interface, the controlling part of the double layer differs. For very dilute electrolytes, the capacitance obeys these magnitudes: $C_D \ll C_H$ and therefore $C \sim C_D$. Hence, the double layer will be solely diffuse as predicted by Gouy-Chapman. In contrast, For very concentrated electrolytes, we can achieve a situation having capacitance of: $C_D \gg C_H$ and $C \sim C_H$ and the Helmholtz layer is the determining criterion of the double layer [8].

2.5.3 Electrode Potentials

The most achievement of the double layer model is its ability to define the electric potential difference between an electrode and its surrounding media like an electrolyte. At the vicinity of the electrolyte surface, where the strength of electrical field $\left(\frac{dV}{dx}\right)$ (*V* is the voltage and x the distance) remains constant, the potential is called *Volta* or the outer potential. After a sudden increase of a potential magnitude namely surface potential, i.e. is a characteristic of metal or solid surface, it reaches a value on the electrode which is called the *Gavlani* potential. The potential difference between the electrode and the bulk electrolyte is the Galvani potential difference, $\Delta\phi$ which is eventually considered as the electrochemical reduction potential of an electrode/electrolyte or a half-cell, *E, known the electrode potential* as well.

For a general electrode reaction such as the electro-oxidation/reduction of a metal:

$$M \rightleftarrows M^{n+} + ne^- \tag{2.33}$$

the electrode potential at equilibrium may be E^0 or zero-current potential. If a potential, *E* rather more reducing or oxidizing than the equilibrium value is applied, there will be a net non-zero current flowing through the electrode. This is called the overpotential, η defined as:

$$\eta = E - E^0 \tag{2.34}$$

The equal and opposite current values for each side of (2.33) implies zero net current. The value of each side is called *exchange current*, I^0 which will be discussed later.

The relationship between the overpotential and current passed through the cell is used as the basis of the experimental examinations of the electrode processes. Figure 2.5 shows a typical current-potential curve recorded for an electrode/electrolyte system. Two types of electrode systems can be classified according to the response to an imposed potential including polarisable and non-polarisable electrodes. This is complied with the reversibility and irreversibility of an electrochemical reaction. Figure 2.5 demonstrates a typical current-potential curve for polarisable electrodes. This is accompanied with an irreversible potential because the rate of attainment of the equilibrium condition is quite slow as indicated on the curve by the application of potential and the current response flowing into electrode. This effect is exhibited by two overpotential magnitudes for anodic (η_a) and cathodic (η_c) parts of the curves. The current value for any overpotential imposed will deviate from the exchange current, $I^{0.}$ Provided that the current response is quite fast, the electrochemical reaction is reversible, and hence the overpotential will become null. A slight slope on the linear parts of the curve corresponds to the mass transfer processes.

For electron transfer processes, it is worth noting to point the electronic levels of electrode and ions. If the electronic levels of electrode and ions become equal the electron transfer occurs. Therefore, any contributions from physical and chemical properties will influence this rate such as molecular arrangement at double layer and its rearrangement under applied potential, and electrical field gradient at the double layer [8].

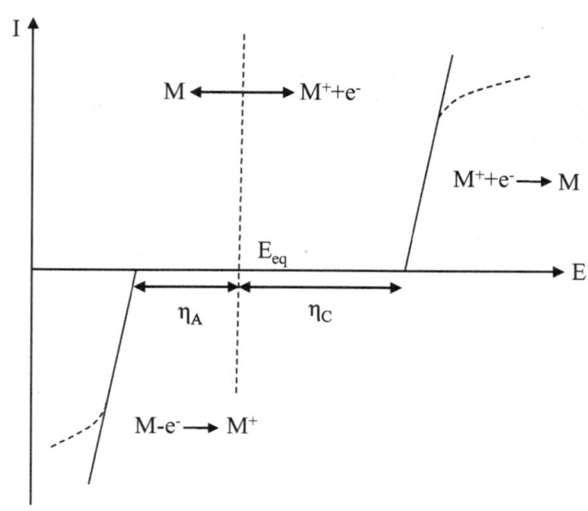

Fig. 2.5 A typical current-potential curve recorded for an electrode/electrolyte system (re-plotted from Crow)

2.5.4 *The Rate of Charge Transfer: Electrode Kinetics and the Butler-Volmer Equation*

The electrode reaction is heterogeneous in nature the rate of which can be expressed as the amount of material produced per unit area of electrode per unit time, i.e. the product flux. For the (2.33), we define the rate of forward, v_a and reverse (v_c) reactions in accordance with the kinetics section in Chap. 1 as:

$$v_a = k_a \cdot \mathrm{e}^{-\frac{\Delta G_a^{\ddagger}}{RT}} \tag{2.35}$$

and

$$v_c = k_c \cdot [\mathrm{M}^+] \cdot \mathrm{e}^{-\frac{\Delta G_c^{\ddagger}}{RT}} \tag{2.36}$$

where k_a and k_c are the rate constant for anodic and cathodic reactions respectively, $[\mathrm{M}^+]$ the metal ion concentration on the electrode surface.

At equilibrium condition, we can write: $v_c = v_a$, thus:

$$k_a \cdot e^{-\frac{\Delta G_a^{\ddagger}}{RT}} = k_c \cdot [M^+] \cdot e^{-\frac{\Delta G_c^{\ddagger}}{RT}} \tag{2.37}$$

For the overall electrochemical reaction, the free Gibbs energy change is: $\Delta G = \Delta G_c - \Delta G_a$. Considering the Nernst equation and on the basis on (2.36) and (2.37), one can obtain:

$$\exp\left[-\frac{\Delta G}{RT}\right] = \frac{k_a}{k_c}.[M^+] = \exp\left[nFERT\right] \tag{2.38}$$

Hence, we can derive the Nernst equation from a kinetics point of view for an electrochemical reaction at the equilibrium state:

$$E = E^0 + \frac{RT}{nF} \ln\left[M^{n+}\right]$$

If an overpotential, η, is imposed to the electrochemical reaction, it deviates from the equilibrium state by amount of the free Gibbs energy change towards anodic and cathodic half-reactions. We consider the reduction reaction for which the free Gibbs energy for both sides of reaction will be changed by a factor of α called the *transfer coefficient:*

$$\Delta G_c^{\ddagger} = \Delta G_c^{\ddagger 0} + \alpha n F \eta \tag{2.39}$$

and

$$\Delta G_a^{\ddagger} = \Delta G_a^{\ddagger 0} - (1 - \alpha)nF\eta \quad (2.40)$$

The rates of anodic and cathodic reactions (2.35) and (2.36) must be re-written according to modified free Gibbs energy functions as:

$$v_a' = k_a \cdot e^{-\frac{\Delta G_a^{\ddagger 0} - (1-\alpha)nF\eta}{RT}} \quad (2.41)$$

and

$$v_c' = k_c \cdot [M^+] \cdot e^{-\frac{\Delta G_c^{\ddagger 0} + \alpha nF\eta}{RT}} \quad (2.42)$$

We now need to use the current density as an approximation of current density for the reaction rates. Since any electrochemical reactions involving a *redox* process at the electrode surface take place by the transfer of one electron per any reaction event, thus the current density, i, produced or consumed by a redox reaction can be equivalent to the reaction rates. We, therefore, write the current densities for anodic and cathodic reactions:

$$v_c' = \frac{i_c}{nF} \quad (2.43)$$

and

$$v_a' = \frac{i_a}{nF} \quad (2.44)$$

Therefore,

$$i_a = nFk_a . e^{-\frac{\Delta G_a^{\ddagger 0}}{RT}} e^{\frac{(1-\alpha)nF\eta}{RT}} = i_0 \, e^{\frac{(1-\alpha)nF\eta}{RT}}$$

and

$$i_c = nFk_c \cdot [M^+] \cdot e^{-\frac{\Delta G_c^{\ddagger 0}}{RT}} e^{-\frac{\alpha nF\eta}{RT}} = i_0 \, e^{-\frac{\alpha nF\eta}{RT}}$$

At equilibrium, the current density of anodic and cathodic reactions will be identical and equal to i_0. Considering an overall electrochemical reaction, the net current density of which is $= i_c - i_a$. Hence:

$$i = i_0 \left\{ e^{-\frac{\alpha nF\eta}{RT}} - e^{\frac{(1-\alpha)nF\eta}{RT}} \right\} \quad (2.45)$$

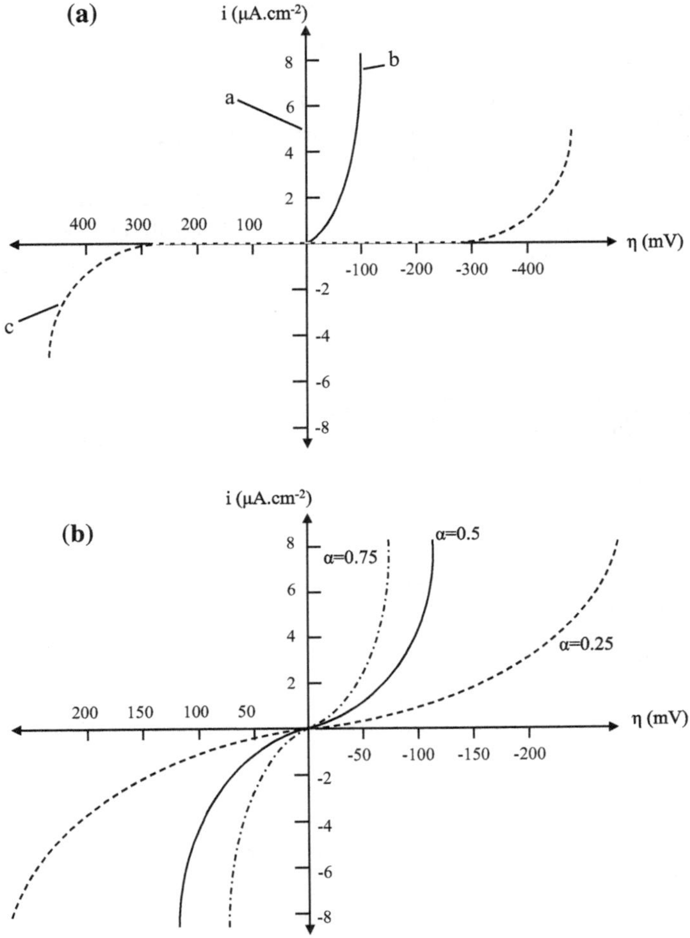

Fig. 2.6 Typical Butler-Volmer plots driven for a one electron electrochemical reaction under **a** different exchange current densities and **b** different transfer coefficients

This is known as the *Bulter-Volmer* equation as the basis of the kinetics examination of electrochemical processes. As an example, assume a general electrochemical reaction with a redox couple. Figure 2.6 plots the current-potential variation for such a system under different exchange current densities and for different transfer coefficients. The transfer coefficient range is between 0 and 1 and as it changes the position of the curve varies.

Two overpotential limits may be applied to the Bulter-Volmer equation. At small overpotentials, the exponential terms of (2.45) are expanded and the other terms are neglected. Thus, the equation can be written:

$$i = \frac{i_0 nF\eta}{RT} \tag{2.46}$$

For higher overpotential limit, which is the most important for applications of the Bultler-Volmer equation in electrochemistry, the cathodic and anodic overpotential must be individually considered.

At larger cathodic overpotentials, the first exponential term and at larger anodic overpotentials the second exponential term is significant. For both anodic and cathodic limits, we reach an identical form of current-potential equation called the *Tafel* equation:

$$\eta = a + b \log i \tag{2.47}$$

a is the and b the Tafel slope. Magnitude of the cathodic and anodic Tafel slopes are $\frac{2.3\mathrm{RT}}{\alpha\mathrm{nF}}$ and $\frac{2.3\mathrm{RT}}{(1-\alpha)\mathrm{nF}}$, respectively. Figure 2.7 shows the extraction of the Tafel equation from I-V curves.

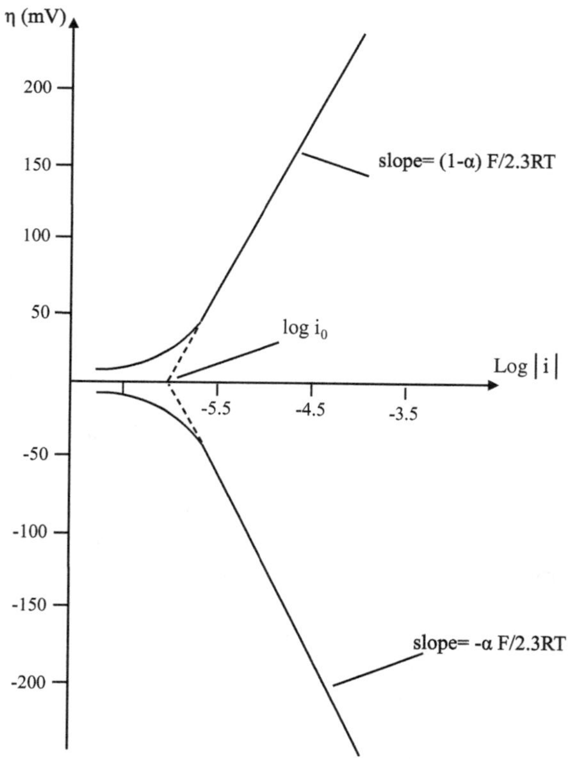

Fig. 2.7 An illustration of the Tafel equation extended from of the Butler-Volmer at high overpotentials

2.5.5 Polarisation

As we discussed before, there are two types of electrodes in terms of polarisation including polarisable and non-polarisibale. The polarisation is defined by a significant and irreversible shift in electrode potential when a current passes through. Provided that the potential of an electrode change only slightly by passing a current, this is called non-polarisable electrode. The best examples for such electrodes are the calomel, silver/silver chloride and H_2/Pt electrodes that exhibit high exchange current density.

Generally, a polarisable electrode undergoes three types of polarisation depending on the electron transfer and mass transport processes. These include: *activation polarisation, concentration polarisation and resistance polarisation.*

The activation polarisation is a consequence of the activation energy of any electrochemical reactions as discussed before in electrode kinetics which can be analyzed by the *Butler-Volmer* equation (2.45) and the *Tafel* equation (2.47). This type of polarisation becomes more important for an electrochemical reaction with a slow electron transfer rate owing to high activation energy. Such electrochemical reactions need to be conducted by establishing a reasonable amount of overpotential known as the *activation* polarisation or overvoltage. Thus, this type of polarisation mainly occurs in the processes with high activation energy consisting of the slow rate determining electron transfer processes.

The concentration polarisation arises due to the accumulation or depletion of ions on an electrode surface. This is called concentration polarisation which induces a change of potential. The change of potential can be readily estimated by the Nernst equation (2.12) in an electrochemical cell consisting of an electrode dipped into one type of electrolyte with two concentrations. Depending on the electrode kinetics, the accumulation or the depletion may occur. For instance, in an electro-reduction reaction, the concentration of ions increases if the rate of mass transport is greater than the rate of electron transfer on the cathode. We will discuss this in more details in next chapter in particular for *electrodeposition* processes.

The resistance polarisation occurs due to the conduction of current through any electrically resistive materials utilized in the electrochemical cell. The resistance polarisation is mainly due to the electrical resistivity of an electrolyte through which the current passes. This leads to an ohmic drop (*IR*) in the potential applied between the two electrodes. It is quite possible to exclude this polarisation by appropriate electrochemical cell design. An example is to employ a *Luggin* capillary to insulate the solution of the reference electrode.

2.6 Electrochemical Instrumentation and Techniques

2.6.1 Electrochemical Cell

The three-electrode cell is the standard laboratory apparatus for the quantitative investigation of most electrochemical properties of materials. It is a refined version of the basic wet electrochemical cell. A typical example is indicated by schematic picture shown in Fig. 2.8. It can be used in many different types of common electrochemical experiments.

First we shall demonstrate the components of the three electrode cell in more detail. The working electrode is the name given to the electrode being investigated. We use the term '*working electrode*' rather than *'anode'* or '*cathode'* because we are not limited to investigations of neither anodic nor cathodic behaviour alone.

The size and dimension of the working electrode depends on the electrochemical test and examination employed. A useful, though not essential, example is to have a surface area of at least 1 square centimeter (1 cm^2) to facilitate the calculation of the current density readily convertible from current measurements. For some other purposes such as semi-industrial and field experiments, the working electrodes, however, are often better performed using larger electrodes. Furthermore, it is very important to keep one peculiar distinct design for every individual examination to avoid any interrupting errors.

Several working electrode designs can be used. Some examples are shown in Fig. 2.9. A simple method is shown in Fig. 2.9a. This design uses mounted small specimens similar to what is used in metallographic specimens, in cold-setting resin to avoid any thermal decay of samples. In addition, an electrical connection must be

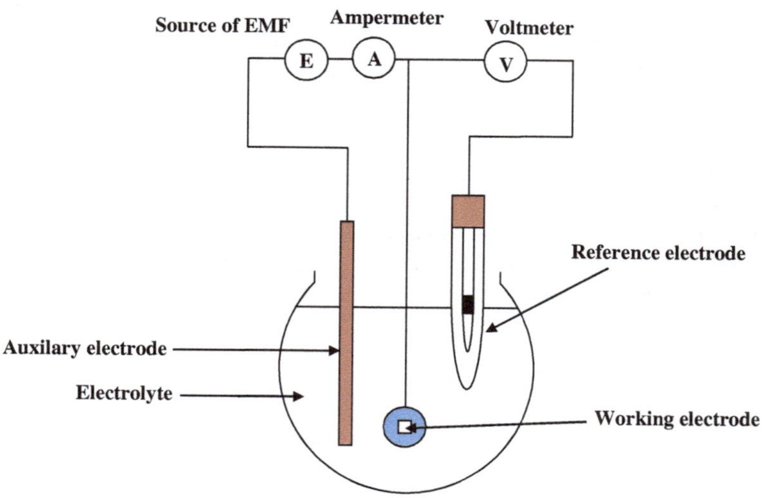

Fig. 2.8 A schematic image of the conventional three electrode cells

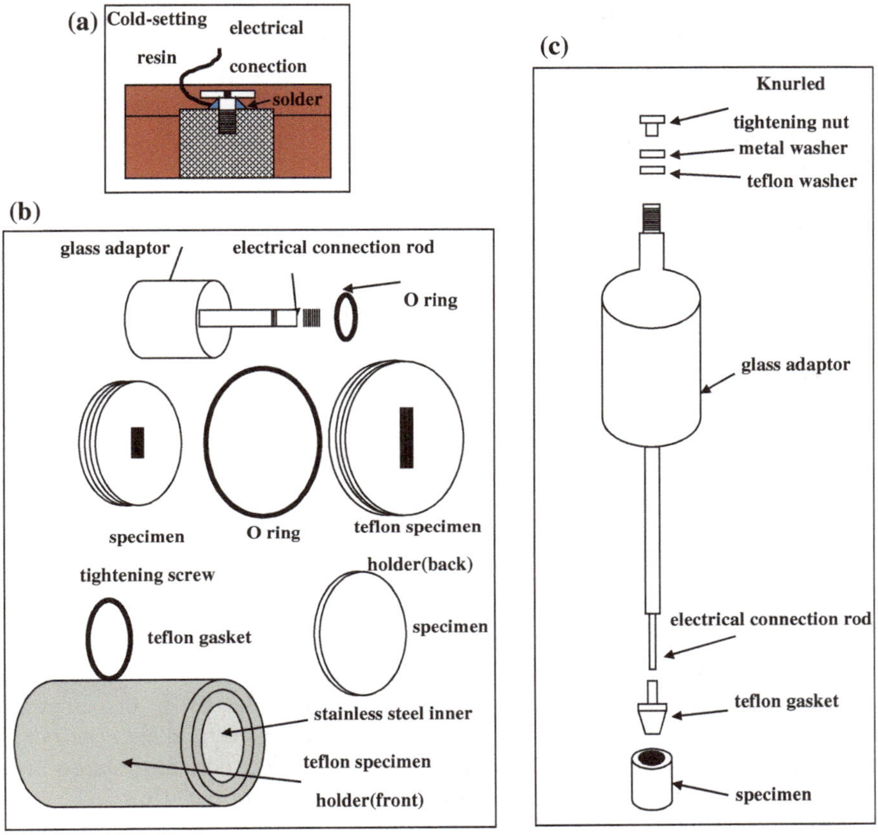

Fig. 2.9 A variety of common designs for working electrodes, **a** mount samples, **b** window-like samples, and **c** rotating disc samples

provided on the back side of the specimen with solder or spot weld before mounting. Each mounted specimen can be frequently used after being ground and polished before each individual experiment. Another common design frequently used in laboratory experiments is to make a window through which the working electrode is exposed to the electrolyte, as shown in Fig. 2.9b. This provides an integrated cell design consisting of the working and reference and sometimes counter electrode for a series of experiment under the same conditions. There is another cell design by which other electrochemical kinetics such as concentration polarisation studies become possible. This is called the rotary disc design, as shown in Fig. 2.9c. Kinetics of processes may be studied by the rotating disc electrodes. The rotation of the electrode sets up a steady hydrodynamic stream of electrolyte and the effect of electrode potentials, current and concentration profile around the surface can be studied.

Table 2.1 Most common reference electrodes and their potentials

Reference electrode	Half reaction	Potential (V vs. SHE) at 25 °C
Hydrogen electrode		
Standard hydrogen electrode, Pt/H_2	$2H^+ + 2e^- \rightleftarrows H_2$	0.0
Calomel or mercury/mercury chloride electrode		
Hg/Hg_2Cl_2, KCl (0.1 M)	$Hg_2Cl_2(s) + 2e^- \rightleftharpoons 2Hg_{(liq)} + 2Cl^-$	0.3337
Hg/Hg_2Cl_2, KCl (1M)	$Hg_2Cl_2(s) + 2e^- \rightleftharpoons 2Hg_{(liq)} + 2Cl^-$	0.2801
Hg/Hg_2Cl_2, KCl (3.5 M)	$Hg_2Cl_2(s) + 2e^- \rightleftharpoons 2Hg_{(liq)} + 2Cl^-$	0.250
SCE (Saturated Calomel)	$Hg_2Cl_2(s) + 2e^- \rightleftharpoons 2Hg_{(liq)} + 2Cl^-$	0.244
Silver/Silver Chloride electrode		
Ag/AgCl, KCl (0.1 M)	$AgCl(s) + e^- \rightleftharpoons Ag(s) + Cl^-$	0.2881
Ag/AgCl, KCl (3 M)	$AgCl(s) + e^- \rightleftharpoons Ag(s) + Cl^-$	0.210
Ag/AgCl, KCl (3.5 M)	$AgCl(s) + e^- \rightleftharpoons Ag(s) + Cl^-$	0.205
Ag/AgCl, KCl (sat'd)	$AgCl(s) + e^- \rightleftharpoons Ag(s) + Cl^-$	0.197

In an electrochemical cell, two electrodes carry the current passing through the cell. The counter or auxiliary electrode is the second electrode. It is not required for measurements of potential. Different materials can be used as the counter electrode such as carbon or graphite, platinum, gold, and titanium.

The third electrode in the conventional three electrode cell is the reference electrode. It is essential to adjust or control the potential of the working electrode at a very stable datum point. As discussed earlier, most of electrodes are polarisable in the sense that their potential changes by the passage of current. However, a few electrodes exist exhibiting a negligible polarisation. These can be employed as the reference electrodes. By far the most convenient reference electrodes to use in electrochemical experiment are standard hydrogen electrode (SHE), saturated calomel electrode or mercury chloride (SCE), silver/silver chloride (Ag/AgCl), etc.

The potential of any reference electrodes is distinct and can be compared with that of the SHE references. It is quite easy to understand, on the basis of the thermodynamics of ions, that the potential of a standard hydrogen electrode is zero. Table 2.1 summarizes the electrochemical half-reactions and corresponding potentials for the most common reference electrodes.

2.6.2 *Electrochemical Tests and Techniques*

Electrochemical test methods are controlled measures of the electrode potential adhering to a known program. The electrode potential may be held constant or changed with time under a peculiar designed program and consequently the current is measured as a function of time or potential. There are separate approaches available to electrochemistry tests namely potential step methods, potential sweep methods, polarography, controlled-current methods and etc. These techniques make possible to understand the kinetics of electrochemical reactions and kinds of electrode processes including *faradaic* and non-*faradaic* reactions. The relationship between electrical charge and mass of consumed or product formed is given by *Faraday's law*; that is, the passage of 96,485.4 C causes 1 equivalent of reaction (e.g., consumption of 1 mol of reactant or production of 1 mol of product in a one-electron reaction). Thus, if reactions are governed by Faraday's law, they are called faradaic processes. Electrodes at which faradaic processes occur are sometimes called *charge transfer* electrodes. Under some conditions, a given electrode-solution interface will show a range of potentials where no charge-transfer reactions occur because such reactions are thermodynamically or kinetically unfavorable. However, processes such as adsorption and desorption can occur, and the structure of the electrode-solution interface can change with changing potential or solution composition. These processes are called nonfaradaic processes. Although charge does not cross the interface, external currents can flow (at least transiently) when the potential, electrode area, or solution composition changes. Both faradaic and nonfaradaic processes occur when electrode reactions take place. Although the faradaic processes are usually of primary interest in the investigation of an electrode reaction (except in studies of the nature of the electrode-solution interface itself), the effects of the nonfaradaic processes must be taken into account in using electrochemical data to obtain information about the charge transfer and associated reactions [9]. Here we give a brief description for the most applied approaches in the electrodeposition of solids, but for detailed discussion about the test, it is recommended to refer to *Bard and Faulkner* [9]. A number of electrochemical techniques which are mostly used for studying the electrodeposition systems are discussed.

Common potential step methods are based on the application of a potential waveform in a step-wise manner from one potential to one or more other potentials, and consequently the current is recorded as a function of time. This kind of experiment is called *chronoamperometry*. A sample two-step potential waveform and corresponding current-time curve are shown in Fig. 2.10a. This technique may define the kinetics of an electrochemical reaction under different mechanisms each of which is stable in a particular potential step. Sharp switching between the potential steps substantially changes the kinetics of a reaction. For instance, we consider a reduction reaction of ions on a solid electrolyte dipped in a solution containing an electroactive species. The potential is switched from a region where faradaic processes do not occur, i.e. E_1 on Fig. 2.10a to a potential E_2 under "mass-transfer-limited" mechanism. The current response is initially large after switching and subsequently decreases to maintain the fully reduced condition at the electrode surface. The initial reduction has created a concentration gradient that in turn produces a continuing flux of the electroactive species to the electrode surface. The reduction of the electroactive species or its current is proportional to the concentration gradient at the electrode surface.

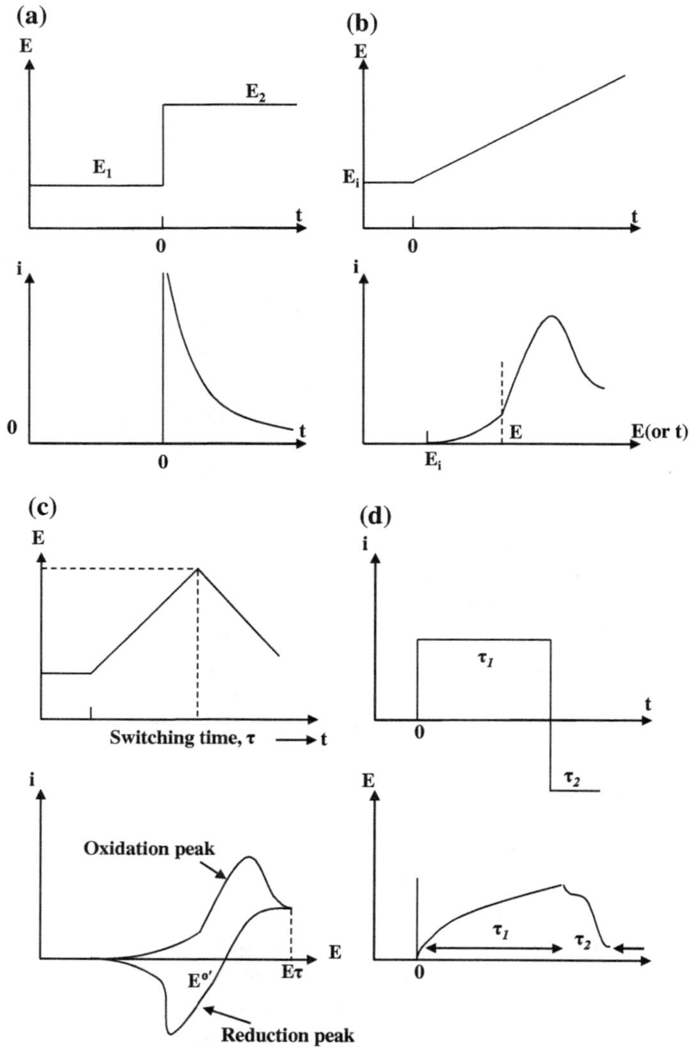

Fig. 2.10 Plots of applied waveform and its response for **a** chronoamperometry, **b** linear sweep voltammetry, **c** cyclic voltammetry, and **d** chronopotentiometry

Thus the slope of the concentration profile at the surface declines with time, and so does the current.

Another most common electrochemical technique is *voltammetry* which commonly determines the kinetics of electrode processes. The voltammetry is based on the *potential step sweep* techniques. In this technique, the current is recorded as the electrode potential is changed. Depending on the steps used to changing the potential, the voltammetry techniques may be divided into two categories: *linear sweep voltammetry* and *cyclic voltammetry*.

Figure 2.10b, c compares the potential sweep waveforms and the corresponding outputs achieved by the two voltammetry techniques. For the linear sweep voltammetry, at lower potentials as is in initial stage of the scan, the current is low owing to sole migration of ions in the electrolyte without any oxidation or reduction reactions. As the potential approaches to the reduction potential of the electroactive solute, the cathodic current rises up to a current peak. Subsequently the current immediately starts to decline due to the lack of the electroactive species limited by mass transfer mechanism. If we add a reverse potential sweep, as shown in Fig. 2.9c, the cyclic voltammetry is performed. The forward sweep related to the cathodic current in similar to the linear sweep voltammetry. In the reverse scan related to the anodic reaction, a rapid change in current occurs which corresponds to the high concentration of oxidisable electroactive species in the electrode surface. Having reached closer to the oxidation potential, an anodic peak appears on the voltammogram until it returns to zero. The shape of the cyclic voltammograms and its variation as a function of scan rate gives more details about the kinetics of the electrode process.

For more than two decades the potential sweep methods such as cyclic voltammetry have been used for a wide range of Systems and mathematical description of these methods are sufficiently developed. So these methods are able to determine the kinetic parameters for a wide range of mechanisms. Cyclic voltammetry method probably the most useful technique in the evaluation mechanism is based on a potential sweep. Cyclic voltammetry because of these capabilities is always a method of choice for primary investigation of kinetics study. When cyclic voltammetry is used for the first time to study a system, it usually provides qualitative analysis on basic information about the system before the semi-quantitative and quantitative studies are performed to calculate the kinetic parameters. In a qualitative study type, voltammograms are recorded in a wide range of potential and scan rates until several peaks appear. The dependence on scan rate to peak amplitude determines the role of adsorption, diffusion, chemical reactions or other processes related to the electrochemical system under study. Figure 2.10c illustrates a typical voltammogram recorded under slow linear sweep of potential for an essentially reversible reaction. Concentration of electroactive species will be in steady state conditions and natural convection will be kept it uniform in a certain distance from the electrode surface. However, in the area near the surface of the electrode the Nernst diffusion layer consists of essentially linear concentration gradient. The Nernst equation can be used. As the potential becomes more negative, surface concentration of reactant species decreases and concentration gradient and flow increases concentration. Finally, the surface concentration of reactants comes to zero and concentration profile in the steady state cannot change. When the scan rate increases, diffusion layer has no enough time to reach equilibrium to be developed into solution and therefore the concentration profiles are not linear anymore. As soon as the potential reaches to the reduction potential of reactant, the surface concentration of dissolved electroactive reactant species is reduced to obey the Nernst equation and then concentration gradients are established. As a result, a current proportional to the magnitude of this gradient is created in the outer electrode surface. Once the concentration gradient becomes zero the flow reaches the maximum point where a peak in current-potential curve appears. The effect of scan rate on the position of peak follows the same argument. Electron or charge transfer rates for reversible systems in all the potentials are significantly greater than the mass transfer rate and there always be an equilibrium state. When the electron transfer rate is not sufficient to maintain equilibrium Nernst electrode surface, the kinetics of reaction is not reversible anymore and form of the voltammogram will change. At low potential sweep rates, electron transfer rate is far beyond the mass transfer rate and cyclic voltammogram recorded will be reversible. At higher scan rates, the high mass transfer rate is comparable with the charge transfer rate. This phenomenon is observed by an increased distance between the reduction and oxidation peaks. One particular characteristic of a cyclic voltammogram for an irreversible system may be accompanied by annihilation or impairing of reverse sweep peak. This is the case for electrodeposition systems. Figure 2.10 illustrates a typical cyclic voltammogram for metal deposition on a foreign substrate. Potential and kinetic parameters for electrochemical reactions can be obtained. The deposition peak is formed under very similar argument stated above. There are significant differences in the reverse scan. The shape of voltammogram is asymmetrical and has an interception with the forward scan. This phenomenon can easily be explained by deposition of materials on the electrode surface which does not intend to dissolve in the reverse scan.

In contrast to potential-controlled techniques, current-controlled techniques have been developed where the current is controlled (mostly is kept constant), and the potential becomes the dependent variable, which is determined as a function of time. These techniques are generally called *chronopotentiometric* techniques, because E is determined as a function of time, or *galvanostatic* techniques, because a small constant current is applied to the working electrode. A constant current or multi-step waveform can be applied and the variation of potential of the electrode is monitored to study the kinetics of the electrochemical processes. Figure 2.10d shows an example current waveform and the corresponding current recorded. The current is reversed after some time (*current reversal chronopotentiometry*). For example, if in the case considered above, the current is suddenly changed to an anodic current of equal magnitude at, or before, the transition time, the component reduced during the forward step will start oxidizing. This reverse pulse has a remarkable effect on the concentration polarisation on the electrode surface This can be an example for pulse electrodeposition techniques. An extended version of this technique is called the *cyclic chronopotentiometry* where the current is successively reversed at each transition (Fig. 2.11).

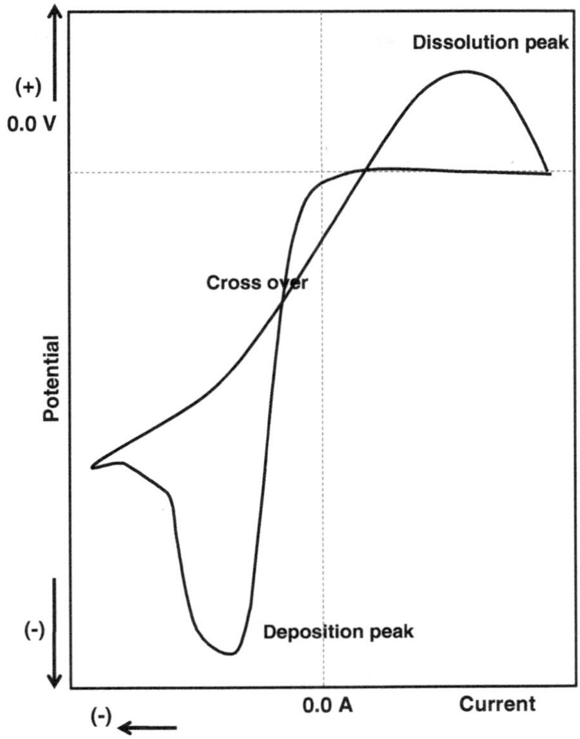

Fig. 2.11 A typical cyclic voltammogram for electrodeposition of a metal on a foreign substrate

References

1. O'.M. Bockris John, K.N. Reddy Amulya, *Modern Electrochemistry*, vol. 1, 2nd edn. (Ionics, Kluwer Academic Publishers, New York, 2002)
2. P. Debye, E. Hückel, Physik. Z. **24**(185), 305 (1923)
3. P.W. Atkins, *Physical Chemistry*, 5th edn. (Oxford University Press, Oxford, 1994)
4. N. Bjerrum, Kgl Danske Videnskab. Selskab, Mat. Fys. Medd. **7**(1) 1926
5. F.J.R.H. Stillinger, R. Lovett, J Chem. Phys. **48**, 3858–3868 (1968)
6. D.R. Crow, *Principles and Applications of Electrochemistry* (CRC Press, 1994)
7. O. Stern, Z. Electrochem **30**, 508 (1924)
8. D.C. Grahame, Chem. Rev. **41**, 441 (1947)
9. J. Bard Allen, R. Faulkner Larry, *Electrochemical methods: fundamental and applications*, 2nd edn. (John Wiley & Sons, Inc, 2001)

Chapter 3
Fundamentals and Principles of Electrode-Position

Abstract In this chapter we explain the fundamentals of electrodeposition starting with the principles of electrolysis, and electrodeposition cells. The electrochemical fundamentals play a vital role in electrodeposition, then these will be discussed in details and modern approaches for the investigation of thermodynamics and kinetics aspects are reviewed. Nucleation and growth of electrodeposition and their modes including overpotential (OPD) and underpotential (UPD) deposition control the microstructure and consequent properties, as will be introduced. Characterisation methods of initial stages of electrodeposition and other stages will be explained using conventional electrochemical techniques and modern microscopical and nanoscopical techniques. The enhancement of electrodeposition process by application of external forces such as magnetic and ultrasound will be explained.

3.1 Introduction

The word *electrodeposition or electrochemical deposition* generally includes the principle electrolysis process in which solids or condensed matters are deposited in different forms and shapes. One historical approach of this process is so called electroplating where it leads to produce mostly metal or alloy coatings for decorative or protective applications. Nowadays, this word covers the deposition of a wide spectrum of solid materials from metals, alloys to semiconductors and polymers in a diverse range of size and dimensions from nano-scale and meso-scale up to macro-sized objects. Nobody can ignore the diversity of electrodeposited materials and their applications in high tech industries such as microelectronic and information storage devices and etc.

Owing to the careful progress in understanding the principles of electrodeposition using electrochemistry, physics and materials science, we now encounter a major field of science and research to exploit novel materials, structures and functions which can be obtained by electrodeposited materials. We in this chapter start with the principles of electrolysis, and electrodeposition cells. The electrochemical fundamentals play a vital role in electrodeposition, then these will be

F. Nasirpouri, *Electrodeposition of Nanostructured Materials*,
Springer Series in Surface Sciences 62, DOI 10.1007/978-3-319-44920-3_3

discussed and modern approaches to investigation of thermodynamics and kinetics are reviewed. Nucleation and growth of electrodeposition and their modes including overpotential (OPD) and underpotential (UPD) deposition control the microstructure and consequent properties, as will be introduced. Characterisation methods of initial stages of electrodeposition and other stages will be explained using conventional electrochemical techniques and modern microscopical and nanoscopical techniques. The enhancement of electrodeposition process by application of external forces such as magnetic and ultrasound will be explained.

3.2 Electrolysis

The general term electrolysis involves the production of solid or gaseous materials under the occurrence of an electrochemical reduction reaction. The production of solids is generally called deposition of solids (or electrodeposition) or metals and the production of gases is called gas evolution. We emphasize on the electrodeposition of solids in general and metals in particular, while the gas evolution is often taken placed during the metal deposition.

3.2.1 Electrolysis Cell

The basic electrolysis cell is shown in Fig. 3.1. This resembles a simple two-electrode electrolytic electrochemical cell. The main components of the cell include: two conducting electrodes namely anode and cathode, an electrolyte containing ions of the metal or solid to be deposited as well as of a supporting

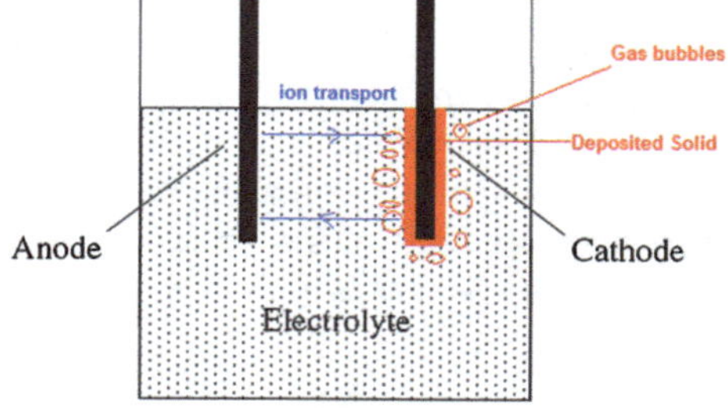

Fig. 3.1 Principle two-electrode cell for electrolysis

solution, and a power supply. In an electrolytic cell, there is no spontaneous electrochemical reaction. In order to conduct a reaction, therefore, a driving force must be supplied by an external source of current or voltage. The reduction reaction including metal deposition or gas evolution take place on the cathode, whereas the oxidation reaction including metal dissolution or gas evolution happen on the anode. Under controlled conditions the electrolysis forms a firm and robust material on the cathode as known an *electrodeposit*.

3.2.2 Electrodeposition Cells and Reactions

We distinct two types of electrodeposition cells namely cells with dissoluble anode and non-dissoluble anode. Former cell consists of metal or alloy anodes which dissolve during the anodic half reaction. The electrodeposition of copper from sulphate solution using a copper anode is a good example for this category. An adequate voltage between the anode and cathode is required to conduct a current through the cell. The anodic half reaction includes the dissolution of copper (3.1) and the cathodic half-reaction is the reduction of copper (3.2):

$$Cu \rightarrow Cu^{2+} + 2e^- \tag{3.1}$$

and

$$Cu^{2+} + 2e^- \rightarrow Cu \tag{3.2}$$

The copper sulphate electrolyte is also dissociated to copper and sulphate ions. The ions will transport in the electrical field between the two electrodes.

For the non-dissoluble anodes, the copper anode is replaced with an inert metal like platinum or carbon electrode. The cathodic half-reaction is similar to (3.2), however, the anodic reaction is not a solid-state reaction. Liquid base auxiliary anodic reactions depend on the electrolyte composition and pH. Some examples are:

$$\begin{aligned} H_2O &\rightarrow 2H^+_{(aq.)} + \frac{1}{2}O_2 + 2e^- \quad \text{for pH} \leq 7 \\ 4OH^-_{(aq.)} &\rightarrow H_2O + \frac{1}{2}O_2 + 4e^- \quad \text{for pH} \geq 7 \end{aligned} \tag{3.3}$$

In above example for electrolysis from copper sulphate solutions, some other auxiliary reactions may take place on the anode as well:

$$\begin{aligned}
&2SO_4{}^{2-}_{(aq.)} \rightarrow S_2S_8{}^{2-}_{(aq.)} + 2e^- \\
&S_2S_8{}^{2-}_{(aq.)} \rightarrow SO_4{}^{2-}_{(aq.)} + SO_3 + \frac{1}{2}O_2 \\
&SO_3 + H_2O \rightarrow H_2SO_4
\end{aligned} \tag{3.4}$$

As is evident, anodic reactions may influence the composition and pH of the electrolyte.

3.2.3 *Electrodeposition Electrolytes*

A typical electrodeposition electrolyte consists of the ions of material to be electrodeposited. Traditionally, the electrodeposition electrolytes of materials are usually based on *aqueous* electrolytes. There are contemporary electrodeposition electrolytes based on *ionic* solutions which have been recently developed for electrodepositing materials in particular nanostructured materials.

3.2.3.1 Aqueous Electrodeposition Electrolytes

In principle, aqueous solutions may be produced from either *simple salts* or *complex* (electrically charged or uncharged) salts comprised of other soluble cationic and anionic compounds. Their primary constituent of both types of aqueous solutions is the salt of the material to be electrodeposited. One component aqueous electrolytes are however rarely used in practice since they do not provide high quality electrodeposits. Therefore, the electrolytes used in practice contain a range of additional compounds, each of which plays a specific role.

The electrical conductivity of the solution is the main characteristics of electrodeposition electrolytes. As we discussed this property in Chap. 2, the conductivity of electrolytes depends on concentration and mobility of the component ions. The conductivity of strong and weak solutions behaves differently due to interactions of the ionic components with one another or with water molecules. The conductivity of electrolytes may be described according to well developed theories, the *Debye-Hückel-Onsager* for the strong electrolytes and Ostwald's dilution law for the weak electrolytes. Since the electrodeposition electrolytes usually consist of various species, it is however very complicated to calculate the conductivity and it is therefore simpler to measure the real conductivity of the electrolytes. The conductivity measurement instrumentation is well explained in [1]. The typical magnitudes of conductivity of different metal electrodeposition electrolytes are in a range of 3–60 $\Omega^{-1}\ m^{-1}$ at room temperature for different metals. It is of course of importance to consider the variation of the conductivity during the deposition process which owes to the presence in the solution of small nonconductive particles or gas bubbles. The effective conductivity value is approximately obtained:

$$\kappa = \kappa_o(1 - \epsilon)^{\frac{1}{2}} \tag{3.5}$$

where κ_o is the conductivity of the pure solution, and ϵ the volume fraction of the nonconductive phase [2].

In order to enhance the quality and performance of electrodeposition electrolyte as well as their conductivity, a range of electrolytes has been developed. The pH of electrolytes is a good measure of the aqueous solutions. Based on the pH, the electrodeposition electrolytes may be divided into *acid, neutral and alkaline* baths, where the pH values are usually <3, =7 and >9, respectively. In the following, examples of each of these three are detailed. In all three cases, it is vital to avoid significant pH changes during the electrodeposition process. For this reason, all the electrolytes consist of additional acid, alkali or buffers.

Acid electrolytes are usually based on simple metal salts such as sulfates or chlorides, less commonly are phosphates or sulphamate salts. A high concentration of the parent acid (i.e. sulphuric acid for nickel sulphate solution, etc.) is used partly to optimise the electrical conductivity, partly to minimise pH changes. The most common buffer used in acidic electrolyte is boric acid which adjusts the pH under the control of this reaction:

$$H_3BO_3 \leftrightarrows H_2BO_3^- + H^+ \tag{3.6}$$

Furthermore, acid electrolytes may be made based on metal ion complexes where the metal is anionically charged such as hexavalent chromic acid (for chromium plating) and acid gold baths, based on the anion $[HCr_2O_7]^-$ and $[AuCN_2]^-$, respectively.

Neutral electrolytes include weakly-acid to weakly alkaline solutions. Due to the poor electrical conducting, this type of electrolytes is not diversely used for electrodeposition. An example for this can be the neutral zinc plating baths, operating in the pH range 7.5–8.8 comprising of complexes of ammonium chloride zincate, $Zn[(NH_3)_2C_{12}]$ or $Zn[(NH_4)_2C_{14}]$ and zinc chelated with polyhydroxycarboxylic acid with an optimum ratio.

The last category of electrolytes is *alkaline solutions*. This may be divided into cyanide-containing and cyanide-free electrolytes. An important issue related with this kind of electrolytes is their reactivity with air. This leads to formation of carbonates in solution which in turn decays the metal deposition process. Cyanide electrolytes have been used for years because they have several advantages such as long durability and stability, high throwing power, uniform films with lesser roughness, and their capability for alloy co-deposition. However, it is presently recommended to replace them with more environmentally friendly electrolyte. Non-cyanide alkaline electrolytes may include zinc plating electrolytes. These electrolytes never encompass the stability of the similar cyanide-containing solutions and are a substitution under critical conditions where cyanide baths are not allowed to be used [3, 4].

For more details about the electrolytes composition and working conditions, refer to [3, 4].

3.2.3.2 Ionic Electrodeposition Electrolytes

Ionic liquids are solvents that are solely composed of ions. The melting points of the ionic liquids are below 100 °C by definition. The first ionic liquid reported was ethylammonium nitrate, $[C_2H_5NH_3]NO_3$ with a melting temperature of 12 °C. The first room temperature molten salt was achieved by thermal mixing of 1-ethylpyridinium chloride with aluminum bromide from which aluminium could be electrodeposited. Great progress in this field has been made recently on organic chloride/aluminium chloride ambient temperature molten salts. Further progress has been achieved by developing water soluble ionic liquids which facilitate their application in atmospheric conditions. Ionic liquids are generally classified into first and second generation liquids [5]; where first generation liquids are the eutectics compositions and second generation have discrete anions. There is another group considered as subdivision of the first generation liquids into separate types depending on the nature of the Lewis or Brønsted acid that forms complexes. The majority of ionic liquids are formed according to:

$$\text{cation} + \text{anion} + \text{complexing agent} \leftrightarrow \text{cation} + \text{complex anion} \quad (3.7)$$

Another possibility is:

$$\text{cation} + \text{anion} + \text{complexing agent} \leftrightarrow \text{complex cation} + \text{anion} \quad (3.8)$$

The equilibrium constant of the reactions mentioned mostly determines the direction of reaction taken place according to the Lewis or Brønsted acid. Different cations and anions of ionic liquids have been introduced in literature [5]. Typical cations are substituted imidazolin ions, like 1- bytyl-3-methylimidazolin, or tetraalkylammonium ions like for example teictyl-methyl-ammonium. In addition important anions are hexafluorophosphate, trifluoromethylsulfonate, bis(trifluoromethylsulfonyl)imide.

Most of the ionic liquids have low vapor pressures even at temperatures of 300 °C and more, whereas they exhibit viscosities similar to water as well as ionic conductivities of up to $0.1\ \Omega^{-1}\ m^{-1}$, and, which makes them interesting for electrochemistry. Furthermore, it is essential to combine the molar conductivity and viscosity to have estimation about the conductivity of ionic liquid electrolytes.

Various ionic liquids composed of transition elements and main group metals have been developed and the electrodeposition of majority of these metals is feasible as includes: Li, Na, Si, Ge, Bi, Ta, Sm as metal, Mn, As, Pt, Nb, etc. as alloy, and Mg, Ti, Cr, Fe, Ni, Cu, Zn, Al and etc. For a complete list of elements, it is referred to [5] as metal and alloy. However, the information of the kinetic controlling processes, morphology evolution for practical application is still not fully attained. There are quite a few reasons that these electrolytes have a good chance to promote in future. Some examples include: (i) the possibility of electrodeposition of metals which are not normally electrodeposited in aqueous solutions like Al, Ti, Ta, Nb, Mo, and W. (ii) enhanced electrodeposition on water-sensitive metal substrates such as Al, Mg and light alloys with good adherence, (iii) elimination of hydrogen evolution in the electrodeposition of materials from ionic liquids, (iv) electrodeposition of alloys which are less feasible by aqueous solutions, (v) electrodeposition of nanostructured materials under peculiar control and finally (vi) more environmental friendly electrodeposition processes [6–8].

3.2.4 Electrodeposition Techniques and Classification

As an electrolytic electrochemical process, the method of electrodeposition is only feasible under application of an electromotive force supplied by an external power supply. Thus, the electrodeposition can be controlled either by applied potential or applied current, as well as by the measured parameters recorded during the process. Hence, on the basis of the applied potential or current, we divide the electrodeposition techniques into two classes: (i) controlled-potential techniques known as *potentiostatic* electrodeposition techniques, and (ii) controlled-current techniques known as *galvanostatic* methods. In the first class, a conventional three electrode cell must be utilised. The potential of the working electrode is kept constant with respect to a reference electrode. In most cases, the potentiostatic techniques are usually preferred owing to their competence in controlling of the process. However, these methods require precise design of the electrodeposition cell which mainly consists of a potentiostats with large output current and voltage capabilities and a stable reference electrode. Care must be taken to minimize uncompensated resistance and the resistance polarisation effects as discussed in Chap. 2. In galvanostatic, controlled-current passing through the cell is applied in constant or pulse modes. The instrumentation of these techniques are almost involve simpler that of the potentiostatic techniques. Two electrode cells are often used. It is convenient to monitor the current passed to ensure 100% current efficiency.

3.2.5 Electrodeposition Kinetics: Potential-Current Relationship

Figure 3.2 depicts a general potential-current curve of a typical electrodeposition of metals from simple salt electrolyte. This has to be conducted in the conventional three electrode cell and the potential of working electrode as the substrate of electrodeposition is measured with respect to a reference electrode. No current passes the cell, until the potential reached the equilibrium potential, E_{eq}. At this potential, the electrodeposition cell works at the equilibrium potential. Applying any potential, $E(I)$, negative than E_{eq} namely cathodic overpotential,

$$\eta = E(I) - E_{eq} \tag{3.9}$$

begins the electrodeposition process of metal. As the magnitude of the overpotential is increased, the cathodic current as a measure of the electrodeposition (or electrolysis) rate increases.

Deposition however may take place at potentials more positive than E_{eq} which is called underpotential deposition (UPD). This will be discussed later in this chapter.

This relationship may be divided into 4 different regions under different kinetics controlled processes. *(1) Activation controlled or charge transfer process, (2) mixed controlled process, and (3) Mass transport or diffusion controlled process.*

3.2.5.1 Activation or Charge Transfer Controlled Electrodeposition

Called sometimes as kinetic control electrodeposition, this kind of electrodeposition processes exhibits an exponential dependence of current versus the overpotential; this behaviour is well described by the Butler-Volmer (2.44). The electrodeposition of metals from simple salt solutions mostly exhibits charge transfer controlled electrodeposition. An example is the electrodeposition of copper from the aqueous solution of 0.15N $CuSO_4$ and 1.0 N H_2SO_4 [4] A typical cathodic linear sweep polarisation (I–V) curve for such electrodeposition processes is shown in Fig. 3.2a. It illustrates that even small increases in overpotential result in large changes in the current density. According to the Tafel (2.46), a linear relationship exists between the logarithm of current density and overpotential for large values of overpotential, $\eta > 100\,\text{mV}$. Figure 3.2 illustrates the Tafel line extrapolated from the linear part if the curve for electrodeposition of a metal under activation control. The intercept of

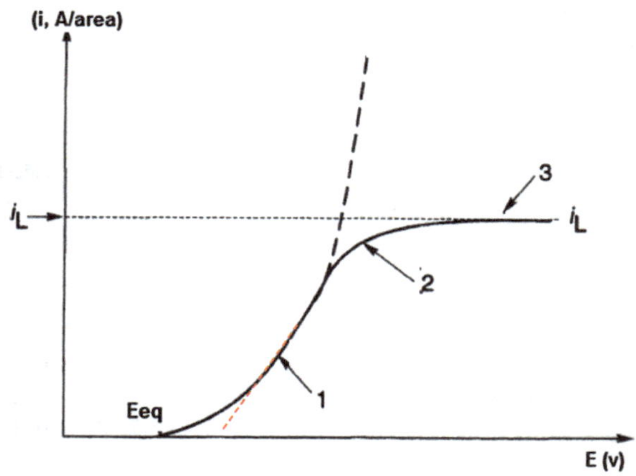

Fig. 3.2 A typical potential-current density *curve* for a typical electrodeposition of a metal exhibiting *(1) Activation controlled or charge transfer process, (2) mixed controlled process, and (3) Mass transport or diffusion controlled process*

the line to x-axis (current density) and the slope may define the exchange current density and the transfer coefficient as described in Chap. 2.

3.2.5.2 Mass Transport or Diffusion Controlled Electrodeposition

One of the basic conditions of the Butler-Volmer equation is the homogeneity of electrolyte at the vicinity of electrode. This condition is not satisfied at high current densities because the migration of ions to and from electrodes is slow and may become rate-determining process. This is shown in Fig. 3.2 on part (3) where further increase of overpotential is not accompanied by any change of current density. This is called the diffusion limited current density.

The behaviour shown in part 3 in Fig. 3.2 is the basis of concentration polarisation and may be a primary controlling step in the reduction of ions on electrodes in most of the metal electrodeposition processes. For instance, industrial electroplating processes demand high speed deposition processes under very high current densities from concentrated electrolytes. Thus is essential to understand and control this type of polarisation in an electrodeposition process.

Let's explain a complete situation where the concentration polarisation occurs at both of anode and cathode. Figure 3.3 illustrates a schematic picture of the concentration profile in a typical two-electrode electrodeposition cell having metal M as the dissoluble anode and an electrolyte comprised of M^{n+} ions. This simplified picture is driven on the basis of this idea that the concentration polarisation dominates all the rate processes. A good example for this is the electrolysis of copper from copper sulphate solutions from concentrated solutions in which copper anodes and copper cathodes are utilised.

A redox couple of M/M^{n+} is assumed and the electrode potentials of the anode (E_a) and the cathode (E_c) are related to the ion activity at the surface of each electrode determined by the Nernst equation.

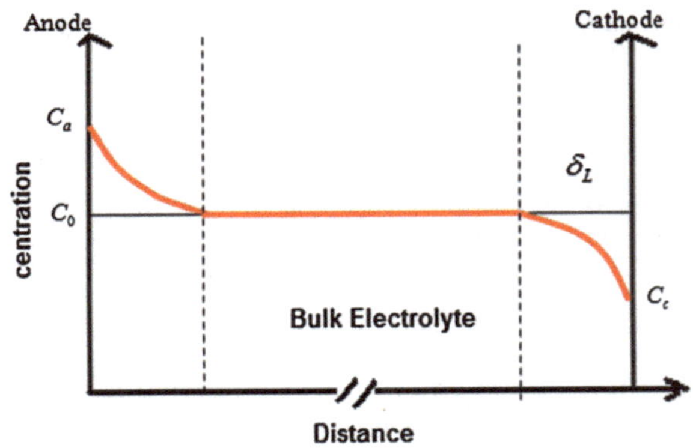

Fig. 3.3 A schematic picture of the concentration profile in a typical two-electrode electrodeposition cell

$$E = E^{\circ} + \frac{RT}{nF} \ln a_{M^+}$$

where a_{M^+} is the activity of the ions.

We assume a constant activity coefficient for an electrolyte comprising of a large excess of supporting electrolyte. Therefore, the electrode potentials before starting the electrolysis at equilibrium or the zero-current potentials will be described as:

$$E = E^{\circ} + \frac{RT}{nF} \ln C_{\circ} \tag{3.10}$$

Once the electrolysis commences, and therefore current flows through the cell, the active ion concentration at the vicinity of electrode just outside the double layer changes to C_a on the anode and C_c on the cathode. The electrode potentials will be written as:

$$E_{(i)} = E^{\circ} + \frac{RT}{nF} \ln C_i \tag{3.11}$$

in which i denotes anode and cathode. Thus, the concentration overpotential (η_c) on the anode and the cathode will be dependent on the concentration change, and is expressed by:

$$\eta_c = \Delta E = E_2 - E_1 = \frac{RT}{nF} \ln \frac{C_i}{C_{\circ}} \tag{3.12}$$

Thus, according to the concentration profile at the vicinity of anode and cathode, one can find the concentration polarisation on the anode and cathode is positive and negative, respectively.

We now attempt to calculate the distance from the electrodes where the electrolyte has its bulk concentration, namely the Nernst diffusion layer, δ_L. It must be noted that the thickness of the Nernst diffusion layer is typically 0.05–0.1 mm and significantly depends on the condition of the hydrodynamic flow on electrolyte. It is remarkably different from the electrical double layer thickness (which is typically less than 1 nm) [1]. We calculate the concentration gradient across the Nernst layer:

$$\frac{dC}{dx} = \frac{C_i - C_o}{\delta_L} \tag{3.13}$$

According to the Fick's law, we can introduce the molar flux, J, in the above equation:

$$J = -D \frac{C_i - C_o}{\delta_L} \tag{3.14}$$

The current density, *i*, is defined as the product of the ion flux and the charge nF per more of ions:

$$i = nFJ = nFD\frac{C_o - C_i}{\delta_L} \tag{3.15}$$

The diffusion of ions through the Nernst layer reaches the maximum at the steepest concentration gradient, which takes place when $C_i = 0$. This concentration gives rise to the maximum current density passing through the cell under diffusion controlled electrodeposition process, called the limiting current density

$$i_L = \frac{nFDC_o}{\delta_L} \tag{3.16}$$

One can derive an equation to calculate the limiting current density from the ionic conductivity, λ, using the Nernst-Einstein (2.24):

$$i_L = \frac{\lambda RTC_o}{nF\delta_L} \tag{3.17}$$

Finally, we find the following equations for the concentration overpotential in terms of current density:

$$\eta_c = \frac{RT}{nF}\ln\left(1 - \frac{i\delta_L}{nFDC_o}\right) \tag{3.18}$$

Or

$$i = \frac{nFDC_o}{\delta_L}(1 - e^{nF\eta_c}) \tag{3.19}$$

The limiting current density is increased by agitating the electrolyte, rising electrolyte temperature and reducing the electrolyte temperature. It is also worth noting that the concentration polarisation is quickly diminished when the electrodeposition current is switch off.

3.2.6 Co-electrodeposition of Various Ions

Let's consider the current density-cathodic potential curves for the electrodeposition of two metals A and B from one electrolyte bath. It is logical from the Farady's law that the current density be considered as a measure of the deposition rate of each constituent in a nominal A-B binary system. Figure 3.4a demonstrates a situation where the current-potential curves do not intersect. The reduction of metal A commences at a formal potential of E_A. At potentials more negative than this value,

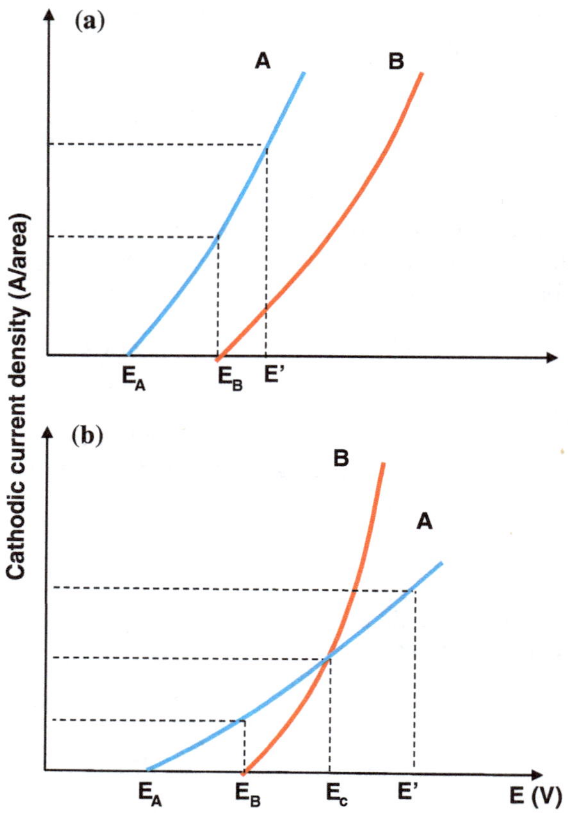

Fig. 3.4 Principle representation of co-electrodeposition of a nominal A-B binary system under two circumstances **a** non-intersecting and **b** intersecting behaviors

solely, the reduction of metal A takes place until a potential E_B. This is accompanied by the reduction of metal B. From this potential forward, e.g. E′, A-B binary system is co-deposited. The more overpotential applied the higher B content of the A-B binary system. Under such circumstance, content of A always exceeds content of B.

Another situation of co-deposition of binary systems is shown in Fig. 3.4b. The co-deposition of A-rich A-B binary system takes place until the potential reaches E_c, where the two current-potential curves intersect. The intersection point exhibit the same current density of electrodepositing of A and B implying that the contents of each constituent of the binary A-B system are the same. Afterward, as the cathodic overpotential increases, the deposition rate of B proceeds, and as a consequence, the binary system will become B-rich. This is a very simple understanding of the co-electrodeposition concept which can be extended to discuss alloy electrodeposition and hydrogen evolution.

3.2.6.1 Electrodeposition of Alloys

By definition alloys are classified into two major groups including homogeneous and non-homogenous. The former may consist of one phase systems comprised of either a solid solution or an intermetallic compound. The later includes two-phase alloys comprised of one or more one phase alloy as well as virtually unalloyed elements. Brenner [9] defines electrodeposited alloys as those cathodic deposits that fall within the scope of the definition given in above. Electrodeposited alloys are similar to thermally processed alloys in structure. However, the cathode product obtained after electrodeposition of two or more metals is not necessarily an alloy and it may be a powder consisting of individual crystals of each metal.

To be able to electrodeposit alloys, it is essential to have close electrode potentials of constituents of atypical alloy, as schematically shown in Fig. 3.5. The definition of close electrode potentials is vast and must be carefully selected. However, as a general rule a one-volt potential window is normal between the electrode potential of alloy constituents for electrodepositing their alloys. However, some methods have been employed to shift their potentials to approach each other for this purpose. We recognize two distinct methods to achieve this.

The first is to manipulate the potential electrode via the Nernst equation, (2.11). The principle factor which changes the electrode potential is the ion concentration of the metal to be electrodeposited. As a rule of thumb, every change of concentration of a doubly charged ion in an order of 100 shifts the electrode potential by 10 mV. Since there is narrow chance of remarkable change of electrolyte concentration, this method is only appropriate to slightly control the deposition current and therefore the alloy content.

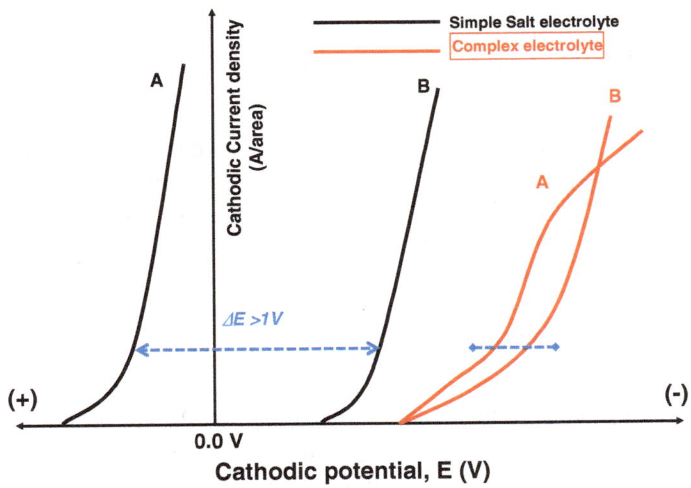

Fig. 3.5 The effect of electrode potentials and type of electrolytes for co-deposition of A-B binary alloy system

The second method is to substantially changing of electrode potential using complex electrolytes for example copper cyanide plating bath. This is schematically shown in Fig. 3.5 for a nominal A-B binary alloy electrodeposited from simple slat and complex electrolytes. When the simple slat electrolyte is used, the electrode potential of metal A and B are far away from each other which do not allow co-electrodeposition. But, the electrode potentials of metals A and B become very close when electrodeposited from a nominal complex bath. One good example for this process is the electrodeposition of Cu–Zn (or brass) alloys in cyanide electrolytes. The electrodeposition of brass from simple salts electrolytes is impossible as the electrode potentials of Cu/Cu^{2+} and Zn/Zn^{2++} redox couples are far beyond the potential window required for co-deposition.

3.2.6.2 Hydrogen Evolution: Mechanism and Overpotential

Hydrogen reduction reaction (HRR) or hydrogen evolution is a well-known phenomenon during electrodeposition of metals from aqueous solutions:

$$2H^+_{aq.} + 2e^- \rightarrow H_2(g) \uparrow \tag{3.20}$$

The electrode potential of HRR can be expressed as:

$$E_{\left(\frac{H^+}{H_2}\right)} = E^\circ_{\left(\frac{H^+}{H_2}\right)} + \frac{RT}{nF}\ln\frac{[H_2]}{[H^+]^2} = -0.0592\,\text{pH} \tag{3.21}$$

And for a general electrochemical reaction, it imposes an overpotential known as the hydrogen overvoltage defined as:

$$\eta_{H_2} = E_{Measured} - E_{\left(\frac{H^+}{H_2}\right)} = E_{Measured} - (-0.0592\,\text{pH}) \tag{3.22}$$

According to the current-potential curves shown earlier in Fig. 3.7, the reduction of less noble is always associated with the reduction of the more noble element provided that the ions of both elements exist in the electrodeposition bath. However, this figure is a schematic view of this process, as for some metals with nominal more noble potentials such as copper the hydrogen evolution takes places [10].

A multi-step mechanism has been suggested for the hydrogen evolution on a cathode by Crow [11]:

(i) Aqueous protons in the hydrated state ($H^+_{(aq.)}$ $or\, H_3O^+$) are transported (i.e. via most probably diffusion or even migration) from the bulk electrolyte right to the vicinity of OHP layer and transferred across the double layer.

(ii) $H^+_{(aq.)}$ *or* H_3O^+ ions will be adsorbed on the cathode by elimination of their hydration layer.

(iii) Charge transfer process takes places on the cathode electrode, M. A possible form of the electron discharge reaction may be:

$$M + H_3O^+ + e^- \rightarrow M - H + H_2O \tag{3.23}$$

(iv) Generation of hydrogen bi-atomic molecules from the hydrogen ad-atoms adsorbed on the cathodes. Two possible HRR reactions have been proposed as:

$$M - H + M - H \rightarrow 2M + H_2 \tag{3.24}$$

Or

$$M - H + H_3O^+ + e^- \rightarrow M + H_2 + H_2O \tag{3.25}$$

(v) desorption of hydrogen gas in form of bubbles.

The rate determining step among them is controversial. Tafel proposed that (3.15) is the rate determining process. He considered a cathodic Tafel slope of 0.03 at 25 °C for HRR for his argument, though such slope is only observable at very low overpotentials on very smooth platinised Pt electrodes. For many electrodeposition systems under practical high overpotentials, the ion discharge reaction would be the slowest stage, where the cathodic Tafel slope is about 0.118 at 25 °C [11]. The (3.22) was derived based on this assumption that the electrochemical HRR is the rate determining stage, (3.25).

3.2.6.3 Current Efficiency

The current efficiency is an important parameter to consider for a typical electrodeposition process. As mentioned, it is quite possible to have two or more faradaic reactions occurring simultaneously at the cathode electrode or the substrate. However, one of them is related to the deposition of materials. Thus, the fraction of the current consumed for solid state deposition to the total current passed through the cell during an electrodeposition process is called the instantaneous current efficiency. The current efficiency is unity (or 100%) when only one reduction reaction occurs during the electrodeposition. A general way to figure out the cathodic current efficiency is to consider the fraction of the total charge involved in the process. This can be readily related to the theoretical mass of deposit using the Faraday's law. Thus, the current efficiency, χ, is written:

$$m = \chi \cdot \left(\frac{M}{n} \cdot \frac{1}{F} \cdot Q \right) \tag{3.26}$$

This is a value between zero and 1, i.e. $0 < \chi < 1$. One good reason for most of electrodeposition processes to have a current efficiency less than 100% is side reactions such as hydrogen evolution, water dissociation, etc.

3.3 Nucleation and Growth of Electrodeposits

3.3.1 Atomistic View

As one of the deposition methods of condensed matter, the nucleation and growth of electrodeposited materials consists of several stages. We may divide the nucleation and growth of an electrodeposit into two major processes including: (I) deposition and (II) electrocrystallisation. The stages of the electrodeposition process are schematically shown in Fig. 3.6.

The first step in the deposition process is the movement of transportation of solvated ions (in aqueous electrolytes hydrated ions) towards the cathode

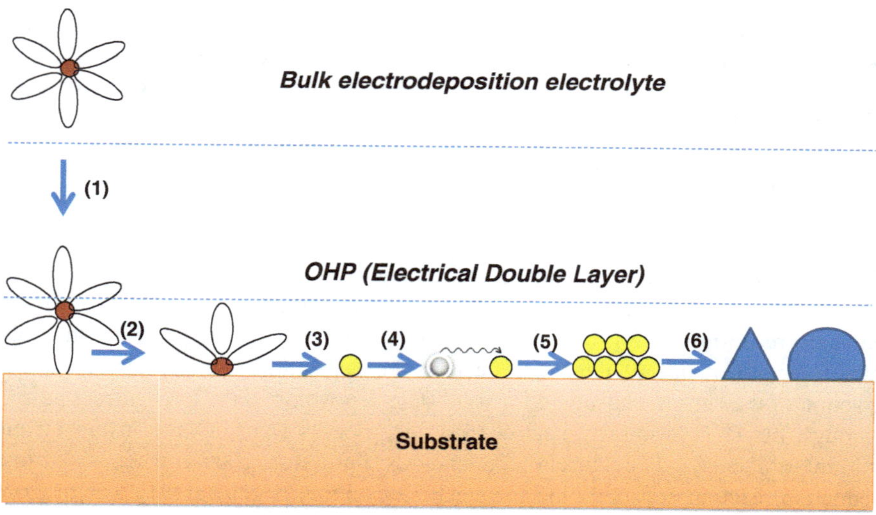

Fig. 3.6 Schematic picture of the nucleation and growth during a typical electrodeposition process from an electrolyte consisting of *(1) ion transport from bulk electrolyte to the OHP, (2) transfer of ions across the electrical double layer, (3) de-solvation (or dehydration) in part or in total, and formation of ad-ions and ad-atoms, respectively (4) surface diffusion of ad-atoms, (5) nucleation of stable atomic clusters and (6) irreversible incorporation of ad-atoms in the atomic lattice and establishment of a peculiar crystallographic texture and morphology or electrocrystallisation*

underwell-known mechanisms of mass transport, i.e. immigration, diffusion and convection. During this stage the hydrated ions approach the border of double layer, i.e. OHP (stage 1 in Fig. 3.6). From this point, the main stage of deposition process begins with crossing the electrifiedinterface. A hydrated ionis waiting at the OHP. Before this ion becomes condensed on the substrate, it has to be released of its sheath of solvating water. This is called the charge-transfer reaction. At this step, the charge transfer is complete and hydrated ions are attached onto the substrate cathode surface. This stage results in the formation of ad-ions and consequently ad-atoms.

The formation of the ad-ions and ad-atoms species follows a complicated pathway mostly depending upon the surface energy of the substarte. We discussed in Chap. 1 the surface growth mechanism and theories, and that there exist several growth sites on a given surface. When an ion will be partly in contact with the substrate surface, first, it will be landed on a peculiar site on the substrate and become a surface *ad-ion*. This ad-ion still has some charge and therefore some molecules of hydration water associated with it. Its charge is less than the ion in solution and as a result must be surrounded by less number of water molecules than that hydrate it in solution. Next, it will be accommodated on the growth sites and the remaining space around it can accommodate water molecules. This is a particular stage during electrodeposition (Stage 4 in Fig. 3.6) called *surface diffusion* which would be the movement of ad-atoms on the substrate surface. A random-walk diffusion process in two dimensions or *surface diffusion* is responsible for the ad-atom movements on the surface. Depending on each category of the growth sites, i.e. flat, kink, edge, etc., the number of water molecules around ions varies. It is the maximum number of the water molecules when an ad-ion is located on the flat surface or terrace. This number progressively decreases as the ad-ion is places in a step, kink, edge vacancy, and hole. Thus, if an ion changes its site on the surface its surrounding will substantially change. Further progress will replace all hydration water molecules by coordinating metal atoms until its full accommodation number. This process is now "zero" charge ion or ad-atom which could be embedded in the lattice. This is called *stepwise replacement of water molecules by metal ions* and formation of *ad-atom* [12]. The surface diffusion takes place for the diffusion of ad-atoms from one site to another for instance from the plane sites to the step sites, and then to the step to a kink site (c.f. Chap. 1, Fig. 1.18), and finally ends up to lattice incorporation. This results in the *nucleation of stable atomic clusters* (Stage 5 in Fig. 3.6).

3.3.2 Thermodynamics of Nucleation

This process may be readily interpreted by considering thermodynamics of nucleation as was explained earlier in Chap. 1. Here we consider a general metal reduction reaction happening during an electrodeposition process:

$$M^{n+} + ne^- \rightarrow M^\circ$$

The Nernst equation defines the electrode potential of the substrate at equilibrium (or zero current electrodeposition potential) by defining the activity of metal ad-ions M^{n+} adhered to the substrate surface $(a_{ad-atoms})$ as follows:

$$E_{eq} = E^\circ + \frac{RT}{nF}\ln(a_{ad-atoms}) \tag{3.27}$$

At this condition there is no driving force to conduct the electrodeposition stages mention in above. Therefore, in order to nucleate on the substrate, the substrate must be under a cathodic overpotential, η more negative than E_{eq} which is the threshold of the initiation of electrodeposition, where current passed through the cell rises (c.f. Fig. 3.2 for a graphical instruction for this concept). Applying such cathodic overpotential will change the activity of ad-ions on the surface to $a^*_{ad-atoms}$. Thus we can write the Nernst equation, (3.19), to obtain the dependency of the overpotential upon the activity of metal ad-ions M^{n+}:

$$\frac{a_{ad-atoms}}{a^*_{ad-atom}} = \exp\left(\frac{-nF\eta}{RT}\right) \tag{3.28}$$

And we have a relationship between the volume Gibbs free energy ΔG_V and overpotential according to the Nernst's equation:

$$\Delta G_V = -\frac{nF\rho\eta}{M} \tag{3.29}$$

where ρ is the mass density and M is the molecular weight of M. If one considers that the formation of spherical nucleus takes place on a substrate, we can obtain the Gibbs free energy of nucleation as (c.f. Chap. 1, (1.27)–(1.33):

$$\Delta G = \Delta G_V + \Delta G_s = \frac{4}{3}\pi r^3\left(\frac{nF\rho\eta}{M}\right) + 4\pi r^2\gamma \tag{3.30}$$

γ is the density of surface energy. Thus, we calculate the barrier energy, ΔG^* and critical radius, r^* as follows:

$$\frac{d\Delta G_{total}}{dr} = 0$$

Therefore

$$\Delta G^* = \frac{16\pi M^2\gamma^3}{3\rho^2 n^2 F\eta^2} \tag{3.31}$$

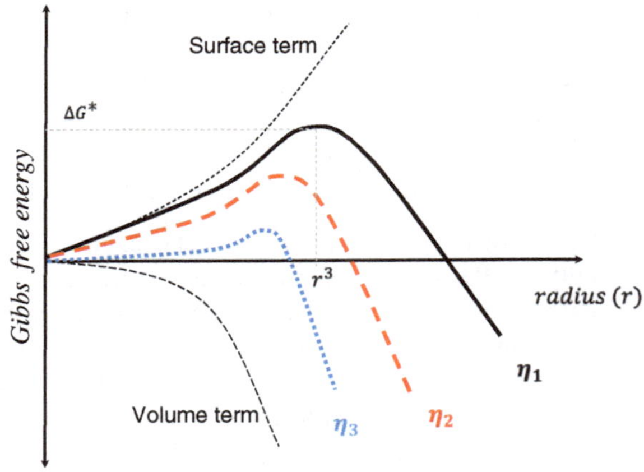

Fig. 3.7 The dependency of Gibbs free energy on the radius of a typical spherical nucleus formed during electrodeposition under different overpotentials: $\eta_1 < \eta_2 < \eta_3$

and

$$r^* = \frac{-2\,M\gamma}{nF\rho\eta} \tag{3.32}$$

This is concluded here that on the basis of a simple assumption the energy barrier of nucleation as well as the radius of an electrodosited sphere significantly depend on the overpotential, as the cathodic overpotential increases the energy barrier and radius of the nuclei are decreased. This is schematically shown in Fig. 3.7.

3.3.3 *Kinetics of Electrocrystallisation*

Electrocrystallisation is generally defined as the nucleation and growth of crystalline materials during electrochemical reactions like electrodeposition. Many people have talked about this topic and proposed different definitions and explanations. There are also several books and monographs dedicated to this astonishing subject [13, 14]. As we talked earlier, the electrodeposition process consists of two major stages namely deposition (i.e. formation of ad-ions, ad-atoms and clusters) and electrocrystallisation. Active sites are essential for electrocrystallisation to begin; therefore, deposition stage consisting of adsorption and nucleation steps is key prerequisite for growth processes.

During electrocrystallisation, two-dimensional (2D) and three-dimensional (3D or bulk) deposits and crystal growth are realised. Several factors affect the mechanisms and modes of 2D or 3D electrocrystallisation which mainly include:

(1) Substrate: native or foreign, microstructure, material, crystallographic texture, etc.,
(2) Substrate/electrolyte interface: double layer arrangement, IHP, OHP, Zeta potential,
(3) Electrolyte phase, and
(4) Kinetics of the mass and charge transfer, and the rate determining step.

The Nernst equation (3.10) may be a good starting point to distinguish the stability of 2D or 3D electrodeposits on the substrate. For 3D or bulk electrodeposits on either native or foreign substrates, the equilibrium potential of a general reduction reaction (e.g. a metal) determines the potential threshold of initiation of phase formation and growth on the substrate as 3D electrodeposits form at any negative potentials or overpotentials with respect to the equilibrium potential. This is called overpotential deposition (OPD). However, the dissolution of 3D phase starts at more positive potentials or overpotential with respect to the equilibrium potential. In contrast, under certain conditions, 2D electrodeposits are formed on a foreign substrate at potentials more positive than the equilibrium potential defined by the Nernst equation. This happens at some particular situations and is called underpotential deposition (UPD).

Therefore, UPD and OPD of a metal correspond to the formation of 2D and 3D electrodeposits, respectively. One important thing to investigate the electrocrystallisation of materials is to understand the influence of electrodeposition kinetics controlled by the charge transfer, mass transport (mainly diffusion), and or both of them. This needs to be discussed here in more details. Further arguments about the 2D and 3D electrodeposits will be explained in Chap. 4.

There is a linear rate of electrocrystallisation mentioned in literature known as V directly related to the current density, i, according to the expression:

$$V = \frac{iV_m}{nF} \tag{3.33}$$

where V_m is the molar volume and F is the Faraday constant. This equation implies that the current density distribution plays an important role in the rate of electrocrystallisation, as the greater the current density applied the higher rate of growth achieved. Thus, a different growth form or mode with possibility of side reactions such as hydrogen evolution is anticipated.

From the thermodynamics point of view, the initiation of electrocrystallisation must be accompanied by imposing a metastable state to the system. This may be achieved by two methods. One is to perturb the equilibrium via the application of a cathodic overpotential which is called *direct electrocrystallisation*. Another way is to alter the temperature or composition of the electrolyte to achieve a supersaturation state which is called *indirect electrocrystallisation*. Under practical conditions, the variation of electrode potential or current density is applied to control the electrocrystallisation processes. Thus, the potential distribution may lead to variation of the electrodeposit morphology and growth rate, or even other side reactions such as hydrogen evolution [15].

The kinetics of electrocrystallisation has been a major topic of research for years [15–17], and there has been development towards understanding of the mechanisms of electrochemical deposition using the electrochemical techniques or recently using other methods such as in situ microscopical techniques, optical or gravimetric methods. In a simple approach, the rate of growth stable centers is assumed to follow first-order kinetics. The number of active growth sites, N_c as a function of growth time t, is concerned with the initial number, N_0, by this expression [18]:

$$N_c = N_0[1 - \exp(-k_n t)] \tag{3.34}$$

where k_n is called the nominal nucleation rate constant.

Based on this equation, two modes of growth during electrodeposition may be mainly recognized namely *instantaneous and progressive* nucleation [19]. The instantaneous nucleation is defined as the number of nuclei formed at the initiation of electrocrystallisation remains constant during the whole electrodeposition. This is the case when $k_n \gg 1$. Then $\exp(-k_n t) \sim 0$ and therefore, $N_c = N_0$. In contrast, the progressive nucleation is devoted to another mode of nucleation where the nucleation is progressively taking place during electrodeposition. Under such circumstance, $k_n t$ has very small value. Thus, this relation can be used as a close estimation: $\exp(-k_n t) \sim -k_n t$. Therefore, $1 - \exp(-k_n t) = k_n t$. We find finally:

$$N_c = N_0 k_n t \tag{3.35}$$

We now interpret the kinetics of electrocrystallisation based on the controlling processes. As mentioned before in Fig. 3.2, at different overpotentials the electrodeposition is controlled by charge transfer (or activation) and mass transport (or diffusion) processes.

Let's first assume that electrocrystallisation is controlled by charge transport process. Under this condition the overall current density of electrodeposition for a total deposition time, t, is written as:

$$i = k \int_0^t A_n(t - \tau_n) v_n d\tau \tag{3.36}$$

where k is a modified nucleation rate constant, A_n is the total area of a single nucleus and $v_n = dN/dt$ is the nucleation rate defined here as the rate of increase in the number of nuclei per deposition time, t. Depending upon the overpotential applied the growth mode would change. The instantaneous nucleation occurs at higher overpotentials. It was shown before that as the overpotential is increased the energy barrier and the critical radius of nuclei decrease, c.f. Fig. 3.7. Thus it is very likely to have an increased number of nuclei formed at high overpotentials, meaning that the nucleation takes place during a narrow period of time and then is suddenly diminished. For the instantaneous nucleation, the overall current density mentioned in (3.36) needs to be modified and is written as:

$$i = kA(t)N_0 \tag{3.37}$$

And for the progressive nucleation:

$$i = kA^2(t)N_0 \tag{3.38}$$

Regardless of the nucleation modes, atomic clusters may form in either two-dimensional (2D) or three dimensional (3D) forms. In both cases, the growth of single nucleus and multiple nucleations can be determined by current-time relationships. For a single cylindrical nucleus with radius r, and height h, the current-time relationship is written for 2D as:

$$i_{2D-cylinder} = \frac{nMF(2\pi h)k_{2D}^2}{\rho}t \tag{3.39}$$

The nucleus in 3D form is a hemisphere with radius r:

$$i_{3D-hemisphere} = \frac{nM^2F(2\pi h)k_{3D}^3}{\rho^2}t^2 \tag{3.40}$$

For the multiple nucleation and when the nuclei do not overlap with each other, the (3.39) and (3.40) will be re-written under two different nucleation modes of instantaneous and progressive. For 2D nuclei and non-overlapping, the equations are:

$$i'^{(instantaneous)}_{2D-cylinder} = \frac{2\,nMF\pi hN_0k'^2_{2D}}{\rho}t \tag{3.41}$$

and

$$i'^{(Progressive)}_{2D-cylinder} = \frac{nMF\pi hAN_0k'^2_{2D}}{\rho}t^2 \tag{3.42}$$

For 3D nucleation and non-overlapping case, the equations are:

$$i'^{(instantaneous)}_{3D-hemisphere} = \frac{2\,nM^2F\pi N_0k'^3_{3D}}{\rho^2}t^2 \tag{3.43}$$

$$i'^{(progressives)}_{3D-hemisphere} = \frac{2nM^2F\pi AN_0k'^3_{3D}}{3\rho^2}t^3 \tag{3.44}$$

The current-time or current transient equations seem to assume the same behaviour under different nucleation modes. As is evident, for example, there exists a parabolic relationship between the current density and time ($i \propto t^2$) for 2D progressive nucleation as well as for 3D instantaneous nucleation. This is misleading if one only relies on the current transient to distinguish a nucleation and growth mechanism. Hence, it is most often essential to make a conclusion about the mechanistic study by investigating the real structures grown using alternative microscopical or other techniques [20].

For those nucleation and growth mechanisms that are associated with overlapping growth centers during electrocrystallisation, the current-time equations are changed. For the nucleation of cylindrical 2D centers under instantaneous and progressive modes we may write:

$$i_{2D-cylinder}^{\prime\prime(instantaneous)} = \frac{2nMF\pi hN_0k_{2D}^{\prime\prime 2}}{\rho} t\, exp\left(-\frac{M^2\pi N_0 k_{2D}^{\prime\prime 2}}{\rho^2} t^2\right) \tag{3.45}$$

And

$$i_{2D-cylinder}^{\prime\prime(Progressive)} = \frac{nMF\pi hAk_{2D}^{\prime\prime 2}}{\rho} t^2 exp\left(-\frac{M^2\pi A k_{2D}^{\prime\prime 2}}{3\rho^2} t^3\right) \tag{3.46}$$

It is obvious that the exponential terms in (3.45) and (3.46) appear due to the presence of the overlap effect. These equations imply that the current initially increases, reaching a maximum current, and then decreases to a plateau. For the progressive nucleation, the current decreases asymptotically to zero. This is shown in Fig. 3.8.

So far, we have discussed the nucleation modes under low overpotentials where the electrocrystallisation occurs under charge transfer regime. Provided that the overpotential increases, the kinetics of electrocrystallisation as well as the

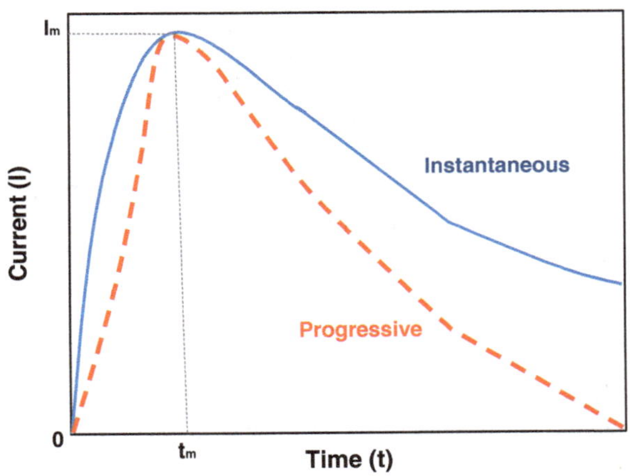

Fig. 3.8 A schematic picture of theoretical current-time *curves* for instantaneous and progressive growth modes under charge transfer control reaction

electrodeposit morphology will change. As the overpotential increases, the current increases according to the Tafel equation until it will be independent of the activation Gibbs energy and thus it the electrodeposition process is dominated by the ion transport rate. This is mainly controlled by diffusion for most of the reduction reactions giving rise to a constant maximum current known as the limiting current density, i_L as discussed earlier in this chapter. The value of the limiting current density depends upon many factors such as electrode geometry, flow conditions, the concentration of metal ions in bulk electrolyte, etc. Another important phenomena associated with the variation of overpotential concerns the change of surface roughness. This will be discussed later in this chapter.

Diffusion-controlled electrodeposition is a wide spread process for many systems such as electrodeposition of metals from aqueous electrolytes or molten salts owing to extremely fast charge-transfer step. Thus, the diffusion of ions towards the cathode plays an important role in determining the nucleation and growth mechanisms.

First, one needs to assume that the electrodeposition process is under diffusion control. There is a model proposed by Delahay and Berzins [21, 22] to investigate the mechanism of electrolysis. This model considers the effect of mass transport based on diffusion on the rate of reduction reaction in a typical current-potential relation normally achieved by cyclic voltammetry which is expressed by:

$$i_{\mathrm{p}} = 2.99 \times 10^5 \alpha^{1/2} A c D^{1/2} v^{1/2} \tag{3.47}$$

in which i_{p} is the current density at reduction peak, A corresponds to electrode area, c is concentration of ions in bulk electrolyte, v the potential scan rate and α is the cathodic charge transfer coefficient (about 0.5 for a typical metal electrodeposition). This relation approves the diffusion-controlled electrodeposition process. However, this relation may be well fitted to experimental data to figure out the diffusion coefficient of ions in a given electrolyte.

In order to explain the growth mode in 3D multiple nucleation diffusion controlled, let's consider the current-time curves. Figure 3.9 schematically compares three modes of diffusion controlled nucleation for non-overlapping and overlapping instantaneous and progressive nucleation. The current–time transient recorded during a typical diffusion controlled electrodeposition process can be divided into three regions, marked as I, II and III on Fig. 3.13. For all the three modes, the first region which is accompanied by a sudden drop at the beginning of electrodeposition is related to s the double layer charging reaction. In the second time interval for all three types of growth the current increases the extent of which depends on the growth mode. For multiple nucleation without overlap this region continues until the end of electrodeposition. This can be instantaneous or progressive. However, the nucleation of single nucleus has been frequently studied [23–26] which shows an increase of current as a function of root square of time. This increase can be due to the growth of either independent nuclei alone or independent nuclei and simultaneous increase in number of nuclei. This is the deposition current

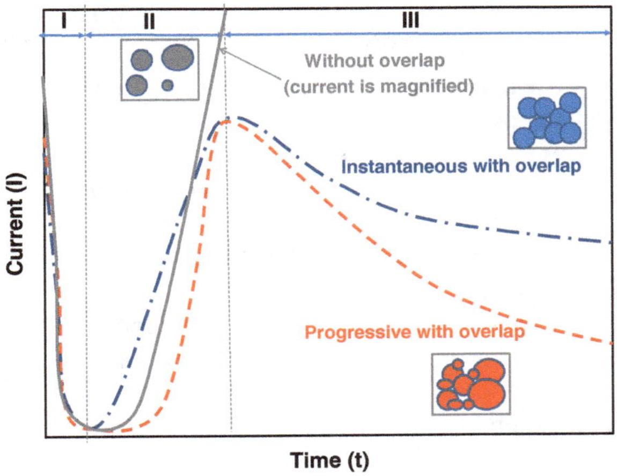

Fig. 3.9 Schematic illustration of different 3D nucleation modes under diffusion controlled electrodeposition

without the overlapping effect. If the nucleation is instantaneous, the current in region II increases linearly with time. If nucleation is progressive, the current of region II increases as a parabolic function of electrodeposition time ($I \propto t^2$). In the third region namely region III there are two opposing effects: growth of independent and overlap.

In earlier studies, Astley et al. considered a simple linear one-dimensional diffusion to figure out the mass transport characteristics [27]. Another explanation for the diffusion of ions was developed by localized spherical diffusion [28]. Scharifker and Hills and coworkers [29, 30] proposed a model assuming the hemispherical diffusion to a hemispherical nucleus, and re-expressing this as an equivalent area of plane surface fed by linear diffusion. The hemispherical diffusion zones may or may not overlap around the nuclei in a typical diffusion-controlled electrodeposition with multiple nucleation. The overlap of diffusion fields was introduced by using of Avrami's theorem [31, 32]. It must be mentioned that at the same time Avrami's argument for considering the overlap between the nucleation hemispheres, it was also derived by Bosco and Rangarajan [33, 34].:

$$\theta = 1 - e^{-\theta_{ex}} \tag{3.48}$$

in which θ is the area on which new nuclei can appear and θ_{ex} is the 'extended area', the theoretical fraction of the area θ which has been nucleated if overlap is ignored. Scharifker and Hills considered a set of hemispherical nuclei randomly distributed on substrate surface growing under diffusion control. The diffusion zones grow and overlap each other. The corresponding flux and growth current was considered in terms of semi-infinite linear diffusion. For instantaneous nucleation of N center simultaneously:

$$\theta_{ex} = N\pi kDt \tag{3.49}$$

A hemispherical diffusion zone emanating from each nucleus with a radius δ defined as:

$$\delta(t) = (kDt)^{1/2} \tag{3.50}$$

in which k is a constant defined for instantaneous nucleation as:

$$k = \left(\frac{8\pi cM}{\rho}\right)^{1/2} \tag{3.51}$$

and for progressive nucleation as:

$$k = \frac{4}{3}\left(\frac{8\pi cM}{\rho}\right)^{\frac{1}{2}} \tag{3.52}$$

Thus, the current-time equations for instantaneous and progressive growth are as follows:

$$I_{instantaneous} = \frac{zFc\theta D^{1/2}}{\pi^{1/2}t^{1/2}}[1 - \exp(-N\pi kDt)] \tag{3.53}$$

and

$$I_{Progressive} = \frac{zFcD^{1/2}}{\pi^{1/2}t^{1/2}}[1 - \exp(-AN_{\infty}\pi kDt)] \tag{3.54}$$

N_{∞} is the number of the density of active sites. As shown in Fig. 3.10 both current transients has an initial rise up to a maximum followed by a fall down to the limiting current density. The current does not become null in contrast with the charge transfer controlled progressive nucleation. The current at peak, I_m and time at peak, t_m, correspond to a point in which the first derivate of the I-t equations becomes zero. In order to facilitate the study of nucleation and growth mechanism,

Fig. 3.10 Non-dimensional plots for **a** instantaneous and **b** progressive nucleation

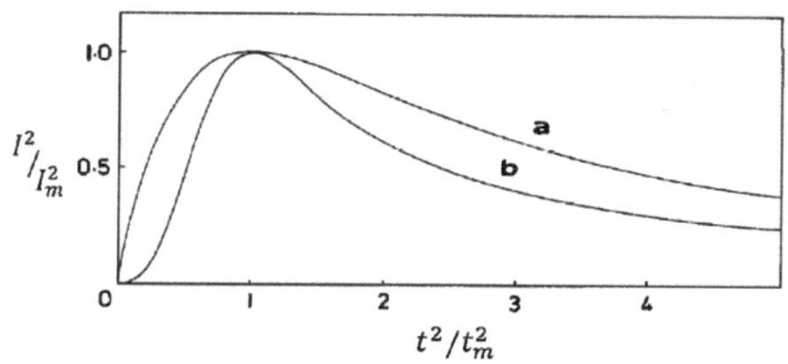

Scharifker and Hills suggested deriving non-dimensional plots I^2/I_m^2 as a function of t^2/t_m^2. Figure 3.10 compares theoretical non-dimensional plots for instantaneous and progressive nucleation as:

$$\left(\frac{I}{I_{\mathrm{m}}}\right)^2 = \frac{1.9542}{t/t_{\mathrm{m}}}\left\{1 - \exp\left[-1.2564\left(\frac{t}{t_{\mathrm{m}}}\right)\right]\right\}^2 \tag{3.55}$$

for instantaneous and

$$\left(\frac{I}{I_{\mathrm{m}}}\right)^2 = \frac{1.2254}{t/t_{\mathrm{m}}}\left\{1 - \exp\left[-2.3367\left(\frac{t}{t_{\mathrm{m}}}\right)^2\right]\right\}^2 \tag{3.56}$$

for progressive nucleation.

The Current transients from the steady-state region of i versus t data (at long times: t_{m}) can be used to calculate the diffusion coefficient at various potentials using the Cottrell's equation [35]:

$$i(t) = \frac{nFcD^{1/2}}{\pi^{1/2}t^{1/2}} \tag{3.57}$$

The depletion of the electroactive species at the vicinity of the surface leads to an inverse $t^{1/2}$ function known as another sign of diffusive control in electrodeposition process. However, in order to utilize this equation in practical measurements of i−t behavior, "Cottrell conditions" must be taken into account. These may include the instrumental and experimental limitations in the actual maximum current and voltage outputs of the potentiostat, time resolution limitations in the recording device in particular during the initial stages of the current transient, recognition between nonfaradaic and faradaic current such as the double layer charging at the beginning of electrodeposition, disturbing of diffusion layer by unwanted convection in longer deposition times, etc.

The model of Scharifker and Hills has been modified later. The model of Scharifker and Mostany [36] introduced a similar value for k (3.51) for either the instantaneous or progressive nucleation as: $k = \left(\frac{8\pi cM}{\rho}\right)^{1/2}$ and consequently an expression for the total current was calculated for circular diffusion zones, having time dependent radii. The initial time was assumed to be the time at which the first nuclei appears and the Avarmi's equation was modified in new form:

$$I(t)_{Scharifker-Mostany} = \frac{1}{\alpha(At)^{1/2}}\left(1 - \exp[-\alpha(At - 1 + e^{-At})]\right) \tag{3.58}$$

in which:

$$\alpha = (2\pi)^{3/2} D \left(\frac{cM}{\rho} \right)^{3/2} \frac{N_0}{A} \tag{3.59}$$

But this modification was later mentioned to be accurate for fitting with the experimental data only at the rising part of the transient [37]. However, it is possible to solve this equation using a genetic algorithm [38]. This model has also been modified for use under conditions of forced convection [39] in which the Cottrell equation is replaced by another expression based on a constant diffusion layer thickness. Other improvement on the nucleation and growth mechanisms have been achieved by Sluyters-Rehbach et al. [40] and Heerman and Tarallo's theory [41, 42] by further assumptions which could be investigate as additional references.

In conclusion, in my view and to the best of our knowledge and based on our research studies, the mechanism of nucleation and growth of electrodeposits must be accompanied by further microscopical or other techniques to reveal a better and more accurate understanding of the mechanism. It is aimed to introduce some of the novel techniques in next sections.

3.3.4 Surface Morphology and Roughness

Surface morphology is defined as the topographic structure of materials on their surfaces. Figure 3.11 shows a typical nano-scale image of a surface which contains hills and valleys over a certain length scale taken by atomic force microscopy. The

Fig. 3.11 A typical surface morphology of a metal surface. The image shows an AFM images of an electropolished Al surface. Reproduced from [43] with the permission of Elsevier

topology has a wide definition owing to its complexity and less degree of numerical predictability as it consists of complex phenomena still under vigorous attention by researchers.

In a peculiar term, surface roughness is a quantitative measure of surface morphology which is of particular scientific interest to understand the growth of thin films surface on a nanometer scale. Surface roughness affects most physical or chemical properties of thin films such as the electrical conductivity [44], the interface dependent magnetism [45–47], the optical properties and interacting waveguides [48], corrosion [49], etc. Further to its applications, this enables understanding of the physics behind the growth and structure of surfaces, though it is challenging due to the complication of growth monitoring as well as complexities of the growth processes.

The first step to determine the surface roughness is to employ some way of quantitatively characterizing tools. Figure 3.12 demonstrates a number of surface characterisation toll with their spatial resolutions. We classify the morphology characterisation techniques into two categories: (i) real-space imaging techniques including atomic force microscopy (AFM), scanning tunneling microscopy (STM), scanning electron microscopy (SEM), surface profilometry instruments, and (ii) wave-matter scattering techniques including high resolution low-energy electron diffraction (HRLEED), reflection high-energy electron diffraction (RHEED), atom diffraction, X-ray diffraction, and light scattering (oblique incident reflectometry, etc.). Each category has advantages, for instance, the imaging techniques provide direct visual interpretation of the surface morphology with capabilities of extraction

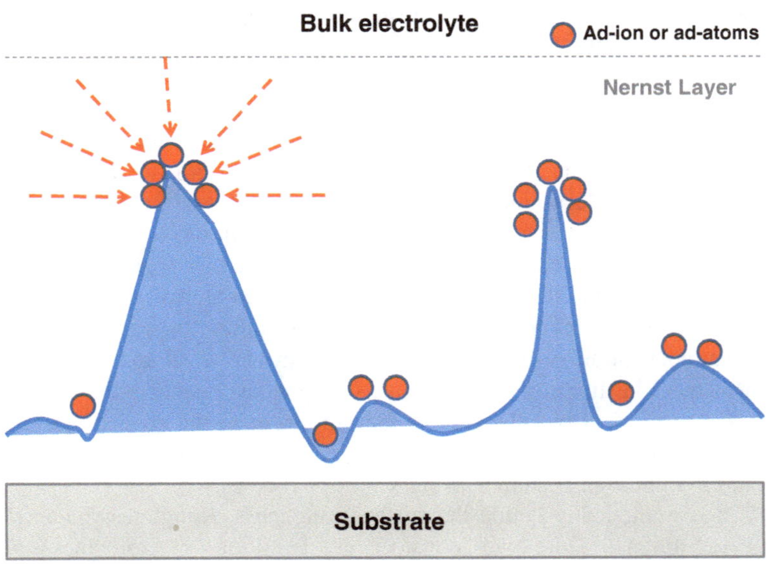

Fig. 3.12 A schematic description of morphological instabilities during a typical diffusion-controlled electrodeposition process

Table 3.1 A comparison between surface imaging and surface scattering analysis techniques

Method	Resolution (Wavelength λ or effect length, L) (Å)	Measurable spatial range (Å)	Measurable height (RMS) range Å
Visible light	$\lambda = 6328$	5×10^4–10^8	10^2–10^4
X-Ray	$\lambda = 1.0$–1.5	10^0–10^4	10^0–2×10^2
HRLEED	$\lambda = 4$	10^0–7×10^4	10^0–4×10^2
RHEED	$\lambda = 0.1$	10^0–10^3	10^0–8×10^2
STM	$L = 10^2$–10^4	7×10^2–10^4	10^0–2×10^3
AFM	$L = 10^3$–10^6	10^0–10^3	10^0–10^4

of surface roughness data statistics, whereas the scattering techniques are non-contact and the data can be extracted from the angular in a short time (Table 3.1).

There have been various mathematical methods developed to analyze the surface roughness and other surface characteristics [50]. Frankly, it has been found that surface morphology or roughness of many thin film surfaces can be interpreted by common *scaling* rules namely *self affine* or *dynamic scaling* [51]. Scaling has been mainly carried out on the surface roughness and the arguments work quite well when the important growth effects in a deposition are "local", or only affect nearby surface heights, an example of which is surface diffusion, where atoms can diffuse to nearby locations depending on deposition conditions such as activation energy and temperature.

Thus, the roughness scaling then could be well used to study the electrodeposition or electrocrystallisation mechanisms as this deposition technique is mainly related with the electrochemical reduction of electroactive species on the surface passing through the electrical double layer. Morphological instabilities during electrodeposition of materials occur. One major reason for such instabilities is the diffusion. During diffusion-controlled electrodeposition any surface irregularities are amplified rather than damped during growth. Referring to previous section, (3.16)–(3.18), the concentration profile of, in the other word, the limiting current density determines the rate of electrodeposition. The question is how this leads to this instability? To answer, we plot a schematic picture of a typical electrode posit with irregularities and distribution of ad-ions or ad-atoms around them. The *Mullins–Sekerka instability* [52] is a close explanation for our argument. It is evident that concentration of ions on the surface around a hill and a valley is quite different. Thus, the limiting current density is greatest near the tip of the protuberance because tip of protuberance is located closer to the edge of the diffusion layer, thus, the rate of electrodeposition is greater. This is an example that necessitates the morphological studies using scaling hypothesis.

Let's first numerically define the surface roughness. Roughness parameter has various definitions. Here we consider a definition of roughness namely scale-dependent roughness which is expressed:

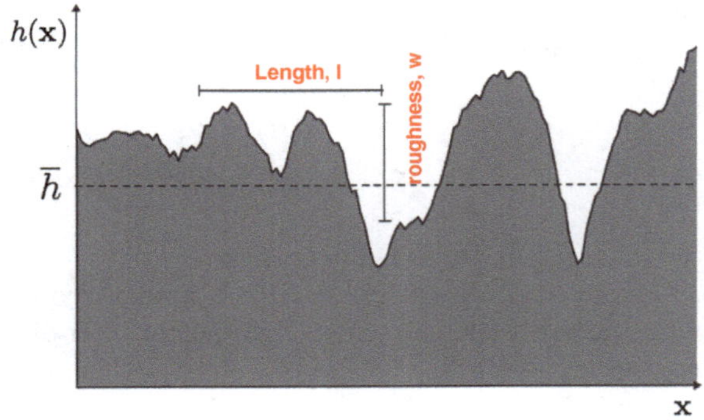

Fig. 3.13 A cross sectional view of surface topography as a measure for roughness

$$w(l) = \sqrt{h - \bar{h}} \tag{3.60}$$

$\bar{h}$ the average height of overall surface as introduced schematically in Fig. 3.13. $w(l)$ is a general term dedicated to roughness which is the root mean square of height difference over the length l. However, one needs to consider the root mean square value for surface roughness, R_q in terms of this expression:

$$R_q = \sqrt{\frac{1}{l}\int_0^l w_{(x)}^2 dx} \tag{3.61}$$

in order to extract the roughness from surface analysis data.

The kinetic roughening of electrodeposited films has been a topic of research for, in particular, recent years. Dynamic scaling has been used to analyze the surface roughness of the films. We simply describe the principles of the concept here, but a more details review paper is given by *W. Schwarzacher* [53]. For investigating the kinetic roughening it is a need to monitor the surface roughness quantity $w(l, t)$ as a function of film thickness, t. Equation (3.26), the relationship between the deposited mass and the charge passed through the cell; the Farday's law simply demonstrates the effect of electrodeposition time, current density, and charge on the film thickness. Therefore, the kinetic roughening could be understood in terms of these parameters. The thickness-dependent RMS of the roughness is, $w(l, t)$ is compared at different film thicknesses in terms of logarithm. Equation in below is used to fit the experimental data as expressed by [54]:

$$w(l, t) = w_{sat}\left\{1 - exp\left[-\left(\frac{l^H}{l_c}\right)\right]\right\} \tag{3.62}$$

$w(l, t)$ is proportional to log(l) below some value l_c namely the cross over length, and saturates above, taking a saturation value, w_{sat}. In this equation H is the so-called Hurst exponent. Accordingly, the kinetic roughening of thin electrodeposited films can be scaled by two models namely: *Normal* scaling and *Anomalous* scaling. A simple power law is used to distinguish the normal and anomalous models as may be expressed by:

$$\begin{aligned} w(l,t) &\propto l^{H} t^{\beta_{loc}} \quad for\, l \ll l_c \\ w(l,t) &\propto l^{H} t^{\beta_{loc}} \quad for\, l \ll l_c \end{aligned} \tag{3.63}$$

in which β_{loc} is the growth exponent.

Family and Vicsek proposed the normal dynamic scaling [55–57] in which $w(l,t)$ behaves take the same values before l_c for different thicknesses, but different w_{sat} values after l_c. In other words, if $\beta_{loc} = 0$ in (3.63) the scaling is normal. This is shown in Fig. 3.14a. The normal scaling may exhibit the simplest model of power law behaviour that w(l, t) could exhibit. However, more complex behaviour has been realised which do not obey this simple model. Thus, another scaling namely Anomalous scaling has been introduced [58–60], where β_{loc} is non-zero. It is different from the normal scaling as the roughness behaves differently before the it saturates for different thicknesses. This model is schematically shown in Fig. 3.14b.

In earlier investigations, the kinetic roughening of the electrodeposited films has been scaled using these equations for copper [61–64], Ni [65–67], Ag [68], Pt [69], and many metals and alloys.

For copper films electrodeposited from simple sulfate solutions, Huo and Schwarzacher have interpreted an anomalous scaling [70]. In addition, more recently Schwarzacher's group has introduced slope analysis as a straightforward complement to scaling analysis for characterizing the morphology of electrodeposited thin films [71]. The surface slope θ, defined as the angle between the local surface normal and the film normal, has been calculated as a function of position for electrodeposited Cu films of increasing thickness. It has been shows that the mean value of θ increases with thickness and demonstrates the close relationship between this observation and the increase in local roughness found in their previous studies from scaling analysis.

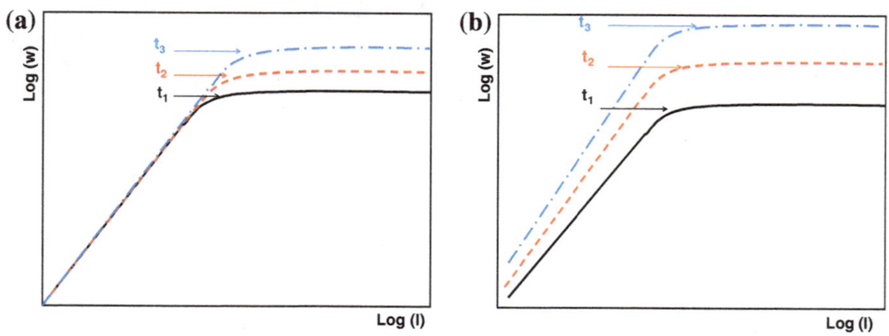

Fig. 3.14 A schematic representation of dynamic scaling models for kinetic roughening of electrodeposited films, **a** normal scaling, **b** anomalous scaling. *Note* t is the film thickness in an order of $t_1 < t_2 < t_3$

3.4 Characterisation of Initial Stages of Electrodeposition Process

Earlier studied using the conventional electrochemical techniques have been unable to precisely describe the initial stages of electrodeposition process. Thanks to advances in materials characterisation techniques, we are now able to detect and monitor the initial stages of electrodeposition as a very useful section of the nanostructuring. The general principles of in situ characterisation methods for the electrochemical deposition of materials on a substrate for a few important techniques are briefly explained in below.

3.4.1 Scanning Probe Microscopy (SPM) Techniques

Scanning probe microscopy is a unique technique to monitor real-time growth of materials by different deposition techniques. This microscopical technique is used in the form of in situ electrochemical mode at the molecular and atomic level. Over the last decades, the electrochemical studies have been gradually expanded on common fundamental areas such as corrosion of metal surfaces, metal deposition, and charge transfer, surface morphology related to corrosion, batteries and semiconductor nanofabrication. Here we briefly explain two experimental modes including scanning tunneling microscopy (STM) and atomic force microscopy (AFM) used electrochemical studies.

3.4.1.1 Electrochemical Scanning Tunneling Microscopy (EC-STM)

Scanning tunneling microscope is a technique to study the structure and properties of materials and thin layers of conductive or partially conductive films on conductive substrates at sub-nanometer scales. STM was invented by Binnig and Rohrer in 1982 [72] and has been shown to be operated in an electrolytic environment [73, 74].

Principles of the measurement system employed in STMs are based on a quantum mechanical effect namely quantum tunneling. STMs are mainly composed of a sharp tip with atomic size radius. When a sharp tip approaches a conductive surface a current is established between the tip and the surface. This phenomenon called "tunneling current" occurs only in the light of quantum mechanics because the electrons are to pass from the tip to the sample surface where the gap (the area between the tip and surface) has an energy level larger than the total energy of the electron potential energy. The amount of electrical current varies as a function of the tip shape and material, geometry and texture and also the voltage difference between the tip and surface. The sharp conductive tip is connected with arms of a piezoelectric material which move it to any level from and position on the surface.

Then it is possible to examine all the surface properties. In determining the properties of different points of surface, STM is used in two modes directly and indirectly. In fact, those properties that are directly obtained from the tunneling current changes under the effect of distance as well as voltage difference applied between tip and surface are called direct mode of STM. The properties of the surface which are determined directly by STM include topography of the surface, work function of different points, density of energy states of different points, surface topography having different vibration levels, and magnetic domains of surfaces and their magnetisation.

In another category of methods, indirect modes, surface properties are investigated under the application of the tunneling current and the change of electronic distribution is recorded from a point of the surface is studied. In this case, some of its characteristics temporarily change so that they can be analysed e surface by other methods of spectroscopy such as Raman spectroscopy or luminescence. For example, detection of luminescence photon, i.e. photons with electrons excited by the return of the tunneling current transferred to the lower levels, energy is considered as a fingerprint of energy levels [75].

STM is used for imaging purposes under two instrumental adjustments: (i) constant current mode where the instrument works at constant electron density, (ii) constant height in which the tunneling current is monitored while the tip is kept at a fixed distance from the substrate surface. The constant-current mode images using negative feedback to control the z-piezo voltage and keep a set tunneling conductance. This mode provides the surface topography on the mesoscopic level but in order to provide atomic level images a correlation must be made between local electron density at the Fermi level and the atomic surface structure. The constant height mode images surface while the tip is located at a fixed height over the surface under the control of current. This mode can be done at higher scan rates provided that the tunneling current fluctuations are controlled by the z-piezoelectric scanner. Topography of flat surfaces is possible using this mode under controlled relationship between tunnel current and tip-substrate separation. Another capability of STM is to characterize the local electronic properties of the surface. This is carried out by manipulating an effective tunneling barrier or decay constant defined from the dependence of the tunnel current on the tip-substrate gap. Tunneling spectroscopy is another STM capability to examine the energy distribution of the density of states under a biased tunneling current at a fixed tip-substrate distance. Normal STM instruments require ultra high vacuum (UHV) to avoid any interrupting factors in the tunnel current or related parameters.

The operation of STM in an electrolyte has been shown a successful method for in situ observation of electrochemical processes such a electrodeposition. Electrochemical STM (EC-STM) in the electrolyte can be used to study the electrochemical corrosion processes or deposition of organic and inorganic systems. The main components of a typical ECSTM are a bi-potentiostat connected to a conventional STM system. A well-defined instrument requires the use of a bipotentiostat to independently control the electrochemical potential of the tip and substrate relative to a reference electrode. This instrumental configuration is different from normal STM in which only the bias is applied between the tip and sample. Here the electrodes need to be specified in the electrochemical environment and the tip electrode is simultaneously placed in the same electrolyte a tunneling probe and an ultramicroelectrode. Consequently, care must be taken to control or eliminate extra possible faradaic reactions happening at the tip itself. These include unwanted redox reactions and deposition and dissolution processes. These reactions if taken place are the basics of another mode of electrochemical SPM called scanning electrochemical microscopy (SECM) which will be discussed later. In constant current mode, the most important practical problems are the reduction of faradic and capacitive current between tip and sample to low levels. So, tip must be insulated using epoxy resin, nail polish, wax or silicone polymers to cover most parts of it except a small area at the tip of the needle. For a more details comprehensive instruction about ECSTM it is referred to [76].

3.4.1.2 Scanning Electrochemical Potential Microscopy (SECPM)

Scanning electrochemical potential microscopy is one of the latest modes of SPM techniques, patented by Veeco company [77]. The SECPM microscope is capable of imaging and mapping the potential of the electrode surface with a nanometer scale resolution. This method is based on the principle that the electrochemical potential at the interface of the electrode/electrolyte or the double layer changes as a function of distance. SECPM detects and images potential changes across the sample surface and provides insights into electrochemical reactions at the liquid/sample interface by mapping the potential distribution of the surface with a resolution comparable to STM. Potential changes between the probe and sample surface immersed in an electrolyte or polar liquid can be detected by potentiometric probe. For further details about the technique it is referred to [78].

3.4.1.3 Scanning Electrochemical Microscopy (SECM)

SECM was introduced by Bard [79] in 1982 as an electrochemical scanning probe technique in which current caused by an electrochemical reaction at the tip as an ultramicroelectrode (UME) is measured as the characteristic parameter to examine the surface and its electrochemical reactions. The overall apparatus for SECM is similar to that used for EC-STM. A bipotentiostat controls the tip and substrate potentials and the tip is transferred to desired positions by piezo controllers. However, the basic working parameters and resulting information obtained by SECM differ from those f EC-STM. As mentioned, the tip is taking part into the electrochemical reactions having substantial contributions into the current. Smaller electrodes usually have a less well-defined geometry and are similar to STM tips produced by etching a wire to a sharp point and then insulating with wax or other coating. In the SECM experiment, the tip and substrate are immersed in a solution containing electrolyte. The cell also contains auxiliary and reference electrodes. The current at the steady state passing through the tip under a potential in far distances from substrate (i_{tip}^{∞}) is expressed by:

$$i_{tip}^{\infty} = ncDFr \tag{3.64}$$

in which r is the radius of the tip electrode. When the tip is moved towards the substrate, at the vicinity of the electrode, the current passed is determined by two processes which develop two SECM modes including: (i) *feedback* mode and (ii) *collection* mode. In the former, two cases are encountered. First, the diffusion of reactive species to the tip is eliminated by the substrate surface which decreases the current. This is generally called positive feedback mode taken place on an conducting substrate. Second, the surface can incorporate into the redox reaction occurring on the tip leading to a larger flux of electroactive species to the tip, which causes the current to increase. This is named negative feedback or blocking mode taken place from an insulating substrate. It is also possible to work in the *collection* mode, where the tip is held close to the substrate at a potential where electroactive products produced at the substrate are detected by the tip.

3.4.1.4 Electrochemical Atomic Force Microscopy

Atomic force microscopy was invented by Binning et al. [80] on the basis of sensing forces between the tip and the sample. Forces that can be detected are in a range of 10^{-13}–10^{-4} N with lateral resolution of the order of nanometer. Figure 3.15 shows a schematic view of main components of an AFM instrument used in electrochemistry. The tip is attached to a piezo cantilever which will be deformed by the forces acting on the tip. The most important advantage of AFM over STM techniques is that there is no need for the samples to be electrically conductive. AFM works in different modes depending on type of the atomic forces. Contact mode is a major category of AFM in which the attractive forces between the tip and substrate dominates. Non-contact mode is another category of imaging modes of AFM where the repulsive forces between the tip and the sample substrate persist. Another more recent technique developed is tapping mode, where the resonance frequency of the cantilever is correlated with the vibrations on the surface in the atomic scale. This technique is quite useful for soft matter and where peculiar investigations require no perturbation on the surface. This enables to record and interpret the amplitude and phase variation of modulating wave for understanding the surface characteristics.

For electrochemical investigation, AFM has become a major in situ examination technique. In situ atomic force microscopy has impressive capabilities with real-time imaging of electrochemical processes on the electrode surface under electrochemical control systems. The sample holder must be designed in order to fit the cantilever along with the conventional three-electrode electrochemical cell connected externally to a potentiostat instrument. This is shown in Fig. 3.15.

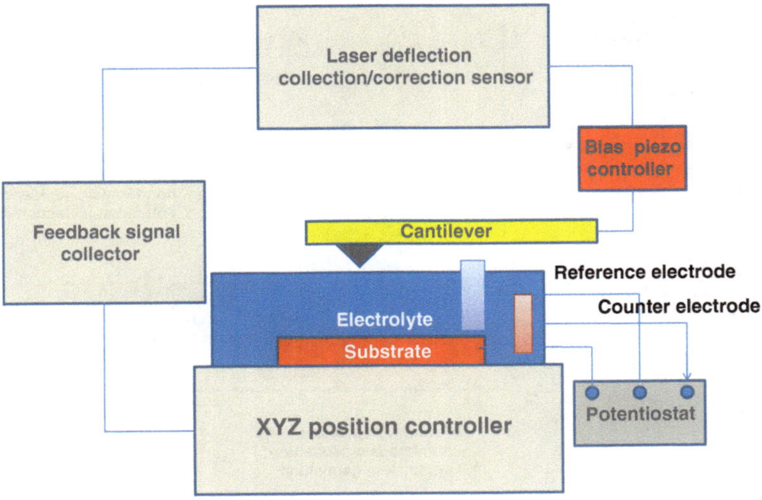

Fig. 3.15 Main components for a typical EC-AFM instrument

3.4.2 *Electrochemical Quartz Crystal Microbalance (EQCM)*

Electrochemical quartz crystal microbalance is known as an electrogravity analyzing technique which is able to distinguish faradaic and non-faradaic reactions during a given electrochemical process. The principle Faraday's law (3.26) gives a simple explanation between the charge passed through the cell and the mass produced during a typical reduction reaction. Conventional voltammogram give details of the redox reactions in terms of current-potential curves. Electrogravity techniques provide useful information about the variation of mass as a function of time and current of the electrochemical reaction such as electrodeposition called massogram. Since the electric currents produced or consumed at an electrode/electrolyte interface involves two faradaic currents and non-faradaic currents, the voltammograms are unable to predict the difference.

Charge transfer to discrete states of species in electrolyte may lead to the faradaic currents. In contrast, non-faradaic currents arise from charge transfer to continuum states of electrodes for example interfacial phenomena. The massograms (variation of mass as a function of potential), we may divide the electric current during a typical electrodeposition process into deposition or non-deposition currents. Deposition currents are associated with the formation of detectable mass, while non-deposition currents are the rest such as hydrogen evolution. The Venn diagram shown in Fig. 3.16 indicates different categories of an electrodeposition process [81].

EQCM is a powerful method in determining the electrochemical reactions involved in an electrodeposition process which can be performed along with other

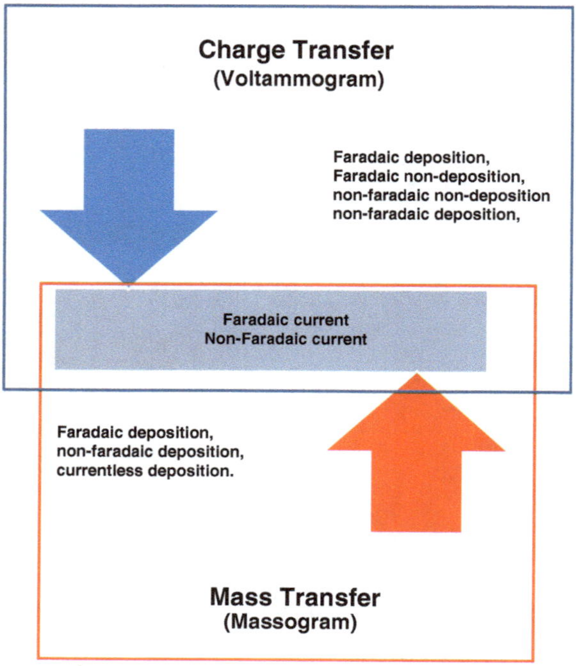

Fig. 3.16 Venn diagram for five distinguishable phenomena in mass transfer and charge transfer processes

conventional electrochemical techniques such as voltammetry, chronoamperometry and etc. This method is based on quartz crystal mass changes (Δm) upon the formation or dissolution of deposits during the electrochemical deposition process. The change of mass is directly proportional to the change of the resonant frequency ($\Delta f = f - f_o$) of the quartz crystal according to the Sauerbrey's equation:

$$\Delta f = f - f_0 = -\left(\frac{f_0}{A x_q \rho_q}\right)\Delta m = -S\Delta m \tag{3.64}$$

in which f_0 is the resonance frequency of quartz crystal before deposition, f the resonance frequency of quartz crystal after mass change, x_q is the thickness of the quartz crystal, ρ_q is the density of the quartz crystal (2.645 g cm^3 at 20 °C), A is the surface area of the quartz crystal, and S is the Sauerbrey constant. $\Delta m/\Delta t$ is measured to figure out the rates of mass transfer directly from the experiments, and then compared with the rates of charge transfer at electrode surfaces.

Also, S is the crystal sensitivity factor (e.g. 56.6 Hz μg^{-1}cm^2 for 5 MHz AT-cut quartz crystal form at room temperature). The Sauerbrey equation shows a linear dependence of the sensitivity factor and in theory does not need to be calibrated for deposition of thin films, rigid and uniformly. However, gas-phase deposition in a vacuum or liquid phase deposition like electrodeposition cannot satisfy all the requirements and therefore will not follow the linear mode and therefore, it needs to be calibrated before experiments [82, 83].

EQCM has been widely used in the studies of nucleation and growth of metal electrodeposits to exploit the UPD and OPD mechanism [82, 83], corrosion studies, and other electrochemical processes.

3.4.3 Oblique Incidence Reflectivity Difference

Oblique incidence reflectivity difference (OI-RD) is one of polarization-modulation ellipsometry (PME) methods. This method is useful to monitor any changes occurring on the surface of a substrate such as electrodeposition of solid phases [84, 85]. One essential condition for this particular method compared to other optical techniques is its instrumental requirement for a Pockel cell and a polariser for nullification of the first and second harmonics of the phase modulated light (note: In the Pockels or Pockels electro-optic effect introduced by F.C.A. Pockels as any changes or production of birefringence in an optical medium is proportional to the electric field applied, whereas in the Kerr effect, the refractive index change or birefringence is proportional to the square of the field). OI-RD has sub-monolayer sensitivity and high detection speed to be used as an effective in situ characterisation method for the early stages of even rapid kinetic electrodeposition reactions.

A typical OI-RD instrument is schematically shown in Fig. 3.17. It is composed of a 15 laser source, a polarizer, a KHz photoelastic modulator between parallel (p) and perpendicular (s) reflection plane, and a Pockels cell. After reflection from the sample, the light passes through an analyzer before being detected by the

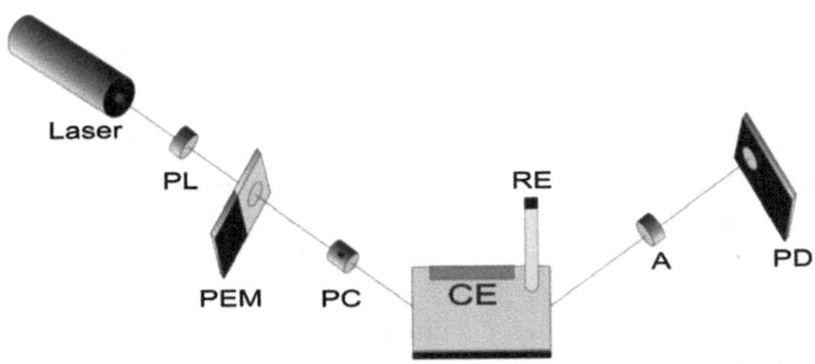

Fig. 3.17 Sketch of the electrochemical cell and optical set up for the oblique incidence reflectivity difference experiment. *Laser*; *PL* Glan-Thompson polarizers; *PEM* Photoelastic modulator; *PC* Pockels cell; *CE* Counter electrode; *RE* Reference electrode; *WE* Working electrode; *A* Analyzer; *PD* Biased silicon photodiode. Reproduced from [86] Reprinted with permission from Elsevier

photodiode. The intensity of light is monitored during the electrodeposition process in terms of various harmonics of the modulation frequency. OI-RD has been used to study the electrodeposition of lead (Pb) [86] and cobalt (Co) [87].

3.5 Modified Electrodeposition Processes Under External Forces

Since electrodeposition is a selective method of material fabrication in the sense that electrode potential must reach a point that enables the electron transfer from electrode to ions in solution, then it is possible to control it via external forces. We specifically point out to distinct methods as will be utilized in this project including: (i) magnetic electrodeposition and (ii) sono-electrochemical deposition or we call sono-electrodeposition.

3.5.1 Magnetic Electrodeposition

Magnetic fields have been used during electrodeposition to channel ions towards the deposition sites and enhance the electrodeposition rate of magnetic ions in solution. This process therefore makes constructive use of magnetohydrodynamics and field-induced convection which creates a transverse flow that disrupts the depletion layer near the electrode. The morphology of radially grown electrodeposits is especially sensitive to magnetic fields when the *Lorentz* force interferes with natural convection [84]. The Lorentz force (f_L) is described as the cross product of the electric current density j and the magnetic induction B:

$$F_L = j \times B \tag{3.66}$$

Due to the small electrical conductivities of aqueous electrolytes ($\sigma \sim 10$ S/m) and assuming small convection velocities (<1 m/s), both induced electric currents and induced magnetic fields can safely be neglected, and the force density is only determined by the external quantities [88]. This main effect in electrochemistry was called the magnetohydrodynamic (MHD) effect and is well-established. This effect is schematically shown in Fig. 3.18 for different configurations of applied magnetic field (***B***) with respect to the ionic flow. Convection generated by the Lorentz force in non-uniform fields has been used to create flow patterns in the electrolyte that would otherwise be unattainable. The convection in the electrolyte induced by magnetic fields superimposed on electrochemical processes increases mass transport indicated by a rise in the limiting current density. In deposition processes, the applied magnetic field produces smoother deposit surfaces [89, 90]. In addition it reduces the hydrogen evolution for non-noble metals and any subsequent hydroxide formation

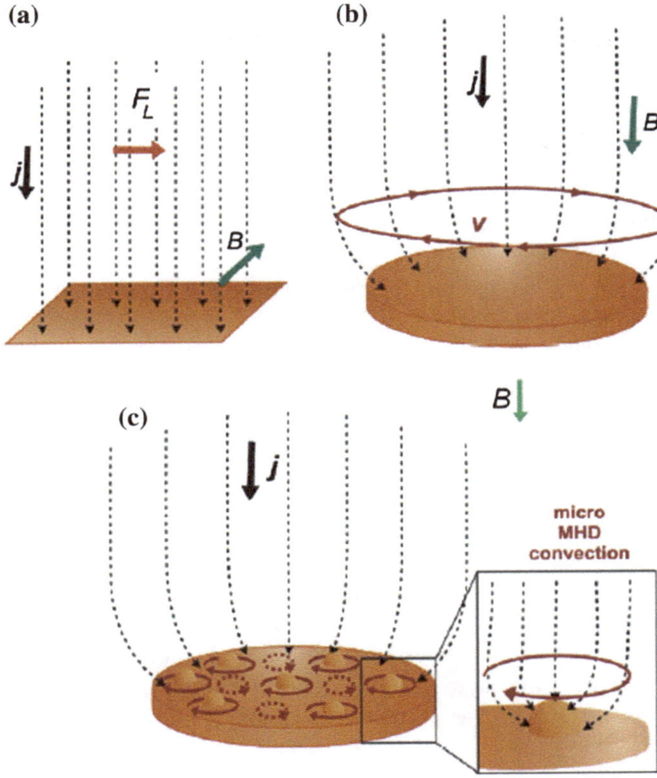

Fig. 3.18 Hydrodynamic flow at an electrode in a uniform magnetic field. **a** Field applied parallel to the electrode surface. The primary MHD flow is parallel to the surface. **b**, **c** Perpendicular to the electrode surface. In (**b**) the primary MHD flow is a vortex around the rim but in (**c**) secondary micro-MHD vortices arise around protuberances on the surface. Reprinted from [95] with the permission of Elsevier

[91–94]. The effect of magnetohydrodynamic flow on the electrode surface during a typical electrode process [95] is schematically shown in Fig. 3.18.

Further to the advantages of homogeneous magnetic fields on electrodeposition, most recently the action of *inhomogeneous magnetic fields* [96] on electrodeposition has been exploited to directly pattern magnetic and non-magnetic metals and alloys, known as the magnetic field gradient force, f_m, which plays an important role in inhomogeneous magnetic fields. f_m reads:

$$f_m = \mu_o \chi \nabla B = \mu_o \chi_{sol} \nabla \frac{B^2}{2}, \tag{3.67}$$

where $\chi_{sol} = \chi_{H_2O} + \sum_i \chi_i^{mol}.c_i$ [11]. Here χ_{H_2O}, χ_i^{mol} and c_i denote the susceptibility of water and the molar susceptibility and the concentration of species i, respectively.

Further to these two types of forces, there exits another type of force in the electrolyte due to the natural convection named the *buoyancy* force. However, the convection can only be driven in regions where gradients of both the magnetic species and the magnetic field do exist. This is in particular important for paramagnetic species [96, 97]. *To describe the ratio of these three forces mentioned, it has been shown that for magnetic elements of small diameter* $d < 1$ mm, *the magnetic gradient force clearly dominates over the Lorentz force in magnitude.*

Most of studies have employed the macro-size magnetic elements to impose inhomogeneous magnetic field to deposit magnetic (Ni, Co, Fe, CoFe, etc.) and non-magnetic (Cu, Zn, etc.) in the presence and absence of homogenous magnetic fields [93, 98–102]. Several features have been explored. For example, the structure (antidots against dots) was reversed for Zn, deposited from diamagnetic Zn^{2+} ions, or Cu with addition of strongly paramagnetic Dy^{3+} or Gd^{3+} ions, which are electrochemically inert. Other techniques have also been used to modify the pattern structures like using pulsed electrodeposition [103], monitoring concentration of ions near electrode surface [104, 105] and three-dimensional fluid flows generated during the electrochemical deposition using astigmatism particle tracking velocimetry measurements [98]. In all these studies, structures created have dimensions down to the micrometer. It was mentioned that the scale of the deposits on large flat surfaces available will be limited by the diffusion layer thickness, which varies inversely with the current density.

3.5.2 Sono-Electrodeposition

In principle, sono-electrochemistry is the processes and phenomena associated with an electrochemical system which is subject to ultrasonic irradiations. Ultrasound is classified for the sound with a frequency beyond human hearing frequency range which is usually above 20 kHz. The ultrasonic frequency range can be divided into two distinct sections: 1. Diagnostic ultrasound with very high frequencies (above 5 MHz) useful for the non-destructive testing of materials as well as fetal scan applications. Owing to their low energy, this category produces no cavitation in the propagation media. 2. Power ultrasonic waves with a frequency mainly ranging 20–100 kHz (and occasionally up to up to 2 MHz) is frequently used in many applications such as chemistry owing to their high energy which can create cavitation and affect the chemistry and processes. Thus, sonoelectrochemistry is related to ultrasonic waves with a frequency up to 2 MHz. Power ultrasound works in two ways: (i) direct mechanical transmission of vibration from the transducer onto a solid to induce vibrations in the surface of the material, and (ii) indirect

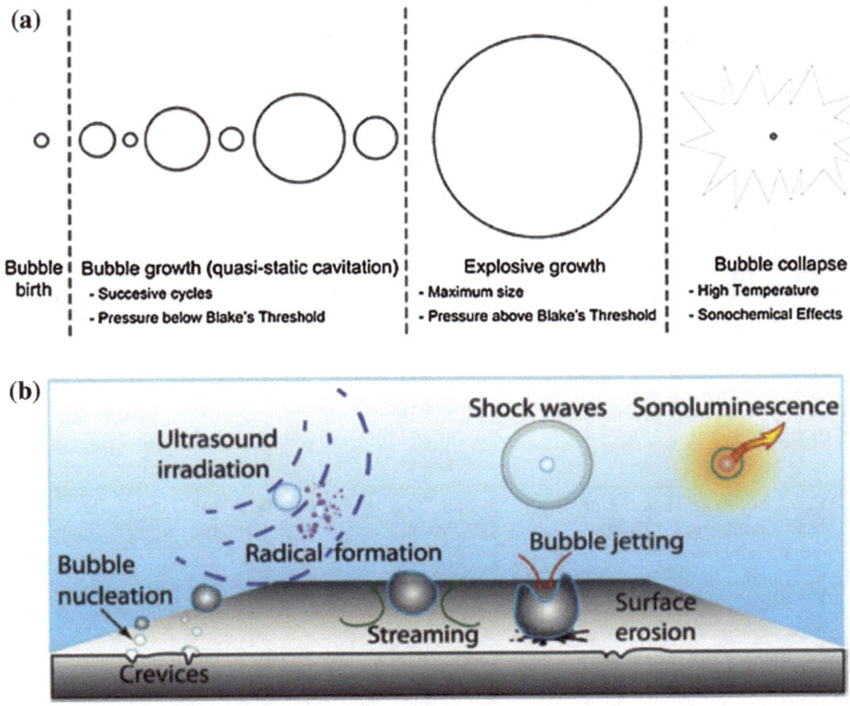

Fig. 3.19 **a** Bubble growth and implosion in a liquid irradiated with ultrasound. **b** Schematic representation of the main effects of cavitation induced by ultrasound irradiation. Reprinted from [109] with the permission of Elsevier

transmission of energy via cavitation induced in a fluid by the transmission of acoustic vibrations [106].

The first application of ultrasound waves in an electrochemical system goes back to 1930 by Moriguchi [107] who showed that the electrolysis of water using platinum electrodes under ultrasound occurred faster at lower voltages. More recently, the positive effect of ultrasonic waves has been discovered in electrodeposition process [108]. It is emphasized here that ultrasound not only affect the electrode/electrolyte interfacial phenomena, but it has significant effects on the electroactive species in the bulk electrolyte. Thus, sono-electrochemistry depends upon the phenomenon of acoustic cavitation. Similar to any soundwave, ultrasound waves are propagated by a series of compression (positive pressure) and rarefaction (negative pressure) waves imposed on the molecules of the medium through which they pass. At sufficiently high power the rarefaction cycle may exceed the attractive forces of the molecules of the liquid and cavitation bubbles will form. It is the fate of these cavities when they collapse in succeeding compression cycles which generates the energy for chemical and mechanical effects [106]. This is a remarkable phenomenon induced throughout the liquid at ultrasonic frequencies that is,

20 kHz or above. The process of formation cavitation and its effect on the electrode process is schematically shown in Fig. 3.19. At adequately high powers, negative pressure generates a cavity or 'bubble' in the liquid during the rarefaction cycle which breaks down the 'attraction' forces of the molecules of the liquid. The bubble becomes larger until at a critical size, it collapses.

Application of ultrasound in the electrodeposition of metals may present many benefits, not only in terms of enhancement of ion mass transfer in diffusion-controlled electrodeposition, charge-transfer improvement, higher cathode current efficiency, but also in terms of the metallurgical characteristics of the electrodeposits such as the grain size refinement, hardness, crystallographic texture and etc. For more details, readers are referred to [109, 110].

References

1. P.W. Atkins, *Physical Chemistry*, 5th edn. (Oxford University Press, Oxford, 1994)
2. Y.D. Gamburg, G. Zangari, *Theory and Practice of Metal Electrodeposition* (Springer, New York, 2011)
3. N. Kanani, *Electroplating- Basic Principles* (Processes and Practice, Elsevier, Oxford, 2004)
4. M. Schlesinger, M. Paunovic (eds.), *Modern Electroplating, 5th edn* (John Wiley & Sons, New Jersey, 2010)
5. F. Endres, D. MacFarlane, A. Abbott (eds.), *Electrodeposition from Ionic Liquids* (Wiley-VCH Verlag GmbH & Co, KGaA, Weinheim, 2008)
6. H. Nattar, M. Bukowski, Hempelmann et al., Z. Phys. Chem. **220**, 1275–1291 (2006)
7. F. Endres, S., Zein El Abedin, Z., Phys. Chem. **221**, 1407–1427
8. F. Endres, S, Z. Phys. Chem. 218, 255–283 (2004)
9. A. Brenner, *Electrodeposition of Alloys* (Principles and Practice, Volumes I, II, III, Elsevier, 1963)
10. S.S. Djokic, *Modern Aspects of Electrochemistry, Electrodeposition Theory and Practice*, No. 48 (Springer, New York, 2010)
11. D.R. Crow, *Principles and Applications of Electrochemistry* (Springer-Science + Business Media, B.V., 1974)
12. J.O'.M. Bockris, A.K.N. Reddy, *Modern Electrochemistry*, 2nd edn. (Kluwer Academic Publishers, New York, 2002)
13. E. Budevski, G. Staikov, W.J. Lorenz, *Electrochemical Phase Formation and Growth* (VCH Verlagsgesellschaft mbH, Weinheim, Germany, 1996)
14. A. Milchev, *Electrocrystallization: Fundamentals of Nucleation and Growth* (Kluwer Academic Publishers, Dordrecht, 2002)
15. F.C. Walsh, M.R. Herron, J. Phys. D Appl. Phys. **24**, 217 (1991)
16. M. Fleischmann (ed.), *Electrocrystallisation, Nucleation and Phase Formation (Faraday Symposia of the Chemical Society 12)* (The Chemical Society, London, 1977)
17. M. Fleischmann, H.R. Thirsk, Electrochim. Acta **2**, 22 (1960)
18. M. Fleischmann, H.R. Thirsk, Ado. Electrochem. Elecfrochem. Eng. **3**, 123 (1963)
19. A. Bewick, M. Fleischmann, H.R. Thirsk, Trans. Faraday Soc. **58**, 2200 (1962)
20. M. Paunovic, M. Schlesinger, *Fundamentals of Electrochemical Deposition*, 2nd edn. (John Wiley & Sons, Hoboken, New Jersey, 2006)

21. P. Delahay, T. Berzins, J. Am. Chem. Soc. **75**(10), 2486–2493 (1953)
22. T. Berzins, P. Delahay, J. Am. Chem. Soc. **75**(17), 4205–4213 (1953)
23. B.R. Scharifker, G.J. Hills, J. Electroanal. Chem. **130**, 9 (1981)
24. C.A. Gunawardena, G.J. Hills, I. Montenegro, Electrochim. Acta **23**, 693 (1978)
25. G.J. Hills, I. Montenegro, B.R. Scharifker, J. Appl. Elecrtrochem. **10**, 807 (1980)
26. G.J. Hills, A. Kaveh Pour, B.R. Scharifker, Electrochim. Acta **28**, 891 (1982)
27. D.J. Astley, J.A. Harrison, H.R. Thirsk, Trans. Faraday Soc. **64**, 172 (1968)
28. G.J. Hills, D.J. Schiffrin, J. Thompson, Electrochim. Acta **19**, 657 (1974)
29. G.A. Gunawardena, G. Hills, I. Montenegro, B. Scharifker, J. Electroanal. Chem. **138**, 225 (1982)
30. B. Scharifker, G. Hills, Electrochim. Acta **28**, 879 (1983)
31. M. Avrami, J. Chem. Phys. **7**, 1103 (1939)
32. M. Avrami, J. Chem. Phys. **8**, 212 (1940)
33. E. Bosco, S.K. Rangarajan, J. Electroanal. Chem. **134**, 213 (1982)
34. E. Bosco, S.K. Rangarajan, J. Electroanal. Chem. **134**, 225 (1982)
35. Bard, A.J., Faulkner, L.R., *Electrochemical methods Fundamentals and Applications*. (John Wiley & Sons, 2001)
36. B. Scharifker, J. Mostany, J. Electroanal. Chem. **177**, 13 (1984)
37. M.E. Hyde, R. Jacobs, R.G. Compton, J. Phys. Chem. Sect. B **106**, 11075 (2002)
38. J. Yu, H. Cao, Y. Chen, L. Kang, H. Yang, J. Electroanal. Chem. **474**, 69 (1999)
39. M.E. Hyde, O.V. Klymenko, R.G. Compton, J. Electroanal. Chem. **534**, 13 (2002)
40. M. Sluyters-Rehbach, J.H.O.J. Wijenberg, E. Bosco, J.H. Sluyters, J. Electroanal. Chem. **236**, 1 (1987)
41. L. Heerman, A. Tarallo, J. Electroanal. Chem. **470**, 70 (1999)
42. L. Heerman, E. Matthijs, S. Langerock, Electrochim. Acta **47**, 905 (2001)
43. M. Ghorbani, F. Nasirpouri, A. Irajizad, A. Saedi, Mater. Des. **27**(10), 983–988 (2006)
44. S.M. Rossnagel, T.S. Kuan, J. Vac. Sci. Technol., B **22**, 240 (2004)
45. F. Nasirpouri, A. Nogaret, D. Atkinson, M Ghorbani. J. Magn. Magn. Mater. **299**(2), 356–361 (2006)
46. V. Ng, J.F. Hu, A.O. Adeyeye, J.P. Wang, T.C. Chong, J. Appl. Phys. **91**, 7206 (2002)
47. J. Gong, S. Riemer, A. Morrone, V. Venkatasamy, M. Kautzky, I. Tabakovic, J. Electrochem. Soc. **159**, D447–D454 (2012)
48. R. Rabady, I. Avrutsky, 2003. J. Opt. Soc. Am. B **20**, 2174 (2003)
49. F. Nasirpouri, M.R. Sanaeian, A.S. Samardak, E.V. Sukovatitsina, A.V. Ognev, Appl. Surf. Sci. **292**, 795–805 (2014)
50. Y.-P. Zhao, G.-C. Wang, T.-M. Lu, *Characterization of Amorphous and Crystalline Rough Surface: Principles and Applications* (Academic Press, New York, 2001)
51. P. Meakin, *Fractals, Scaling and Growth far From Equilibrium* (Cambridge University Press, Cambridge, 1998)
52. W.W. Mullins, R.F. Sekerka, J. Appl. Phys. **35**, 444 (1964)
53. W. Schwarzacher, J. Phys. Cond. Matter. **16**, R859–R880 (2004)
54. F.P. Nasirpourim, Ionics, **19**(3), 535–542 (2013)
55. F. Family, J. Phys. A: Math. Gen. **19**, L441 (1986)
56. F. Family, T. Vicsek, *Dynamics of Fractal Surfaces* (World Scientific Publishing, Co., New Jersey, 1991)
57. F. Family, T. Vicsek, J. Phys. A: Math. Gen. **18**, L75 (1985)
58. M. Schroeder, M. Siegert, D.E. Wolf, J.D. Shore, M. Plischke, Europhys. Lett. **24**, 563 (1993)
59. S. Das Sarma, S.V. Ghaisas, J.M. Kim, Phys. Rev. E **49**, 122 (1994)
60. J.M. Lopez, Phys. Rev. Lett. **83**, 4594 (1999)

61. G.L.M.K.S. Kahanda, X.-Q. Zou, R. Farrell, P.-Z. Wong, Phys. Rev. Lett. **68**, 3741 (1992)
62. J.M. Pastor, M.A. Rubio, Physica D **96**, 384 (1996)
63. A. Iwamoto, T. Yoshinobu, H. Iwasaki, Phys. Rev. Lett. **72**, 4025 (1994)
64. S. Huo, W. Schwarzacher, Phys. Rev. Lett. **86**, 256 (2001)
65. J. Morales, S.M. Krijer, P. Esparza, S. Gonzalez, L. Vazquez, R.C. Salvarezza, A.J. Arvia, Langmuir **12**, 1068 (1996)
66. M. Saitou, A. Makabe, T. Tomoyose, Surf. Sci. **459**, L462 (2000)
67. M. Saitou, W. Oshikawa, M. Mori, A. Makabe, J. Electrochem. Soc. **148**, C780 (2001)
68. Y. Shapir, S. Raychaudhuri, D.G. Foster, J. Jorne, Phys. Rev. Lett. **84**, 3029 (2000)
69. G. Nabiyouni, Jalali Farahani B. Appl. Surf. Sci. **256**, 674–682 (2009)
70. W. Schwarzacher, S. Huo, Phys. Rev. Lett. **91**, 119601 (2003)
71. L. Liu, W. Schwarzacher, Electrochem. Commun. **29**, 52–54 (2013)
72. G. Binnig, H. Rohrer, Helv. Phys. Acta **55**, 726 (1982)
73. R. Sonnenfeld, P.K. Hansma, Science **232**, 211 (1986)
74. H.-Y. Liu, F.-R.F. Fan, C.W. Lin, A.J. Bard, J. Am. Chem. Soc. **108**, 3838 (1986)
75. H. Siegenthaler, STM in electrochemistry, in *Scanning Tunneling Microscopy II*, ed. by R. Wiesendanger, H.J. Gunterodt (Springer, New York, 1992)
76. T.P. Moffat, *Electroanalytic Chemistry: A Series of Advances*, ed. by A.J. Bard, I. Rubinstein, vol. 1 (New York, M. Dekker, 1999)
77. C. Li, K.J. Kjoller, Veeco Instruments Inc., 2007. Scanning electrochemical potential microscope US Patent US7156965 B1
78. M. Herpich, J. Friedl, U. Stimming, *Surface Science Tools for Nanomaterials Characterization,* ed. by C.S.S.R. Kumar. (Springer, 2015), pp. 1–67, 2015
79. A. Bard, K.L. Fan, Anal. Chem. **61**(2), 132–138
80. G. Binnig, C.F. Quate, C. Gerber, Phys. Rev. Lett. **56**, 930 (1985)
81. G.A. Snook, A.M. Bond, S. Fletcher, J. Electroanal. Chem. **526**, 1–9 (2002)
82. F. Nasirpouri, Ionics **17**(4), 331–337 (2011)
83. F. Nasirpouri, Prot. Met. Phys. Chem. Surf. **47**(4), 534–539 (2011)
84. A. Wong, X.D. Zhu, Appl. Phys. A **63**, 1 (1996)
85. P. Thomas, E. Nabighian, M.C. Bartelt, C.Y. Fong, X.D. Zhu, Appl. Phys. A **79**, 131 (2004)
86. G.Y. Wu, S.E. Bae, A.A. Gewirth et al., Surf. Sci. **60**, 1886 (2007)
87. W. Schwarzacher, J. Gray, X.D. Zhu, Electrochem. Solid-State Lett. **6**, C73 (2003)
88. J.M.D. Coey, G. Hinds, J. Alloy. Compd. **326**, 238–245 (2001)
89. M. Uhlemann et al., Eur. Phys. J. Special Topics **220**, 287–302 (2013)
90. O. Devos et al., J. Electrochem. Soc. **145**, 401 (1998)
91. H. Matsushima et al., Electrochim. Acta **53**, 161 (2007)
92. M. Uhlemann et al., J. Electrochem. Soc. **152**, C817 (2005)
93. J. Koza et al., Electrochim. Acta **56**, 2665 (2011)
94. A. Krause et al., Thin Solid Films **515**, 1694 (2006)
95. L.M.A. Monzon, J.M.D. Coey, Electrochem. Commun. **42**, 38–41 (2014)
96. A. Nogaret, F. Nasirpouri, J.C. Portal, H.E. Beere, D.A. Ritchie, A.T. Hindmarch, EPL (Europhys. Lett.) **94**(2), 28001 (2011)
97. G. Mutschke et al., Phys. Rev. Lett. **109**, 229401 (2012)
98. O. Gorobets et al., J. Phys. Chem. C **112**, 3373 (2008)
99. P. Dunne et al., Phys. Rev. Lett. **107**, 024501 (2011)
100. P. Dunne, J.M.D. Coey, Phys. Rev. B **85**, 224411 (2012)
101. P. Dunne, R. Soucaille, K. Ackland, J.M.D. Coey, J. Appl. Phys. **111**, 07B915 (2012)
102. J. Koza et al., Electrochem. Comm. **11**, 425 (2009)
103. K. Tschulik et al., Electrochim. Acta **56**, 5174 (2011)
104. R. O'Brien, K. Santhanam, Electrochim. Acta **32**, 679 (1987)
105. K. Tschulik et al., Technisches Messen 78, 232, and 2012. Analyt. Chem. **84**, 2328 (2012)
106. T.J. Mason, V.S. Bernal, B.G. Pollet (eds.), *Power Ultrasound in Electrochemistry From Versatile Laboratory Tool to Engineering Solution* (John Wiley & Sons, West Sussex, United Kingdom, 2012)

107. N. Moriguchi, The effect of supersonic waves on chemical phenomena (III). The effect on the concentration polarization. J. Chem. Soc. Jpn. **55**, 749–750 (1934)
108. R. Walker, Ultrasound and electroplating. Chem. Br. **26**, 251–254 (1990)
109. I. Tudela, Y. Zhang, M. Pal, I. Kerr, A. Cobley, Surf. Coat. Technol. **259**, 363–373 (2014)
110. J.Y. Hihn, F. Touyeras, M.-L. Doche, C. Costa, B.G. Pollet, *Power Ultrasound in Electrochemistry From Versatile Laboratory Tool to Engineering Solution*, ed. by B.G. Pollet, John Wiley & Sons, West Sussex, United Kingdom

Chapter 4
Electrodeposition of 2D and 3D Meso and Nanostructures

Abstract This chapter deals with electrodeposited two dimensional (2D) and three dimensional (3D) meso and nanostructures. Using the fundamental arguments made in previous chapters, the underpotential deposition (UPD) and overpotential deposition (OPD) of electrodeposited systems will be discussed. This is consistent with the nucleation and growth theories. Next, 3D nucleation and growth will be explained in two different views: one as free standing metal or alloy centres on a substrate, and two multiple 3D nuclei and consequent coalescence for the formation of either thin (less than one micrometer in thickness) or thick films (more than one micrometer in thickness). Several examples of recent progresses of each topic will be addressed.

4.1 Introduction

In this chapter we will focus on the development of electrodeposition on two dimensional (2D) and three dimensional (3D) structures. We first consider in this chapter that underpotential deposition (UPD) leads to 2D structures whereas overpotential deposition (OPD) develops three dimensional structures. In the former it is generally considered that a monolayer of an electrodeposit forms on a foreign substrate but in the later the OPD leads to formation of 3D single nuclei (generally on an microelectrode), multiple 3D nuclei as free standing metal or alloy centres on a substrate, and multiple 3D nuclei and consequent coalescence for the formation of either thin (less than one micrometer in thickness) or thick films (more than one micrometer in thickness). We will explain some example for each category, though there are more in the literature.

F. Nasirpouri, *Electrodeposition of Nanostructured Materials*,
Springer Series in Surface Sciences 62, DOI 10.1007/978-3-319-44920-3_4

4.2 UPD and OPD: General Consideration

We first overview the principle mechanisms of UPD and OPD during a typical electrochemical deposition process of a metal (M) on a foreign substrate. For simplicity, we talk about the electrodeposition of a metal (M) on substrate (S). Formation of a solid phase M on a substrate S depends on (i) the binding energy (U) between the metal ad-atoms ($M_{\text{ad-atoms}}$) and substrate atoms (S), (ii) the crystallographic mismatch factor between M and S expressed by [1]:

$$f = \frac{a_S - a_M}{M}. \tag{4.1}$$

a_S and a_M are lattice parameters of substrate and metal bulk crystals in their relaxed state, respectively. Several factors influence the lattice mismatch factor such as temperature, thickness of the deposit, M, strain-induced effects during growth and etc.

For the electrodeposition purpose, we consider a process occurring at about equilibrium conditions where the supersaturation is small and kinetics is not perturbing the deposition. Under such conditions, for growth of a distinct M phase on S without formation of an alloy between them, we may distinguish two modes:

(i) The binding energy between $M_{\text{ad-atoms}}$ and S is smaller than the binding energy between $M_{\text{ad-atoms}}$ themselves or in other word between $M_{\text{ad-atoms}}$ on native substrate (M):

$$U_{\text{M}_{\text{ad}}-\text{S}} \ll U_{\text{M}_{\text{ad}}-\text{M}_{\text{ad}}} \tag{4.2}$$

This binding energy leads to OPD of a 3D M phase forming on S. This is well described according to the "Volmer-Weber" or island growth mode (Chap. 1, Fig. 1.21). In this case, the crystallographic M-S lattice mismatch has no effect on the phase formation.

(ii) The binding energy between $M_{\text{ad-atoms}}$ and S is greater than the binding energy between $M_{\text{ad-atoms}}$ themselves or in other word between $M_{\text{ad-atoms}}$ on native substrate (M):

$$U_{\text{M}_{\text{ad}}-\text{S}} \gg U_{\text{M}_{\text{ad}}-\text{M}_{\text{ad}}} \tag{4.2}$$

This energy state favors the formation of 2D phase formation mechanism (c.f. Chap. 1, Fig. 1.22). In electrochemical deposition this is named UPD. However, the morphology and shape of 2D film changes depending on the surface concentration of $M_{\text{ad-atoms}}$ at the equilibrium potential of reduction of M (Nernst' equation), If the lattice mismatch between M and S atoms is negligibly small, 2D UPD layers are followed by 3D OPD crystallites as epitaxial thin film according to the "Frank-van der Merwe" or layer-by-layer growth. Thus, no boundary may be seen between UPD and OPD films. Otherwise, in systems exhibiting large M-S lattice misfit, the 2D UPD initially forms on S which is subsequently followed by 3D OPD bulk phase. The UPD under this mode is composed of a strained microstructure. 3D phase then grows according the "Stranski-Krastanov" growth mode. There is a sharp boundary between the UPD and OPD phases due to the formation of misfit dislocations [2].

4.2.1 Underpotential Deposition (UPD)

In principle, UPD processes are classified as one of the two major divisions of a typical electrodeposition process in which 2D metal phase, 2D M-S surface alloy phase and 3D M-S bulk alloy phases form in the underpotential range (i.e., defined as a range of potentials more positive than the formal potential of a given reduction reaction of M). Under such potential range named under potential range, there are only sorption phenomena of electrolyte constituent such as electroactive species occurring on the substrate. Therefore, no faradaic reactions such as reduction reaction or co-reduction of M-S tend to occur. The interfacial phenomena associated with general electrochemical reaction would significantly be changed in order to provide feasible conditions for the sorption of species on the surface. This is then re-named as an *interphase* region which must provide desirable conditions for the formation of M UPD on substrate [2].

In order to understand the UPD mechanism, the chemical potential of M ad-atoms, $\mu_{\mathrm{M_{ad}}}$ is defined as a function of their surface concentration and electrode potential, E. We distinguish two step mechanism for the formation of M UPD.

(1) Formation of ad-ions on the substrate surface which is expressed by:

$$M^{n^+}_{\mathrm{sol}} \rightarrow M^{n^+}_{\mathrm{ad}} \tag{4.3}$$

in which solvated M ions from the electrolyte phase are partially desolvated and enter the substrate phase as ad-ions.

(2) Total or partial discharge of M ad-ions on the surface influenced by a charge transfer coefficient factor, λ as shown by:

$$M^{n^+}_{\mathrm{ad}} + \lambda z e^- \rightarrow M^{n(1-\lambda)^+}_{ad}. \tag{4.4}$$

The transfer coefficient ranges $0 < \lambda < 1$ and depends on the M-S bond strength and type. $\lambda = 0$ for completely ionic bond and $\lambda = 1$ for completely covalent bond. Theoretical calculations have exploited λ to possess values close or equal to unity for the formation of M on native substrate or in systems with relatively small differences of electronegativities of Me and S [3–5].

At equilibrium condition, the chemical potential of completely discharged M ad-atoms ($\mu_{M_{ad}}$) on S assuming $\lambda = 1$ is calculated by:

$$\mu_{M_{ad}^{n+}} = \mu_{M_{ad}} - z\mu_{e^{-1}}. \tag{4.5}$$

This condition is applied for the formation of a monolayer of M on S, known as UPD layer. For formation of 3D bulk phase, the Nernst equation is applied. Any negative overpotential applied will form OPD whose nucleation and growth was completely studies in Chap. 2.

The UPD phenomena have been discovered for many metal deposits on polycrystalline and single crystalline substrates from aqueous and ionic electrolytes. The presence of UPD is easily detectable in cathodic and anodic peaks appearing in cyclic voltammograms. In situ microscopical (STM, AFM, SECM), optical wave reflection (Oblique incident reflectometry) and electrogravimetric (EQCM) methods provide useful information on the UPD formation. We here provide two example research works from our group on the formation and UPD and OPD of lead (Pb) on two different metal substrates: polycrystalline copper (Cu) [6, 7] and gold (Au) [8, 9]. This is an instructive explanation to provide information on understanding of UPD and OPD of a given metal on substrate using either conventional electrochemical techniques or in situ techniques.

4.2.2 UPD and OPD in Electrodeposited Lead (Pb) on Copper (Cu)

One of the relevant studied M-S systems to explore the existence of the UPD is Pb on copper substrates deposited with either high vacuum [10] or electrochemical techniques [11]. It has been generally shown that two types of Pb deposits form. First, a monolayer of Pb and second, three-dimensional islands of Pb grow on copper. In the electrodeposition of Pb on Cu, both of the UPD as the initial monolayer and the OPD as the bulk deposited potential are distinctly formed. The structure and properties of the UPD monolayer have been studied extensively. However, the OPD phase has attracted much interest recently.

The mechanism of lead underpotential deposition has been frequently studied using different techniques including conventional electrochemical techniques, i.e., cyclic voltammetry and chronoamperometry [12–14], single crystal rotating ring disk with ex situ low-energy electron diffraction and scanning tunnelling microscopy [15–17] and also nanogravimetric electrochemical quartz crystal measurements [18]. Pb UPD on Cu in a chloride-containing electrolyte is accompanied by the incorporation of chloride ions into the deposition mechanism through desorption. In addition, surface alloy formation during the formation of the UPD on copper has been observed [11].

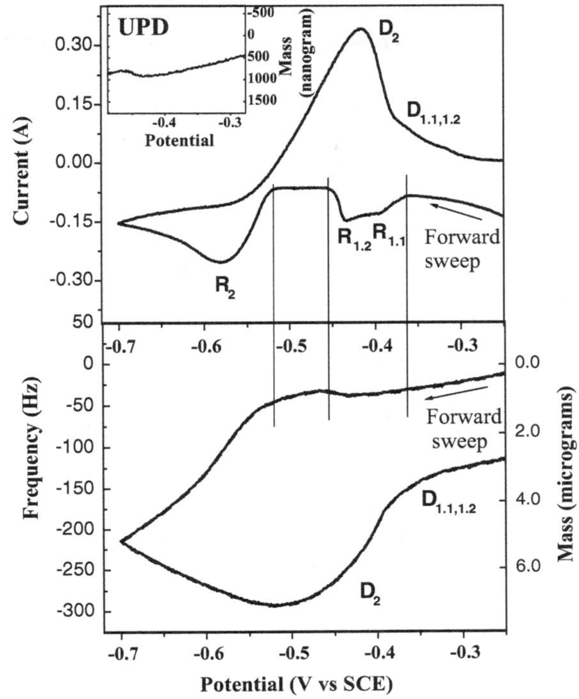

Fig. 4.1 A typical first sweep cyclic voltammogram (CV) (the *upper panel*) taken from 1 mM lead(II) perchlorate trihydrate and 10 mM HCl at room temperature on polycrystalline copper substrate at 10 mV/s scan rate. The *lower panel* illustrates the corresponding in situ measurement of frequency-potential or cyclic massogram. The *inset to upper panel* shows a magnified part of the UPD peak from the massogram. Reprinted from [6] with kind permission of Springer

Nasirpouri has investigated the UPD and OPD of Pb on copper film electrodeposited on a gold coated quartz crystal [6, 7]. Figure 4.1 shows a typical first sweep cyclic voltammogram (CV) (the upper panel) taken from 1 mM Pb $(ClO_4)_2{\cdot}3H_2O$ and 10 mM HCl at room temperature on polycrystalline copper substrate at 10 mV/s scan rate. The lower panel illustrates the corresponding in situ measurement of frequency against potential. As shown in the figure, a double peak, labelled R1.1 and R1.2 on the curve, occurs at $E_{1/2}|_1 = -0.38$ and $E_{1/2}|_2 = -0.43$ V versus SCE. The formation of a double-peak structure of UPD is in good agreement with other studies on Cu single crystals [15, 16] as well as on polycrystalline copper [19]. Wu et al. have used OI-OR method to explore the formation of UPD of Pb on copper. Accordingly, it is likely due to the different binding energies of copper sites with different geometries. Furthermore, the adsorption of chloride ions has an important effect on the splitting the UPD peak in the sense that its presence enhances the kinetics of Pb UPD deposition.

A small drop in frequency can be seen on the corresponding frequency-potential curve with a starting point arbitrarily mentioned by the lines on the curve, shown in Fig. 4.2. According to the frequency-potential curve, the height of the peak in the range of −0.38 to −0.43 V is measured about 10 Hz on the cyclic voltammogram, as shown on the inset. Figure 4.3 shows the variation of resistance recorded versus time at the mentioned potential range at above. The variation of resistance is considerable when a rigid solid film forms on the quartz crystal depending upon any change of the viscoelasticity of the crystal [20].

Slight increase in resistance versus time at potentials from −0.39 to −0.42 V SCE are clearly observed exhibiting the formation of a solid film. If the film is assumed to be solid, rigid and smooth, using Sauerbrey's equation (Chap. 3, 3.64), the corresponding mass of the layer deposited on copper (with an area of 1.37 cm^2), is achieved 242.04 ng (nanograms). Assuming the deposit is made of lead (II) with a density of 11.35 g/cm^3, this amount of mass leads to a thickness of 1.58 Å for the deposit which can be correlated to the thickness of one monolayer (ML) of lead on such surface area.

This value qualitatively can explain the formation of underpotential deposition of lead as a monolayer which forms below the formal deposition of lead (from the solution is −0.457 V (SCE) [21]) and is consistent with the previous reports either on single crystalline or polycrystalline copper substrates (Fig. 4.2). The anodic

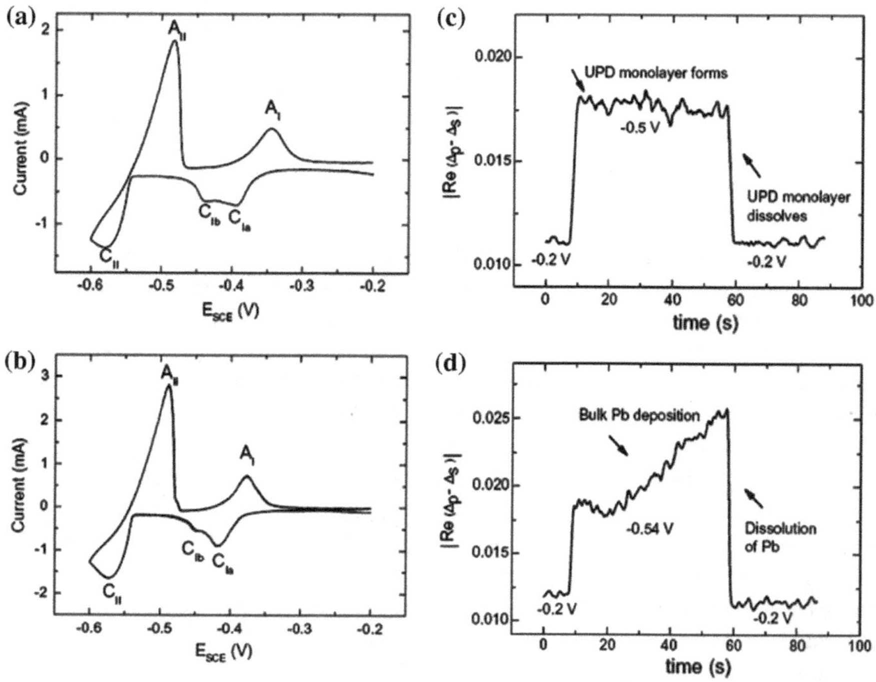

Fig. 4.2 Cyclic voltammograms for a (111)-textured polycrystalline Cu substrate measured at a scan rate of 10 mVs^{-1} in: **a** 100 mM $HClO_4$ + 1 mM $Pb(ClO_4)_2$; **b** 100 mM $HClO_4$ + 1 mM Pb $(ClO_4)_2$ + 20 mM KCl. Potentials are quoted with respect to SCE. OI-RD signal recorded when stepping the Cu substrate potential **c** from E = −0.2 V to E = −0.5 V for t = 50 s (underpotential deposition—UPD) and **d** from E = −0.2 V to E = −0.54 V for t = 50 s (overpotential deposition—OPD). The electrolyte was 100 mM $HClO_4$ + 1 mM $Pb(ClO_4)_2$. *A* and *C* denotes anodic and cathodic peaks on CVs. From [19] with permission from Elsevier

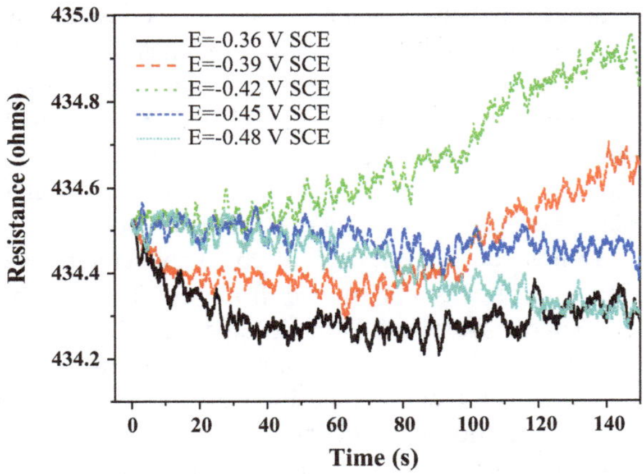

Fig. 4.3 Variation of resistance versus time in a range of potential corresponding to the Pb UPD peak on copper

peaks in the back scan contains a shoulder which can arbitrarily separated to two peaks $D_{1.1}$ and $D_{1.2}$ on Fig. 4.1, demonstrating that the UPD of Pb is dissolving.

Once the deposition potential approaches below the formal potential of lead, the bulk film of lead starts to form. Only very few reports have studied the formation of bulk lead films. The results obtained by oblique incident oblique-incidence optical reflectivity difference (OI-RD) and in situ atomic force microscope (AFM) have demonstrated the effect of chloride ion on the properties of the bulk film over the UPD range. The OPD of Pb reveals the effect of halide ions on the morphology and growth mechanism of bulk Pb on copper, as shown before in Fig. 4.2 [19].

Figure 4.1 shows that at potentials more negative values in respect to the formal potential of lead, the curve reveals a greater peak corresponding to overpotential deposition of lead. OPD of Pb, labelled R_2, commences from ~ -0.52 V and the corresponding change in frequency (or mass) is remarkable compared to the UPD of lead. This implies that a more dense material (bulk phase) is formed on the substrate. The corresponding dissolution peak on the anodic curve (D_2) has also a greater peak current than of the UPD dissolution peak.

In Nasirpouri's experiments there is an offset between the starting and ending points of scans, as mentioned by other authors previously [22–24]. It has been demonstrated in the literature that when experiments involve only relative frequency shifts which are measured in a fixed solution, the offset caused by the viscous loading of the liquid, has negligible effect on the accuracy of the Sauerbrey's equation for the determination of small mass changes in rigid deposits. The drift in the data presented here is not readily explainable, even it has been observed before by other authors though. During the cathodic part of the scans, solid films should form causing a decrease in the frequency and an increase in resistance due to the deviation from complete reversible scans. Some factors have been mentioned to cause such effect including surface roughness, liquid entrapment and the formation of stable solid films. Nasirpouri's best speculation was that the small changes in frequency are in response to roughening of the deposit due to the hydrogen evolution which will be discussed later and also viscous changes caused by depletion and regeneration of metallic ions or chloride ions during the voltammetric scan. Furthermore, hydrogen evolution plays a critical role on roughening of the deposit making some spaces for the entrapment of liquid. However, several interfacial processes are known to occur during electrosorption, including adsorption/desorption of solvent and changes in the position of interfacial plane between the electrode and the solution.

Deposition of overpotential Pb on copper substrate was further investigated. The OPD range was determined from cyclic voltammetric and frequency-potential (massograms) curves. Potentiostatic experiments (chronoamperometry along with EQCM) were carried out in a range of potential from −0.48 to −0.64 V (SCE) to produce overpotential deposition of lead (Pb) from the perchlorate solution.

The variation of frequency in Hz (or mass per micrograms) and resistance in ohms of quartz crystal during potentiostatic measurements (electrodeposition) of OPD of lead are shown in Fig. 4.4. Both frequency and resistance do not exhibit any considerable change until the potential approaches −0.50 V (SCE). For simplicity, the curves recorded more positive than this potential are not inserted in these figures. The variation of the frequency and resistance recorded at −0.52 V (SCE) start to be slightly increasing with deposition time signifying that a rigid

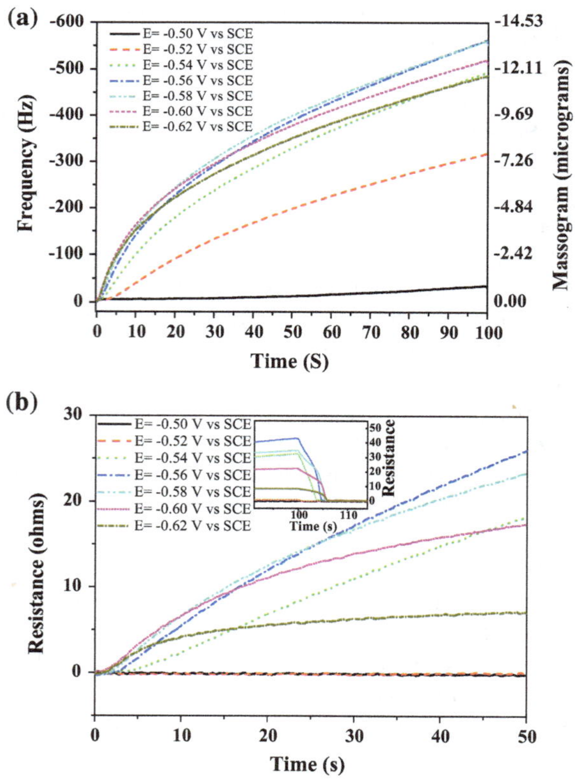

Fig. 4.4 **a** Variation of frequency in Hz (or mass per micrograms) of the copper/gold plated quartz crystal recorded at different potentials during potentiostatic measurements (electrodeposition) of OPD of Pb, **b** variation of resistance in ohms recorded at different potentials during potentiostatic electrodeposition of OPD of lead from 1 mM lead(II) perchlorate trihydrate and 10 mM HCl at room temperature. The *inset* shows the fall of resistance to its initial value after switching the potential to −0.2 V SCE

deposit starts to grow on the substrate. This behaviour is seen for all other curves recorded at more negative potentials or higher overpotentials. It is, too, worth noting that at the beginning of the formation of the OPD of lead, the deposit grows at a higher rate. As the deposition time is extended, the deposition rate intends to be dramatically reduced.

The variation of resistance versus time during electrodeposition from perchlorate solution is consistent with massograms, signifying that a bulk deposit forms on the substrate. As long as this parameter reflects the viscoelasticity of deposit on the quartz crystal, then any increase of resistance exhibits the formation of a rigid film during deposition. After switching the potential to a value of −0.20 V SCE (in the range of dissolution of the deposit), the resistance returns to its initial value. Both frequency and resistance plots are normalized to zero by subtracting the initial values of both parameters.

The change of deposition rate after a short period of time can be understood by looking at the current transients recorded during potentiostatic deposition, as shown in Fig. 4.5a. Each transient shows a peak at the beginning, followed by a drastic fall until it reaches a plateau. This is a typical diffusion controlled nucleation and subsequent growth of metallic electrodeposits, well described by Scharifker and Hills, Chap. 3. The peak corresponds to the nucleation of metallic sites on the surface which is followed by the reduction of current exhibiting three-dimensional (3D) growth. The rise in current in each transient reflects the electroactive area increases because of either each independent nucleus grows in size and/or the number of nuclei increases. At this stage a diffusion controlled process directs the growth of deposit and then the current falls at final stage. We have compared our results with Scharifker-Hills predictions based on non-dimensional plots I^2/I_m^2 versus t/t_m (Chap. 3, 3.55 and 3.56). Non-dimensional plots are constructed from the I–t transients, where the current I_m and the time t_m correspond to the maximum current peak. Typical plots together with theoretical curves derived from Scharifker equations for the instantaneous and progressive nucleation are shown in Fig. 4.5b for electrodeposition of Pb from perchlorate solution. The experimental data are found to best describe the 3D instantaneous, diffusion-controlled nucleation and growth mechanism of Pb from this solution. A typical SEM image is shown as inset to Fig. 4.5b from Pb electrodeposited on copper for 20 s at −0.60 V SCE. All the mature nuclei grown at the initial stage of electrodeposition have almost the same size.

Figure 4.6 illustrates the variation of frequency (or mass) against the total charge passed through the cell recorded for different OPDs. The slope of these curves demonstrates the correlation between the actual deposited mass obtained by gravimetric method and the total charge passed through the cell. The equation relating the variation of frequency to charge (3.64) can be used to elucidate which species are involved in the mass changes that accompany electron transfer reactions and hence is very useful tool for characterising the mechanisms of such reactions. Therefore, comparison of the change of frequency with charge passed for 100 s makes one able to calculate the ratio of molecular weight to the number of electrons

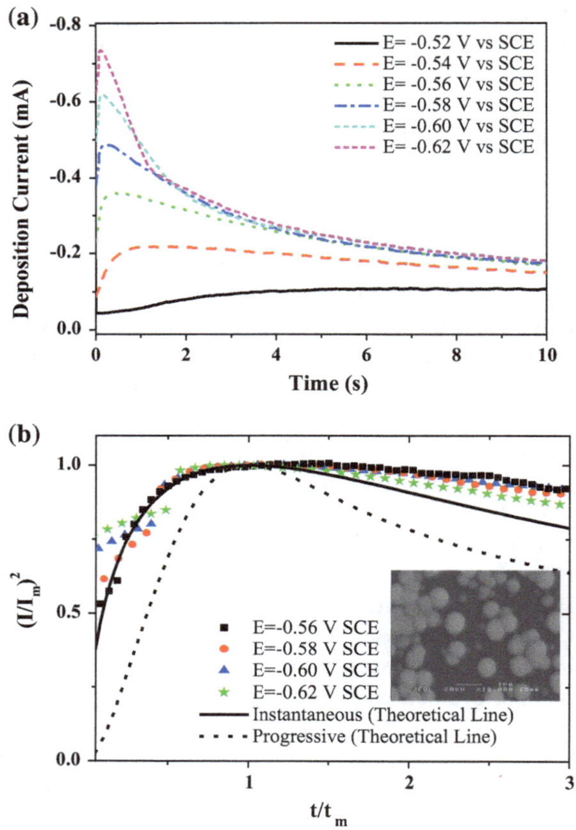

Fig. 4.5 **a** Current transients recorded at different potentials during potentiostatic electrodeposition of OPD of Pb on Cu, **b** non-dimensional plots ($(I/I_m)^2$ versus $(t/t_m)^2$) for electrodeposited lead at different overpotentials from 1 mM lead(II) perchlorate trihydrate and 10 mM HCl at room temperature. *Inset* shows a typical scanning electron micrograph taken from Pb electrodeposited on copper for 20 s at −0.60 V SCE. Reproduced from [6] with the permission of Springer

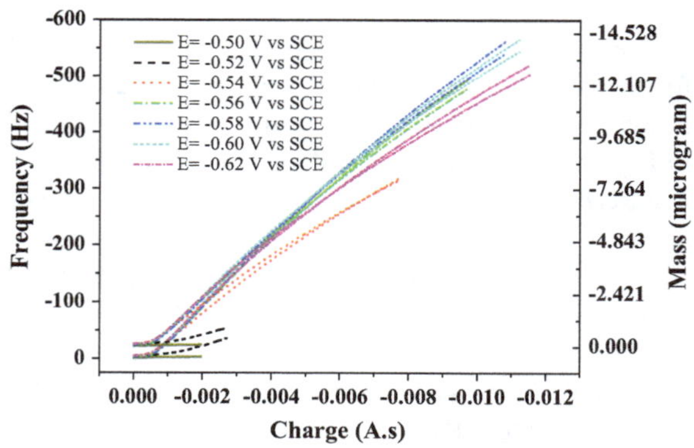

Fig. 4.6 Variation of frequency (or mass) versus charge passed through the cell recorded at different potentials during potentiostatic electrodeposition of OPD of lead from 1 mM lead(II) perchlorate trihydrate and 10 mM HCl at room temperature. From [6] with the permission of Springer

transferred in the reactions. Calculated mass (M) of the deposit to the number (n) of electrons transferred gives M/n ratio for each OPD, which are compared at different potentials with the theoretical ratio of Pb(II) and mentioned in Fig. 4.7. The

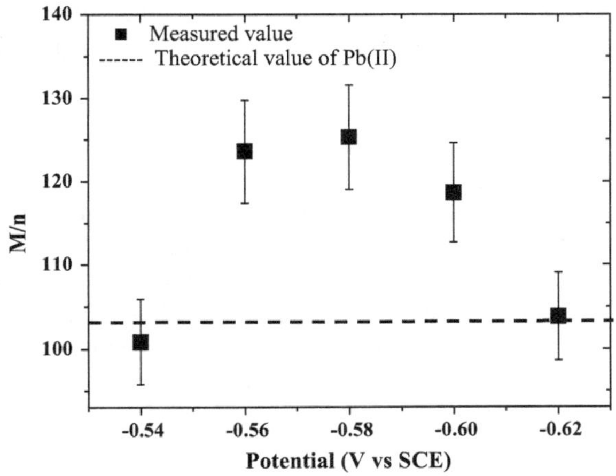

Fig. 4.7 Calculated ratio of molecular mass of the deposit to the number of electron transferred (M/n) for each OPD potentials

theoretical mass change per mole of electrons (or electrochemical equivalent) defined here as M/n ratio is approximately consistent with incorporation of Pb (II) in the formation of the Pb(0) deposit. At very low overpotentials, i.e., less than −0.52 V SCE, M/n is too low, notifying that the bulk deposit does not form. At an applied potential of −0.54 V SCE, the measured ratio is equal to the theoretical electrochemical equivalent of Pb(II)/Pb, i.e., 103.6 g/mole. It means that as soon as the OPD of lead starts to form, the magnitude of M/n approaches the ratio of Pb (II). But, at higher potentials up to −0.58 V SCE, the mass equivalent of the deposit is remarkably higher than of Pb (II). Any increase in the overpotential after −0.58 leads to the reduction of the measured M/n ratio with EQCM.

The behaviour observed for the variation of M/n observed using EQCM with incorporation of side reactions with the deposition of Pb. Most of the mechanisms proposed for the formation of lead UPD is based on the reaction [25, 26]:

$$[PbCl_4]^{2-}_{ads.} + 2e^- \rightarrow Pb_{UPD} + 4Cl^- \tag{4.6}$$

However, Moffat [11] earlier pointed out that the transformation from the halide to the lead ad-layer in the UPD results in the formation of vacancies and adatoms which condense to form holes and islands, respectively. These features may be rationalized by the formation of an alloy phase.

Mechanism of OPD of Pb has not been as extensively investigated in detail. It was previously mentioned by Gray et al. [27] that electrodeposition of lead (II) from a perchlorate electrolyte on single crystal copper can be associated with different compositions. Using the OI-RD results, it was proposed that immediately following the formation of the UPD of the first Pb monolayer, two different types of bulk-phase films grow depending upon the magnitude of overpotential.

At low overpotentials a bulk-phase Pb film grows on top of the UPD monolayer, while at high overpotentials either a PbO and PbCl2 compound or a rough Pb bulk-phase layer may grow on top of the UPD layer. Furthermore, just recently Schwarzacher group has mentioned that in the presence of halids like Cl^-, a low density of large Pb islands is formed on the surface, because chloride ions adsorb on the surface and blocks the nucleation sites in the OPD range [28]. However, the same group has newly mentioned the effect of surface alloying of Pb with gold and copper substrates [29]. According to our results in here, the higher mass change per mole of electrons is observed at low overpotential. In the sense that the formation of an alloy can take place on the surface [27], here we ascribe any rise in the electrochemical equivalent to the formation of a compound between reduced lead and an alloying element such as adsorbed chlorine, as long as the formation of an oxide layer is less probable but not impossible as the solution was exposed to air during experiments. Assuming chlorine as the alloying element, the mass change per mole of electrons of a chlorine atom is 34.45 g/mole. Considering $PbCl_2$ as the composition of the compound to form, the change of mass per mole of electrons of such compound in pure state is equal to 174.5 g/mole. The increase of the M/n ratio to higher than the theoretical electrochemical equivalent of Pb can be ascribed to the formation of a mixture of Pb and $PbCl_2$. For the deposits obtained at even higher overpotentials more negative than −0.58 V SCE, it is seen that the mass-charge (M/n) ratio decreases signifying the presence of gas phase reaction like hydrogen evolution. The deviation of the measured M/n from the theoretical value takes place, when both the formation of compound and the hydrogen evolution occur. However, the hydrogen evolution accompanying the deposition of OPD of Pb has a strong effect on the morphology evolution in some metallic films. This possible effect can lead to roughen the Pb deposit as proposed previously by OI-RD leading to the change of mass per mole of electrons [19].

4.2.3 UPD and OPD in Electrodeposited Lead (Pb) on Gold (Cu)

Nasirpouri has also investigated the formation and mechanism of Pb UPD and OPD on gold substrate [8, 9] using cyclic voltammograms and massograms. Table 4.1 summarizes the potential range of formation of Pb on Au coated on a quartz crystal. Figure 4.8 shows a typical first sweep cyclic voltammogram (CV) and the frequency-potential curve (massogram) taken from 1 mM $Pb(ClO_4)_2 \cdot 3H_2O$ and 10 mM HCl at room temperature on a gold substrate at 10 mV/s scan rate.

Two distinct reduction and their corresponding dissolution (or oxidation) peaks are clearly observed on each side of potential sweep directions. The first reduction peak which is labelled UPD is a reduction peak occurring in the range of −0.32 to −0.50 V versus SCE. This reduction peak does not correspond to the equilibrium potential of 1 mM Pb(II)/Pb (about −0.457 V SCE [15]), occurring at more positive potentials. A small drop in frequency can be seen on the corresponding frequency-potential curve which has a height of the peak of about 10 Hz on the

Table 4.1 Data analysis carried out on the first cyclic voltammogram and massogram (frequency-potential curve) taken from 1 mM Pb(ClO4)·3H2O + 10 mM HCl on a fresh gold-coated quartz crystal from [8] with permission from Springer

Reaction type	Peak current (mA)	Peak potential V versus SCE	Peak start potential V versus SCE	Peak end potential V versus SCE
UPD reduction	0.050	−0.390	−0.320	−0.500
OPD reduction	–	–	−0.520	–
UPD dissolution	0.059	−0.460	−0.490	−0.440
OPD Dissolution	0.0267	−0.380	−0.430	−0.340

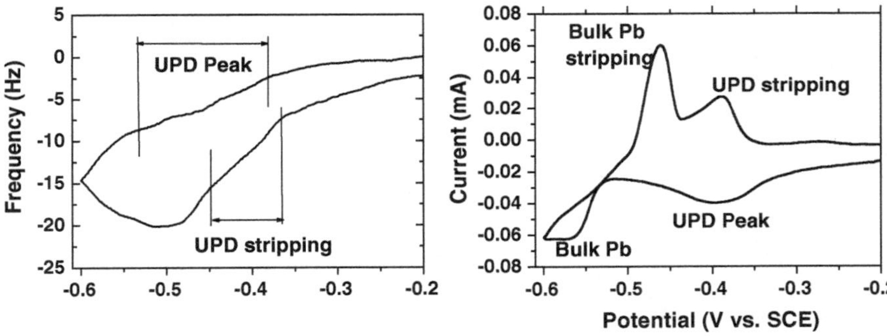

Fig. 4.8 *Upper panel* insitu measurement of frequency-potential or cyclic massogram with recognizable regions of UPD and OPD and *lower panel* a typical first sweep cyclic voltammogram (CV) taken from 1 mM $Pb(ClO_4)_2 \cdot 3H_2O$ and 10 mM HCl at room temperature on gold substrate at 10 mV/s scan rate. Reprinted from [8] with permission from Springer

cyclic voltammogram, as shown on the inset. With the same argument presented in [6] Nasirpouri has demonstrated that this value qualitatively can explain the formation of underpotential deposition of lead as a monolayer which forms below the formal deposition of lead. There is a shoulder on the backward potential sweep direction on Fig. 4.8, demonstrating that the UPD of Pb is dissolving.

At more negative potentials than the observed UPD peak in a range more negative than the formal potential of lead, a greater reduction peak is observed corresponding to overpotential deposition of lead (OPD). OPD peak starts from ~−520 mV and the corresponding change in frequency (or mass) is notable compared to the UPD of lead. This implies that a more dense material (bulk phase) is formed on the substrate. The corresponding dissolution peak on the anodic curve has also a greater peak current than of the UPD dissolution peak.

Figure 4.9 shows the effect of potential sweeping rate on the cyclic voltammograms and massograms taken from the perchlorate solution on the gold substrate

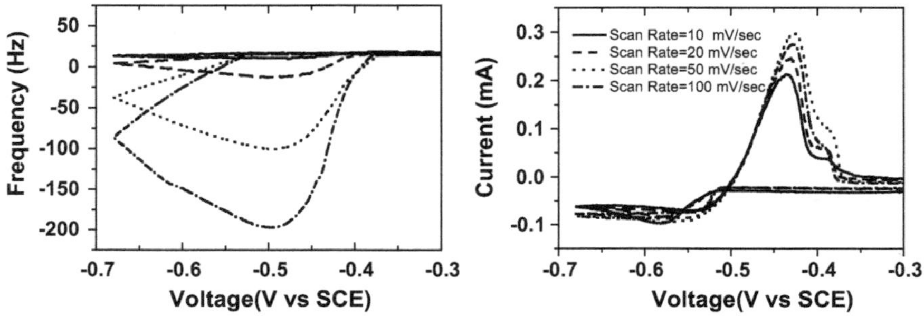

Fig. 4.9 Cyclic massograms (the *upper panel*) and voltammogram (the *lower panel*) taken from 1 mM $Pb(ClO_4)_2{\cdot}3H_2O$ and 10 mM HCl at room temperature on gold substrate at different (10–100 mV/s) scan rate. Reprinted from Ref [8] with permission from Springer

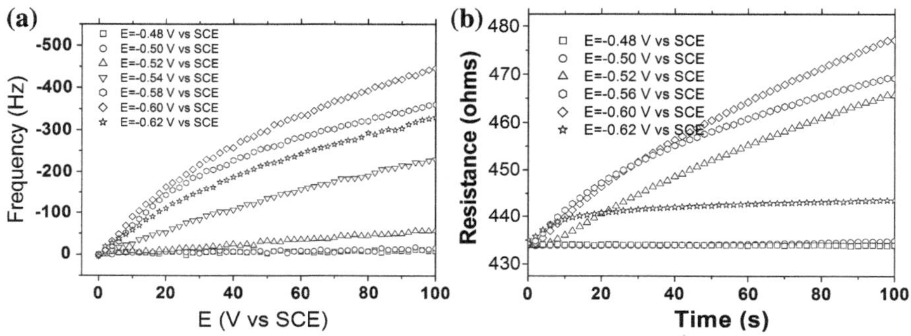

Fig. 4.10 Variation of **a** frequency and **b** resistance of gold plated quartz crystal recorded at different potentials during potentiostatic measurements (electrodeposition) of OPD of lead on gold from 1 mM $Pb(ClO_4)_2{\cdot}3H_2O$ and 10 mM HCl at room temperature. Reprinted from [8] with permission from Springer

at room temperature. Both UPD and OPD peaks are clearly observed, while the peaks are shifted towards more negative potentials at higher currents or masses. Increasing scan rate appears to be associated with significant increase in peak currents exhibiting a diffusion controlled process.

Electrochemical deposition of lead OPD Pb on the gold substrate was mentioned to have interesting features when the potentiostatic current transients and the corresponding massograms were investigated. Potentiostatic experiments were conducted in a range of potential from −0.54 to −0.64 V (SCE) to produce overpotential deposition of lead (Pb) at room temperature. Frequency-time and resistance-time curves recorded at different OPD potentials on the quartz crystal during potentiostatic measurements are shown in Fig. 4.10. Both frequency and resistance do not exhibit any considerable change until the potential approaches −0.50 V (SCE). The variation of the frequency and resistance recorded at −0.52 V (SCE) start to be slightly increasing with deposition time notifying that a rigid

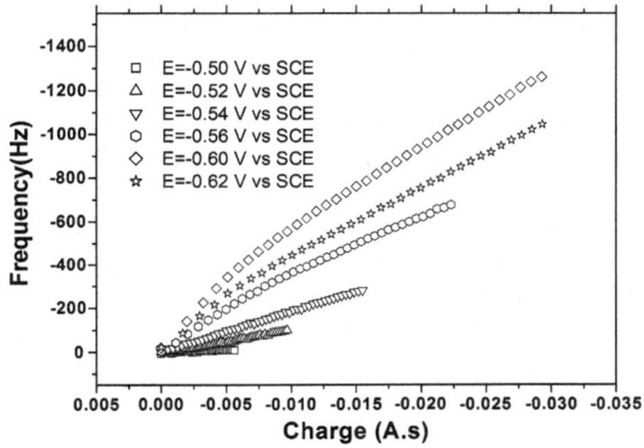

Fig. 4.11 Variation of frequency (or mass) versus charge passed through the cell recorded at different potentials during potentiostatic electrodeposition of OPD of lead on gold from 1 mM Pb $(ClO_4)_2{\cdot}3H_2O$ and 10 mM HCl at room temperature. Reprinted from [8] with permission from Springer

deposit grows on the substrate. This behaviour is seen for all other curves recorded at more negative potentials or higher overpotentials. It is, too, worth noting that at the beginning of the formation of the OPD of lead, the deposit grows at a higher rate. As the deposition time is extended, the deposition rate intends to be dramatically reduced. The variation of resistance versus time during electrodeposition from perchlorate solution is consistent with massograms, signifying that a bulk deposit forms on the substrate. As long as this parameter reflects the viscoelasticity of deposit on the quartz crystal [20], then any increase of resistance exhibits the formation of a rigid film during deposition.

Figure 4.11 illustrates the variation of frequency (or mass) versus the total charge passed through the cell recorded at different OPD potentials. The slope of these curves demonstrates the correlation between the actual deposited mass obtained by gravimetric method and the total charge passed through the cell, signifying the sum of faradic and non-faradic. At very low overpotentials, i.e., less than −0.52 V SCE, M/n is too low, notifying that the bulk deposit does not form. Afterwards the curve slope obtained at potentials up to −0.60 V SCE is increasing. Any increase in the overpotential after −0.60 leads to reduce the curve slope.

We interpret the behaviour observed for the variation of the curve slope which corresponds to the electrochemical equivalent based on Sauerbrey equation with incorporation or side reactions with deposition of Pb. It has been previously mentioned that electrodeposition of lead (II) from perchlorate electrolytes on copper is associated with including different compositions whit changing the OPD potential. The other side reaction can the hydrogen evolution which makes the deposit rough changing the viscoelastic behaviour of the quartz crystal. It has been pointed out that hydrogen evolution has a strong effect on the morphology evolution in some metallic films like nickel [30].

For the electrodeposits obtained at higher cathodic overpotentials, the mass-charge ratio decreases signifying two major phenomena associated with deposition of lead: (i) incorporation of a non-metallic element, which exists in the electrolyte, with lower molecular mass and greater number of electrons taken part into the reduction reactions such chlorine and oxygen, (ii) the hydrogen evolution. The deviation of the real mass from the theoretical mass of deposit reveals that hydrogen evolution (gas phase reactions) takes place at higher overpotentials.

Nasirpouri has also studied the nucleation and growth mechanism of Pb on Au based on Scharifker-Hills theory. Figure 4.12a shows the current transients recorded during potentiostatic deposition. Typical potentiostatic current transient (I-t) curves demonstrate a diffusion control deposition behaviour with the characteristic peak corresponding to the nucleation-and-growth process and for longer times, which is caused by a diffusion controlled process and described by the Cottrell equation. According to the I-t curves, each transient has one well-defined recognizable current maximum as a clear first peak followed by a sharp fall and posterior growth. The I-t transients have a normal dependence to overpotentials, whereas the current density increases with increasing of overpotential. The peak corresponds to the nucleation of metallic sites on the surface which is followed by the reduction of current exhibiting three-dimensional (3D) growth. Increasing the peak current at higher overpotentials means that the number of sites (nucleation rate) increases owing a higher nucleation rate. Typical plots together with theoretical curves derived from Scharifker-Hills equations for the instantaneous and progressive nucleation are shown in Fig. 4.12b. The experimental data are found to best describe the 3D instantaneous, diffusion-controlled nucleation and growth mechanism of Pb from this solution. This mechanism was confirmed by ex situ examination using SEM taken from the initial stage of electrodeposition which exhibited almost the same size nuclei formed at the same time and the growth afterward.

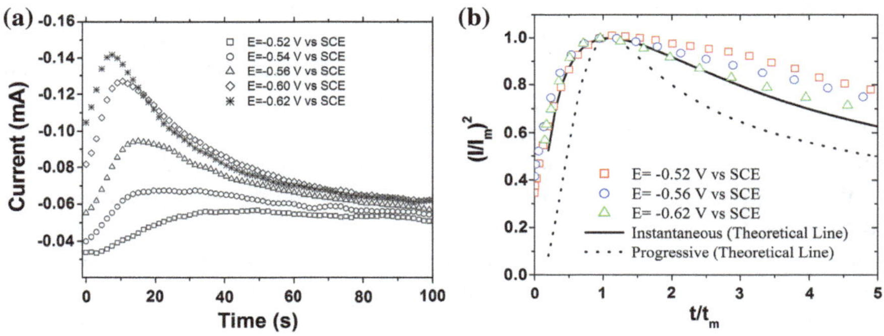

Fig. 4.12 **a** Current transients recorded at different potentials during potentiostatic electrodeposition of OPD of lead on gold nd **b** non-dimensional plots (I^2/I_m^2 vs. t/t_m) for electrodeposited lead on gold coated quartz crystal at different overpotentials from 1 mM $Pb(ClO_4)_2{\cdot}3H_2O$ and 10 mM HCl at room temperature. Reprinted from [8] with permission from Springer

4.3 Three Dimensional (3D) Growth of Electrodeposits

3D nucleation and growth of metals or alloys have been of particular importance for either fundamental science or technological applications for nanostructure or film growth. Electrodeposition of 3D structure obeys the Volmer–Weber island growth mechanism as explained earlier in this chapter. The most important factor to control the island Volmer–Weber growth in electrocrystallisation is the cathodic overpotential employed. This is defined as the difference between the applied electrode potential and the potential at which the nucleation of nuclei commence on a foreign substrate. The nucleation potential is generally referred to the formal electrode potential of typical redox couple in a typical reduction reaction, e.g., M/M^{2+}. This is similar to OPD of metals or alloys on various substrates. However, the magnitude of the cathodic overpotential significantly affects the nucleation and growth stages associated with t island growth mechanism. Figure 4.13 demonstrates three nucleation regimes for the island growth during a typical electrocrystallisation process including: (a) Island growth is kinetically controlled regime such that the island shape is dictated by the energy of the different facets. (b) Island growth is under mixed kinetic/diffusion control, usually resulting in the formation of hemispherical islands. (c) Island growth is under diffusion control resulting in the formation of diffusion limited aggregates or dendrites. Based on the author's experience, we encounter two of the modes (a and b) mentioned are of major applications and importance as we talk about them in next sections. However, mode (c) may be interesting for fundamental aspects such as analysis of fractal growth [31].

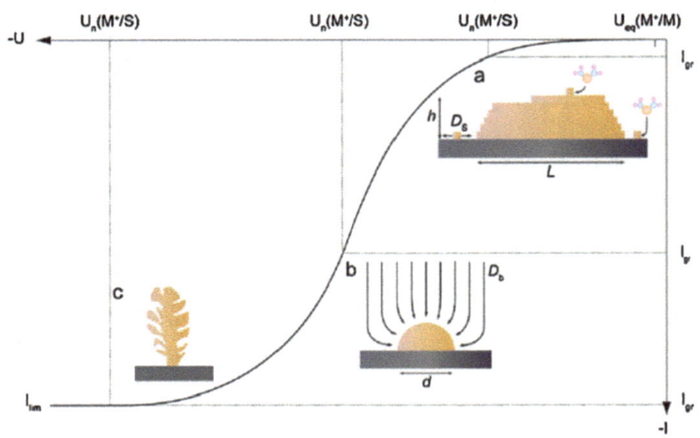

Fig. 4.13 Steady state current–voltage curve for the electrodeposition of material M on substrate S. The nucleation potential for deposition of material M on substrate S is indicated for **a** kinetic control, **b** mixed kinetic/diffusion control, and **c** diffusion control. Reprinted from [31] with the permission of Elsevier

4.3.1 3D Nuclei as Free Standing Meso- and Nano-Structures

Under small negative overpotential, the kinetic (or charge transfer) control island growth dominates for electrocrystallisation with sufficiently slow rate. This allows ad-atoms to either directly grow on an island or indirectly attach first to substrate and then diffuse on the surface to find a growth site. The former leads to vertical growth, while the later results in island lateral growth. P. Searson and co-workers have explained the deposition rates of ad-atoms on a growing island and the substrate in terms of the probability [32, 33] of electron transfer. It is obvious that the island growth of individual centres dominates when the probability of a metal ion is reduced on an island is greater than that of a metal ion is reduced on the substrate. Furthermore, a faceted island is generated under slow direct attachment on the facets namely layer-by-layer (or spiral) growth. It is complex growth mode where it is island mode on the substrate while is layer-by-layer on the growing island. This is valid only for small negative overpotentials because at higher overpotentials the growth rate is so high that faceting is no longer the case [32, 33]. This will turn into diffusion controlled hemispherical island growth which will be discussed later. It is therefore important to explain this area as the development of regular and faceted islands by electrodeposition.

4.3.1.1 3D Mesoparticles, Nanoparticles and Nanowires

Electrodeposition of nanoparticles and nanowires on flat substrates has become a field of rigorous research. Behm and co-workers have demonstrated periodic arrays of monosized metal islands decorated on defects of the reconstructed gold substrate surface [34–37]. Kolb et al. have used atomic force microscopy to transfer electrodeposited metal particles with narrow size distribution from the AFM tip on a surface [38–41]. Other works also were performed on the electrodeposition of metal particles on silicon [42, 43]. More recently, R. Penner and co-workers have enhanced the use of electrochemical deposition to fabricate well distributed metal nanoparticles and nanowires on carbon substrates [44]. In his critical review, many aspects of electrodeposition of nanoparticles and nanowires and their distribution on the surface have been discussed in details using experimental and theoretical calculations. Broad size distributions of nanostructured islands are achieved under different techniques developed such as direct pulse electrodeposition, dual pulse, and hydrogen evolution assisted technique. This broad size distribution has been found to be due to the overlapped diffusional zones discussed before in Chap. 3. This is even more pronounced when the metal nanostructure centers nucleate on the substrate surface. Using the Brownian dynamics computer simulations of nanostructure growth it is interestingly found possible controlling the diffusional overlap between nanostructures using either of two growth strategies which will be described in this section on nickel nanoparticles and nanowires for efficient production of the size-selective electrodeposition of metal nanoparticles and nanowires that are narrowly distributed in diameter.

Here we focus on some recent works on the electrodeposition of nickel nanoparticles and nanowires. The electrodeposition of nickel from sulphate solutions on highly oriented pyrolytic graphite (HOPG) at a distinct potential reveals that particles and wires of nickel are formed at step edges and terraces, but this is not a reliable method for the production of individual island mesostructures. The morphology evolution of nickel mesostructures was investigated by changing the electrodeposition conditions in two ways including: dual-step potentiostatic deposition and lowering the pH value leading to co-evolution of hydrogen [45, 46].

Electrodeposition of Nickel Nanowires

Figure 4.14a, b shows nickel nanowires electrodeposited from Watts bath on HOPG using a dual pulse technique [45]. In the dual-step potentiostatic deposition method, two potentiostatic pulses are applied in rapid succession. The first step is

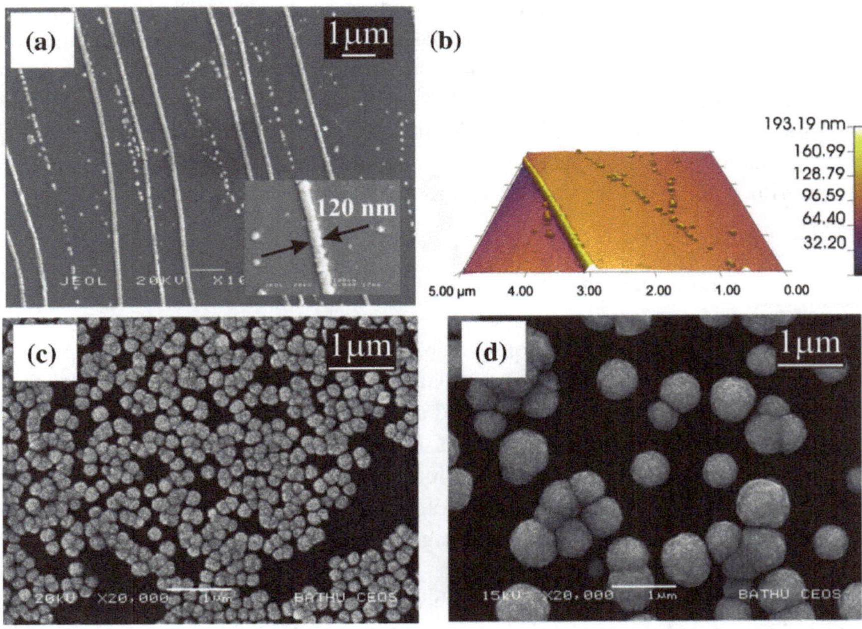

Fig. 4.14 Effect of dual-pulse plating and pH value on the morphology of nickel mesostructures. **a** SEM image of nickel nanowires electrodeposited at −850 mV versus Ag/AgCl for 5 ms and −580 mV versus Ag/AgCl for 400 s. *Inset* to **a** shows the precise diameter of a typical nickel nanowire in a high magnification image. **b** Three dimensional AFM image shows the growth of one nickel nanowire with diameter of 100 nm. **c, d** SEM images of nickel nanoparticles electrodeposited at −1.1 and −1.2 V versus Ag/AgCl for 50 s. Reprinted from [45] with permission from Elsevier

dedicated to a nucleation pulse lasting just 5 ms and having a deposition potential of −850 mV versus Ag/AgCl. This pulse produces "seed" nanoparticles pseudo-randomly on the graphite surface. Immediately following the nucleation pulse, a growth voltage pulse with a deposition potential of −580 mV versus Ag/AgCl is applied lasting up to a minute [47]. The potential of this growth pulse is at the beginning of the nickel reduction peak in the cyclic voltammogram acquired on freshly cleaved HOPG. At this potential, the rate of nucleation is exceedingly slow, so the dual pulse has the added benefit of enforcing a separation between nucleation and growth in time. If the nucleation pulses are omitted, the nucleation density along step edges would be low on average and metal nanowires are not obtained. The selection of the growth potential is critical and involves a compromise because values of this potential that are too negative cause the nucleation and growth of metal particles on terraces, whereas values that are too positive can cause the rate of nanowire growth to be depressed to the point where several hours are required. The optimization of nucleation and growth for each metal, and each metal plating solution, involves considerable trial and error [48–50].

This dual pulse method produces a high nucleation density along step edges. With continued growth, this dense linear array of metal nuclei evolves into a continuous nanowire with a diameter of 50 nm or so. Figure 4.14 shows SEM and AFM images of nickel nanowires, electrodeposited with the dual-pulse method, indicating that nanowires are deposited on step edges of HOPG. At the current conditions of electrodeposition, nickel is deposited in the form of a nanowire having a diameter of about 120 (±3.45) nm, shown in the inset to Fig. 4.14a. The length of the nanowires produced with this technique was observed to be more than 20 μm with a uniform diameter along the wire axis.

Electrodeposition of Nickel Nanoparticles

Another electrodeposition technique has been used by Nasirpouri et al. to produce Ni nanoparticles by exploiting the co-evolution of H_2 during reduction of Ni^{2+} from a low pH Watts solution. Figure 4.14c, d shows the nickel nanoparticles produced using this techniques. This method has been introduced earlier by Penner and co-workers [48] using nickel nitrate solution at high pH values which make an appreciable contribution of hydrogen evolution during nickel electrodeposition. Ni electrodeposition from sulfate supporting electrolytes at pH 2 and 4 has been investigated [51] and a strong influence of H_2 evolution and its effect on the morphology of Ni electrodeposits have been observed in this electrolyte at conditions indicating excess of H_3O^+ over Ni^{2+}, e.g., at pH 2. In the work carried by Nasirpouri et al. results clearly indicate the presence of hydrogen evolution in a Watts bath with a low pH value of 2. The electrode in this experiment is partially covered with hydrogen bubbles visible to the eye. In the cyclic voltammograms acquired in the negative direction, separate peaks for nickel deposition and H_2 evolution were not observed because a significant barrier to the nucleation of nickel exists on graphite. The contribution of hydrogen evolution to the current passed through the cell is higher when the deposition potential approaches more negative values. The deposition potential, and hence the H_2 coevolution rate, strongly influences the morphology of electrodeposited nickel structures on HOPG. This effect is seen in the scanning electron microscopy (SEM) images of Fig. 4.14c, d, which were obtained for nickel nanoparticles prepared using deposition potentials of −1.1 and −1.2 V versus Ag/AgCl from solution 3. The size of nickel particles measured from SEM images was 191.8 (±13.1) nm and 414.38 (±28.4) nm for the particles electrodeposited at −1.1 and −1.2 versus Ag/AgCl, respectively. It is also interesting to observe the narrow size distribution of nanoparticles electrodeposited due to hydrogen evolution. This has been described using Brownian dynamics simulations incorporating inhomogeneous diffusion-controlled nucleation and growth without interparticle coupling [48].

As is evident in the images, we clearly find a complete absence of crystalline facets on nickel mesostructures prepared using either dual-pulse deposition or low pH methods. The cause of this has been attributed to fact that these mesostructures are nanocrystalline. The formation of nanocrystalline nickel deposits with dual-pulse plating under high current densities has been well studied for many years [52]. The nanocrystalline nature of the nickel particles prepared from a nitrate solution has been characterized using electron microscopy data.

4.3.1.2 3D Faceted Mesocrystals

As explained earlier, the island growth mode would lead to faceted crystals provided that the growth rate is so slow which is achieved under very low overpotentials or close to equilibrium. The shape or morphology evolution of islands formed during an appropriate electrocrystallisation process can be explained by the differences in surface energy of different crystallographic planes [53]. We have explained the growth rates of different facets of crystals in Chap. 1 using the periodic bond chain (PBC) theorem. The accommodation factor is the lowest for those facets having the highest surface energy. Thus, growth rates are faster on facets with higher energies which will be occupied faster than facets with lower surface energy. Thus, it is crucially important to control the growth process to keep meso or nanocrystals with high index facets in which number of steps and kinks possess high densities. One important application for these nanocrystals is in high performance catalysts. Figure 4.15 shows a triangular diagram that nicely correlates convex polyhedrons of an fcc metal enclosed by different crystallographic planes [11]. The vertices of this diagram correspond to the three polyhedral crystals enclosed by low-index facets, that is, cubes by {100} facets, rhombic dodecahedrons by {110} facets, and octahedrons by {111} facets.

The stereographic models on the sides and interior of the triangle represent four types of polyhedrons bounded by high-index facets: trioctahedron (TOH), tetrahexahedron (THH), trapezohedron (TPH), and hexoctahedron (HOH). For instance, the surface free energies of a nanocrystal made of an fcc metal, increase in the order of $\{111\} < \{100\} < \{110\}$ [54].

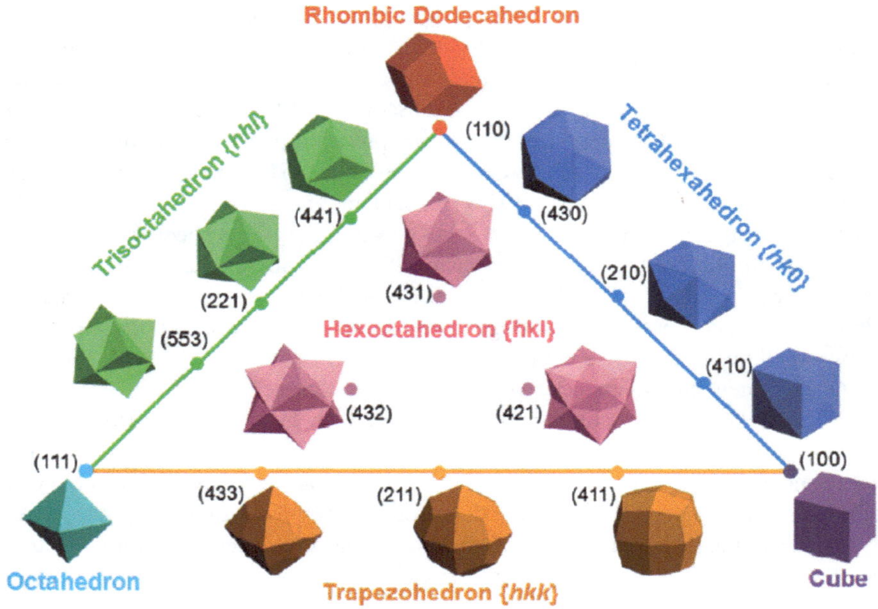

Fig. 4.15 An example of possible crystallographic structures based on a fcc lattice. *Triangular* diagram showing fcc metal polyhedrons bounded by different crystallographic facets. Reproduced from [54] with permission from Copyright (2016) American Chemical Society

In electrocrystallisation process, several factors influence the crystal morphology and shape including the nature of the metal to be electrocrystallised, the nucleation overpotential, direct current or pulse current techniques, electrolyte type, etc. We will have a brief overview on the developments in this field over past decade.

Electrodeposition using this technique has been effectively used for growing mesocrystals with a range of sizes, from nano to macro, and to realise regularly shaped crystals of a few metals like Au, Pb, Ag, Sb and Pt. Indeed, under specific deposition conditions, one can flexibly manipulate the size and the shape of the electrodeposited architectures. However, some metals like Ni have not revealed faceted mesocrystals due to the incorporation of side reactions such as hydrogen evolution which has a pronounced influence on the deposition mechanism and the produced structures. This will be discussed in next few sections.

Electrodeposition of Silver Mesocrystals

Electrocrystallisation of silver has been a subject of study for years [55, 56]. Nasirpouri and co-workers have recently worked on the electrodeposition of silver mesocrystals on HOPG surface from an aqueous electrolyte [57]. In order to obtain a quick overview of the electrodeposition system $AgNO_3$, cyclic voltammetry was

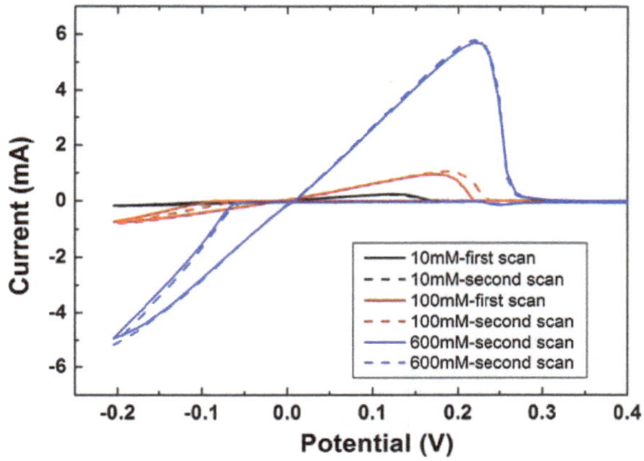

Fig. 4.16 Typical cyclic voltammograms obtained for Ag on HOPG when depositing from $AgNO_3$ solution in $NaNO_3$ as the supporting electrolyte with different concentrations [57]

performed on HOPG at different electrolyte concentrations. Figure 4.16 illustrates typical cyclic voltammograms obtained for silver on a HOPG working electrode which displays the characteristic features of nucleation, namely the large peak separation and the crossover on the cathodic branch. For comparison a cyclic voltammogram for the second sweep taken for each $AgNO_3$ concentration is shown by the dashed line. A negligible difference is observed between the first and second sweeps and any difference is likely reflected by the irreversible nature of the solid state deposition of silver on HOPG. It seems likely that the deposited silver is not readily anodically stripped and therefore promotes the nucleation and deposition of silver in the following cathodic cycle.

The reduction process occurs mainly in one step revealed by a cathodic reduction peak negative with respect to the formal potential for silver reduction. It has been frequently mentioned [58, 59] that the electrodeposition of silver obeys a diffusion controlled mechanism on carbon substrates. Here we also observe that the reduction peak current increases with increasing the electrolyte concentration and it is reasonable to assume that this corresponds to a mass transport regime.

Chronoamperometric measurements give further details of nucleation and growth of silver on HOPG. Several investigations have discussed this using current transients recorded during the potentiostatic deposition [60, 61]. Generally for the current transients for this electrodeposition system, the reaction current rises after a minimum has been reached. The higher the cathodic overpotential applied, the higher is the initial value of the current, the faster its decay, shifting the minima of the current transient curves to shorter times and making the following increase steeper. Figure 4.17a, b show the current transients recorded during electrodeposition of silver from 10 and 600 mM $AgNO_3$ solutions, respectively, at different deposition potentials.

In general, current transients for metal electrodeposition is explained as follows: the rising part of the curves is generally linear due to the growth of silver nuclei, where charge transfer takes place directly at the growth sites and newly formed silver atoms are immediately incorporated in the crystal lattice. The increase of the

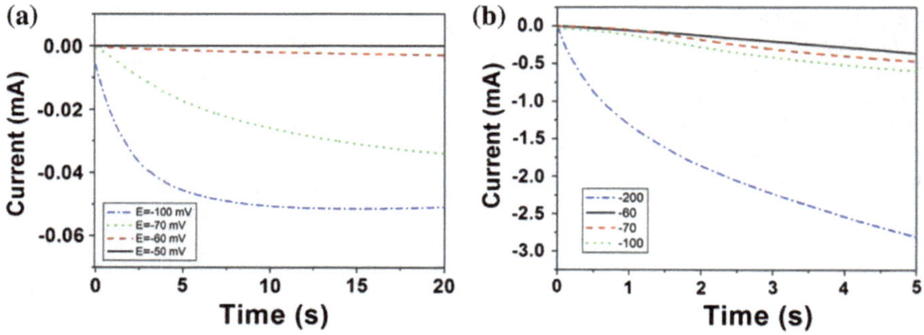

Fig. 4.17 Current transients recorded during electrodeposition of silver from **a** 10 mM $AgNO_3$ solution at different deposition potentials and **b** from 600 mM $AgNO_3$ solution at different deposition potentials [57]

current, corresponding to an increase of the reaction surface, lasts until a maximum value is reached. At that time the whole surface is covered with growing silver crystallites [60], and no new active growth sites are formed. The current transients were stopped before the current starts to drop. This was done since the purpose of our work was to create individual silver centres. The currents corresponding to the rising parts of the current transients are linear in $I^{1/2}$, indicative of the mass transfer controlled growth of hemispherical nucleation as described by the equation:

$$I = nF\pi N\frac{(2Dc)^{3/2}M^{1/2}t^{1/2}}{\rho^{1/2}}, \quad (4.7)$$

where nF is the molar charge of the electrodepositing species, N is the total number of nuclei, D is the diffusion coefficient, c is the bulk electrolyte concentration, M and ρ is the molecular weight and density of the electrodeposit, and t is the time. According to the equation, the magnitude of the current, and hence the nucleation rate or N itself, is dependent on various factors. Indeed, in our study we find a strong effect of the deposition potential and also concentration on the current and nucleation rate.

Figure 4.18 shows an SEM image of typical silver on HOPG electrodeposited at a potential of −0.1 V for 60 s. It is seen that the mechanism of electrodeposition is progressive with the formation of three dimensional silver centres with different sizes and also irregular shapes. It is surprising that surface has not been fully covered by a continuous silver film, instead the nuclei formed grow and become bigger. It has been mentioned [61] that the number of nuclei does not change after a sufficiently long time. A simple explanation of this effect is that the constant number of clusters is determined from the direct exhaustion of active sites on the substrate. Therefore, the distribution and the number of supercritical clusters reflects the number of active sites. However, this might not be the case, for instance, due to the formation of nucleation exclusion zones arising around growing stable centres.

Fig. 4.18 SEM image of silver electrodeposited on HOPG at a potential of −0.1 V for 60 s reflecting the progressive mechanism of deposition with irregular crystal shapes and sizes

The overlapping of these zones may interrupt the nucleation and thus lead to a saturation nucleus density [61].

On the basis of these findings, Nasirpouri et al. [57] studied the change of morphology of silver electrodeposits by changing the deposition potential, the electrolyte concentration and also the deposition time. Under specific electrodeposition conditions, one can realise highly faceted silver mesocrystals with well defined shapes and size. Figure 4.19 demonstrates some typical silver mesocrystals realized by various electrodeposition conditions. Several shapes and morphologies from modified cube to octahedron have been obtained under different electrocrystallisation conditions. However, some of crystals form non-equilibrium shapes which are not anticipated by theory [62].

As explained before the surface or facet energy may define crystal morphologies and some crystal habits appear more frequently than others. Silver crystallizes in an fcc-type lattice. Before, different possible shapes and morphologies of a fcc cube crystal were illustrated in Fig. 4.15. Provided that the cubic system symmetry rule is considered, then some of the features observed are interpretable. For instance, if one corner is replaced by triangular or hexagonal faces, then the shape will be modified accordingly. The possible modifications of a cube system are shown in Fig. 4.20a. Radmilovic et al. have studied the crystallographic aspects of silver mesocrystals electrodeposited under almost the same conditions [63] It has been shown that final form that a given crystal will take obviously depends on the relative development of cubic and octahedral form faces. However, if a given substance is electrocrystallised under different conditions, it might exhibit the same crystallographic structure but different habits. A crystal is assumed to form a shape with minimum surface and consequently minimum total energy which is determined by the sum of the energies of all unsaturated bonds on the crystal surface. For fcc silver, the surface energy will be dependent on the ratio between the number of unsaturated bonds and the coordination number 12.

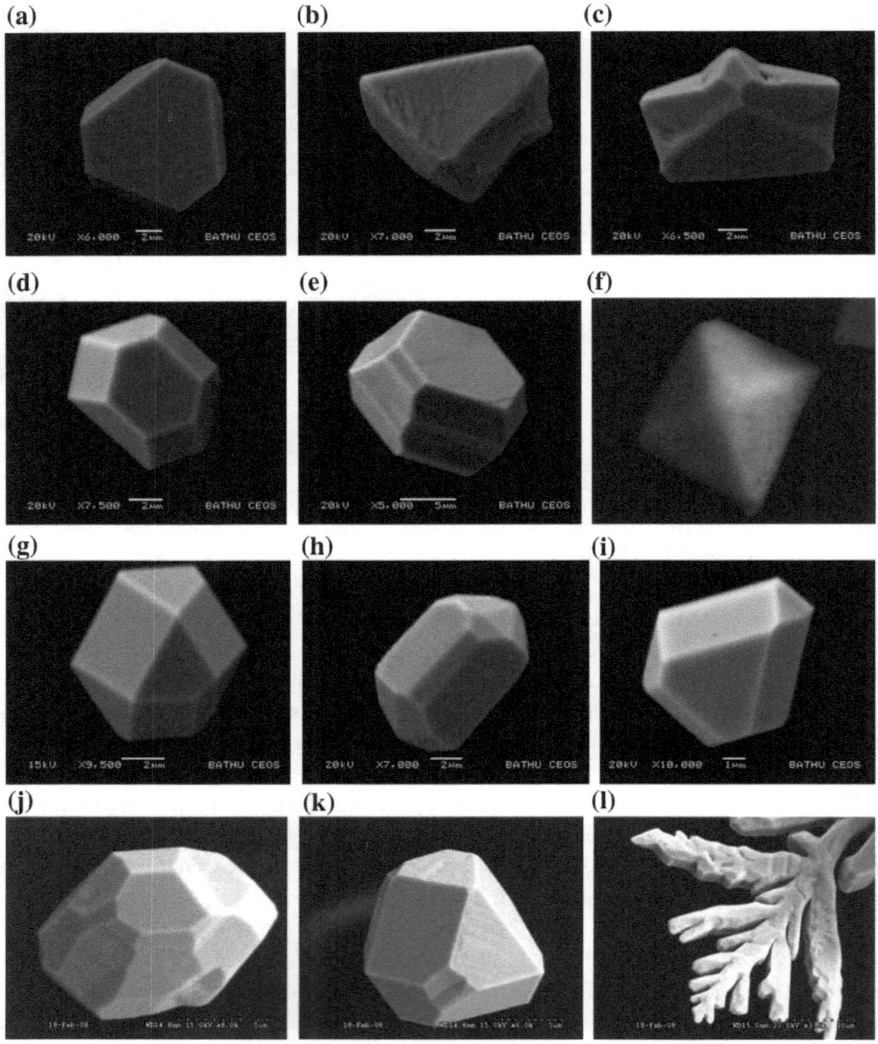

Fig. 4.19 Different shapes of silver mesocrystals electrodeposited at **a–c** 10 mM $AgNO_3$ for 60 s, **b** 100 mM $AgNO_3$ for 30 s, **c** 250 mM $AgNO_3$ for 10 s and **j–l** 600 mM $AgNO_3$ for 5 s. Electrodeposition potential has been in a range of −60 to 200 mV. The increase of overpotential is from *left* to *right* in each concentration

Previous studies have shown that silver mesocrystals electrodeposited exhibit the cubic crystal with all kinds of different habits which may include {111} and {100}, {110}, {120}, {112},{310}, etc. under equilibrium conditions which are schematically shown in Fig. 4.20b. For those that are not approved by the equilibrium crystallisation, i.e., the nonequilibrium shapes, are named multiply-twinned particles

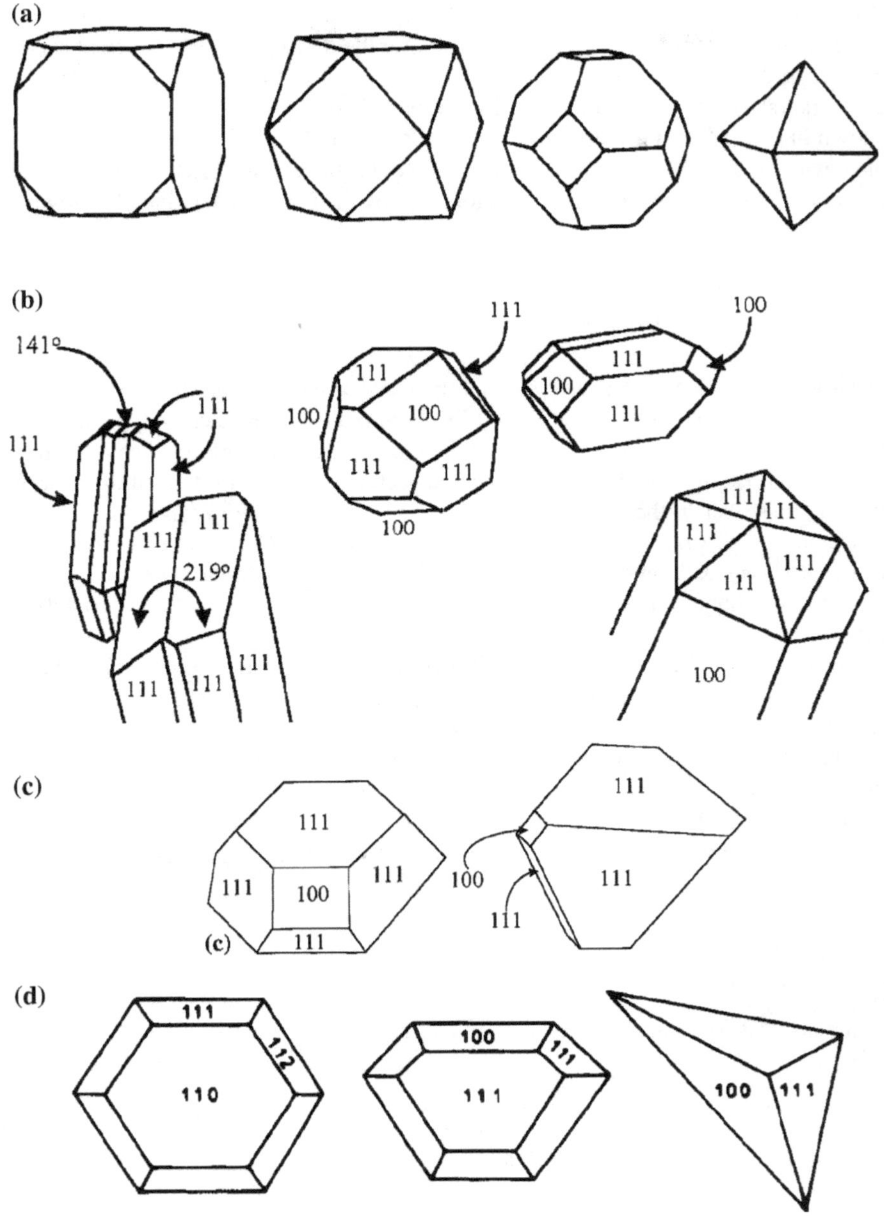

Fig. 4.20 **a** Possible modifications of a cube crystal for silver mesocrystal electrodeposition, schematic description of crystal habits correlated with those observed in reality for **b** equilibrium silver mesocrystals, **c** non-equilibrium silver mesocrystals, and **d** silver mesocrystals grown under kinetics growth influences. Adapted from [63] with the permission of Springer

(MTP's) such as decahedral, icosahedral, etc. In addition, surface impurities such oxygen, nitrogen, etc. may be another reason for the discrepancies between observed mesocrystals and equilibrium shapes of small particles. For instance, Michlev and co-workers [64] have reported the formation of Ag_7NO_{11} crystals, schematically shown in Fig. 4.20c. Furthermore, the presence of {111} close-packed plane steps is also evident. This has been explained by the contribution of growth kinetic parameters rather than the surface energy factors in growth. This is schematically shown in Fig. 4.20d.

Electrodeposition of Copper Mesocrystals

Nasirpouri and co-workers have studied the electrochemical growth of copper from a simple sulphate bath on HOPG [57]. Figure 4.21 illustrates a typical cyclic voltammogram taken for copper deposited on a HOPG working electrode from 2 mM $CuSO_4$ solution. It reveals a general behavior of electrodepositing material with a cross over in the cathodic region and separated reduction/dissolution peaks. There is one distinct reduction peak at E = −0.4 V Ag/AgCl and two oxidation peaks at E = +0.29 and +0.47 V Ag/AgCl. The two oxidation peaks seen are assigned to copper stripping (+0.29 V) and oxidation of Cu^o to CuO (+0.470 V) [48].

Nasirpouri et al. investigated the electrodeposition of copper onto HOPG based on the potential-current characteristics achieved during the cyclic voltammetry. Electrochemical step-edge decoration has been investigated in details for several different methods to produce nanoparticles and nanowires. The controllable production of metal electrodeposits has been shown to be possible using a multi-pulse plating method. Two different plating pulses ere used including single pulse and dual pulse. Figure 4.22a shows SEM and AFM images of copper electrodeposited

Fig. 4.21 Typical cyclic voltammograms obtained for Cu on HOPG when deposited from 2 mM CuSO4 solution [57]

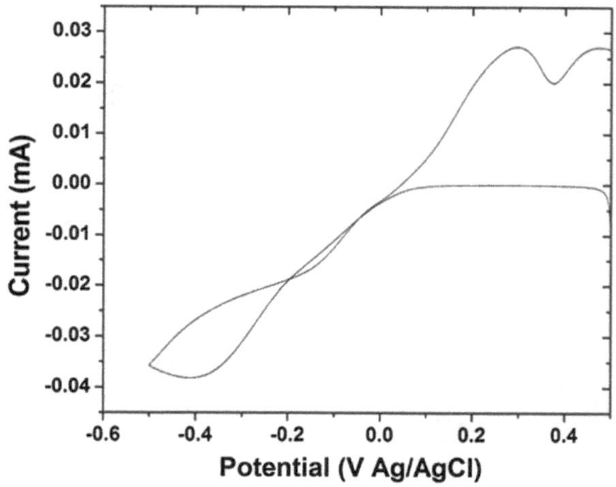

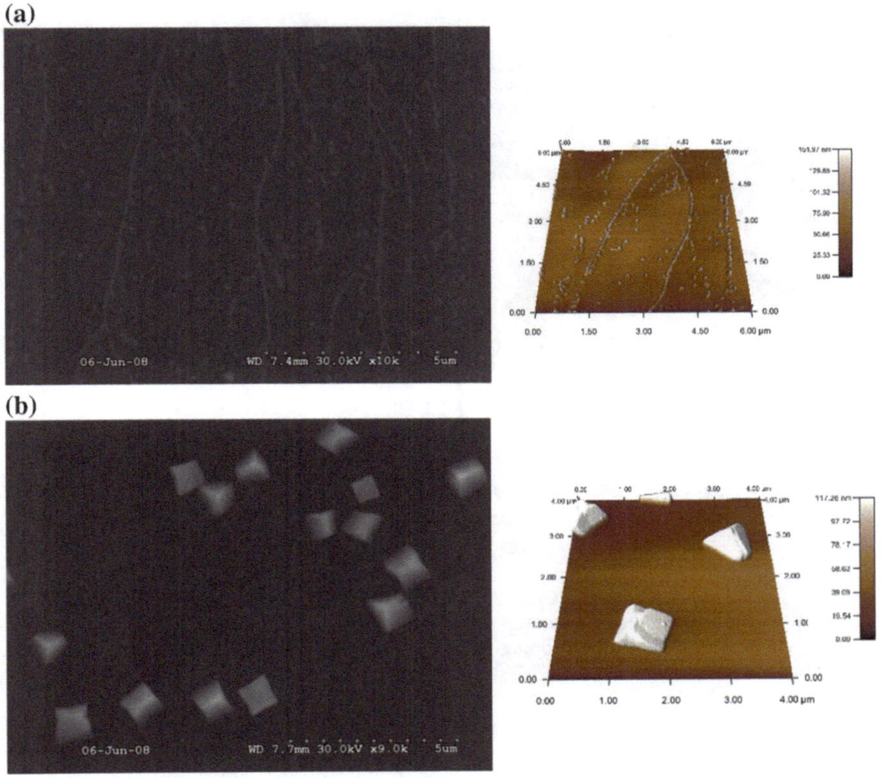

Fig. 4.22 SEM and AFM images of copper electrodeposited **a** from 2 mM on HOPG at a distinct deposition potential of −0.1 V Ag/AgCl for 60 s, and **b** pulse electrodeposited **a** from 2 mM on HOPG at −0.1 $V_{Ag/AgCl}$ for 5 ms and −0.055 Ag/AgCl for 600 s [57]

on HOPG at a distinct deposition potential of −0.1 V versus Ag/AgCl. It is seen that copper is electrodeposited in the form of particles and also wires on the HOPG surface. The general form of particles is faceted without great regularity across the whole sample. When the dual pulse technique is employed, three dimensional copper tetragons are formed with same shape and size as shown in Fig. 4.22b. The dual pulse technique permits the electroactive species to add to the metallic centre and grow in a regular form. The first short pulse corresponds to the nucleation stage and the second long pulse corresponds to the growth of three dimensional copper mesocrystals.

Wen-Yin Ko et al. [65, 66] have electrodeposited copper nanoparticles from an aqueous solution containing copper(II) tetrafluoroborate hydrate ($Cu(BF_4)_2{\cdot}xH_2O$) in the presence of organic surfactants including dodecylbenzene sulfonic acid sodium salt (DBSA), and poly(vinylpyrrolidone) (PVP) on a (111) textured gold substrate and have obtained copper particles with different architectures such as pyramid, cube, and multipod. Figure 4.23 shows different morphologies of copper

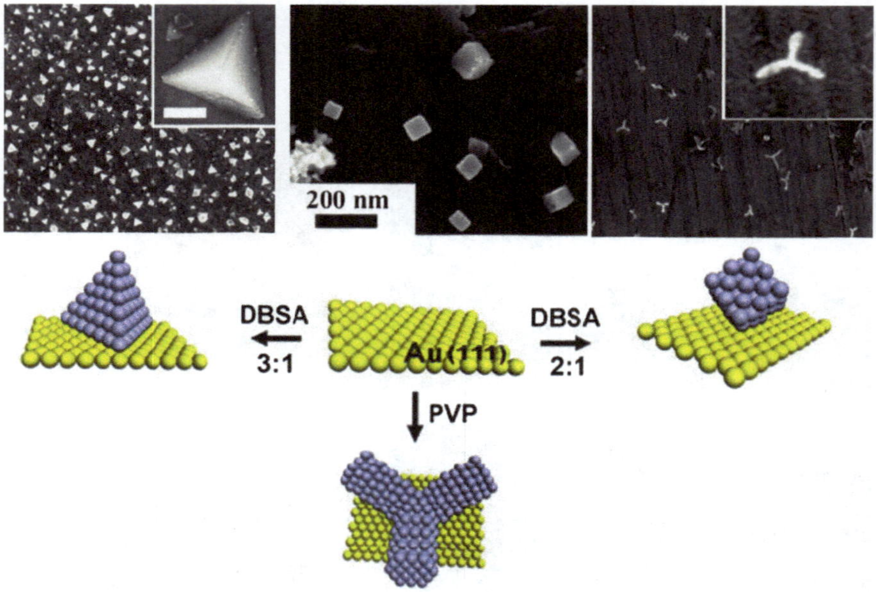

Fig. 4.23 Pyramidal, cubical, and multipod Cu mesocrystals electrodeposited on gold substrates by using DBSA and PVP as the capping reagents. Under different surfactant concentration, the morphology of mesocrystal change as shown in the graphical scheme in the *bottom panel*. Reproduced from [66] with permission of Springer

nanocrystals obtained in this study. It has been claimed that at the initial stage of the deposition, Cu atoms epitaxially electrodeposit onto the (111) domains of Au surfaces, where DBSA does not participate in the electrochemical process but adsorbs onto some surfaces of the deposited Cu, thereby forming the first (111) plane of pyramid-shaped nanocrystals. Subsequently, the surface-regulating capping agent of DBSA plays a critical role as the reaction proceeds, which results in stronger interactions with the atoms on the {111} facets than those on the {100} facets, leading to the preferential addition of Cu atoms onto the {100} facets. This leads to the enhancement of the growth rate along the [100] direction and/or reduction in the growth rate along the [111] direction, and ultimately, in the formation of pyramid-shaped Cu NPs. Multipod copper mesocrystals were unexpectedly found when the PVP polymer was used in place of DBSA, as shown in Fig. 4.23.

PVP is believed to bind preferentially to the {100} and {111} planes, which induces the selective interaction and/or interaction strengths between PVP and the crystallographic facets of a Cu lattice [65, 66].

Electrodeposition of Other Metal Mesocrystals

Xiao et al. [67] have demonstrated the realisation of various forms of 3D architectures of lead (Pb) mesoscrystals. Different morphologies were shown ranging from nanowires with cylindrical and square cross sections to mesoparticles with triangular, hexagonal, octahedral, decahedral, and icosahedral shapes and complicated structures such as multipods, porous nanowires, nanobrushes, and snowflakes. Their approach was based on the manipulation of electrodeposition potential and changing the aqueous solution composition. In general, mesocrystals (icosahedron, decahedron, and octahedron in Fig. 4.24a–c; icosahedron, decahedron, hexagon, and triangle in Fig. 4.24g–i, respectively) form at applied voltages close to the thermodynamic equilibrium potential, while nanowires (Fig. 4.22d, e), nanobrushes (Fig. 4.24f), and monopods and multipods (Fig. 4.24k, l) were demonstrated at high cathodic overpotentials.

The starting nuclei formed were shown to be responsible for different morphologies. Furthermore, the shape evolution associated with increasing reduction potentials was speculated in analogy with colloidal synthesis in which high monomer concentrations persist. With increasing reduction potential, structures with higher chemical potential and higher aspect ratio such as nanowires are expected to form on the cathode. In fact, the crystal habit or shape of mesocrystals follows the sequence of their specific surface energies that determine their chemical potentials. The electrodeposited multipods have this property as shown in Fig. 4.24g–l.

The novel shapes of icosahedron, decahedron, and octahedron lead mesostructures obtained at applied voltages close to the equilibrium potential has interpreted by a model based on a modified Wulff construction [68, 69] for shape formation. According to this model, which minimizes the total surface and twin-boundary energies of an assembly of single-crystal units at constant volume, the polyhedracan is described as a collection of single-crystal tetrahedra that are twin-related at their adjoining faces. For examples, five tetrahedra arranged symmetrically about an axis form a decahedron, and 20 tetrahedra in a three-dimensional configuration assembled about a common vertex produce an icosahedron. A strong evidence for this interpretation is the observation of the theoretically predicted notches at the twin boundaries of a decahedron, as shown in Fig. 4.24b. The formation mechanisms for porous nanowires and nanobrushes are still under investigation. Possible scenarios can be hydrogen evolution, the capping effect at high growth rate, or an oscillatory behaviour in the nucleation kinetics that can result in branched structures. Other electrodeposition parameters can also affect the architecture of the mesostructures. For example, snowflake shaped mesostructures were observed in the presence of ethanol truncated mesoparticles appeared when nitric anions were partially replaced with acetic anions. High yields of these mesostructures can be produced, though more effort is needed to achieve monodispersive architectures [67].

Other researchers from a joint work between Simon Bending's and Laurence Peter's groups have demonstrated the possibility of tuning architectures of tin

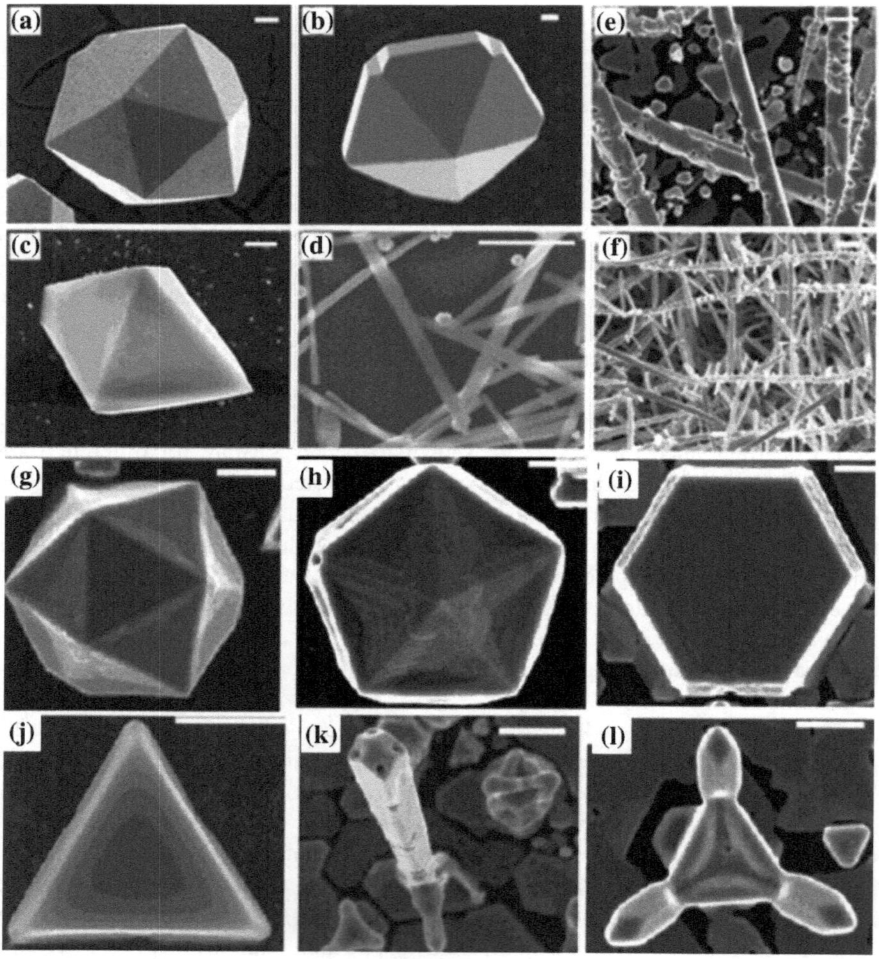

Fig. 4.24 Pb mesocrystals electrodeposited **a–f** from solution containing 5 mM lead nitrate and 0.1 M boric acid under a constant potential of −0.8, −0.95, −1.1, −1.2, −1.7, and −2.0 V, for **a–f** respectively, and **g–l** from solution containing 5 mM lead acetate and 0.1 M boric acid under a potential of −0.65, −0.8, −1.0, −1.1, −1.3, −1.4 V for **g–l**, respectively. The scale bar length is 500 nm. The growth time is 60 s. Adapted with permission from [67]. Copyright (2016) American Chemical Society

(Sn) and bismuth (Bi) mesocrystals by electrochemical deposition. The electrodeposition mechanism of Sn from tin(II) tetrafluoroborate solution and fluoroboric acid solution on polycrystalline boron-doped diamond (BDD) was studied indicating that the growth process is not fully diffusion limited as assumed by the Scharifker–Hills model and there is a significant kinetic component. Normal electrocrystallisation behaviour was speculated for Sn which is consistent with the

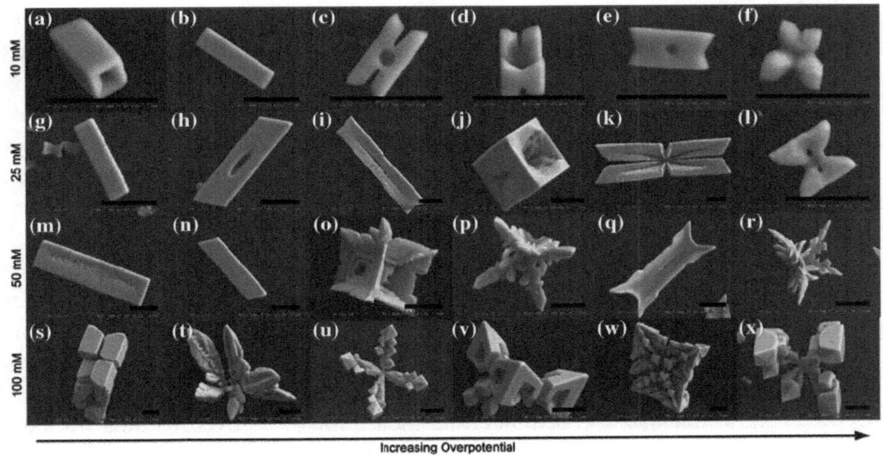

Fig. 4.25 SEM images of Sn crystals grown under different conditions from solutions with a range of concentration 10–100 mM and electrodeposition potential ranging from −740 to −900 mV. The scale bar is 5 mm in all cases. Reproduced from [70] with kind permission from Simon Bending and Laurence Peter as well as from the Royal Society of Chemistry

argument presented earlier in this chapter Fig. 4.13. At low overpotentials the growth is accompanied by a small number of simple shaped, faceted crystals such as square cuboids with low surface energies, whereas higher overpotentials lead to a higher density of crystals and much more complex shapes such as dendrites with very high surface energies. Figure 4.25 shows a range of Sn mesocrystals electrodeposited on BDD substrates. Figure 4.25a, b show relatively large square cuboid rods, while the crystal in (Fig. 4.25a) seems to be hollow. For higher potentials (Fig. 4.25c–f), the crystals look like the outer skeleton of a defected square cuboid. In 25 mM cuboid shape mesocrystals are seen. At low overpotentials (Fig. 4.25g–i) l ordered square cuboid and at higher overpotentials and longer growth times (Fig. 4.25j–l) hollow cuboid shape was shown. At higher concentrations there a range of rather complete, solid square cuboids to very complex, dendritic crystals (Fig. 4.25o–r, t–x) were demonstrated.

It has been concluded that the long axis of the cuboids corresponds to the [001] direction, while the two (equal) short lengths are along the two equivalent directions [100] and [010]. This is the shape expected for slow kinetically limited growth when the energetically favoured high atomic density surfaces are selected. For higher concentrations and overpotentials the growth mechanism of Sn mesocrystals was shown to change. The fastest growth happens along the body diagonal [111] direction (e.g., Fig. 4.25o, p) while the faces of the cuboid are only partially developed. For the highest concentration (Fig. 4.25, bottom row) the crystals grow in a dendritic, fern-like manner. The crystals in Fig. 4.25u, x show, in particular, that the structures still grow as single crystals, albeit multiply twinned ones. The main branches are of quite irregular shapes but at the end caps the cuboid parts are arranged along identical axes. Spherical diffusion layers are formed around more acute pointed parts of the crystal, resulting in more efficient 3D diffusion. This leads to faster growth at these locations and incident atoms have little time to diffuse on the surface, resulting in a highly non-equilibrium surface structure. This explains the formation of the skeletons, since the fastest growth occurs at the corners rather than the faces of the crystal. This is also the origin of the dendritic growth when, even further away from equilibrium, the crystal grows preferentially at the crystal corners rather than at edges or on faces [70].

In another study, the same group have performed real-time atomic force microscopy (AFM) imaging of bismuth mesocrystals that were grown under electrochemical control at low overpotentials to ensure a slow growth rate and allow in situ observation of the growth. A two-step potential pulse method was applied to a boron-doped diamond (BDD) working electrode with a short high overpotential, −0.4 V (2 s), to nucleate the bismuth, and then a long low overpotential for slow growth, −0.32 V (4.4 h). Growth rates of individual crystals and detailed growth mechanisms could be followed in real time because of the slow crystal growth under the experimental conditions employed in the study, shown in Fig. 4.26.

The close proximity of AFM tip and tip holder to the working electrode appears to hinder the diffusion of bismuth to the BDD substrate, therefore slowing down the growth process. This must be noted in such experimental research. However, in situ AFM is capable to produce useful information of growth of a crystal during electrodeposition. The detailed evolution of growth of an individual bismuth crystal has been shown in Fig. 4.26a. Figure 4.26a–d, clearly show mobile steps on the surface of the crystal indicating a step-flow growth mechanism. Numerical calculations of growth regions (Fig. 4.26i–iii) show that rapid growth is clearly associated with sharp corners of the crystal and step edges.

Application of a constant potential at long time scales and the motion of the AFM moving across the surface of the crystal creates convective transport of Bi^{3+} to the active area. In practice, this motion was minimal and not enough to significantly increase the growth rate. Figure 4.26b shows several line scans taken across the front portion of the crystal during its growth. The first line scan taken shows flat crystal faces and a well-defined rhombohedral structure. Growth of such a crystal then follows a cyclic step-flow growth mechanism. Steps are highly reactive centers because of their high surface area and therefore preferential site for bismuth nucleation. The difference images of Fig. 4.26 indicate that the growth of all the older steps is interrupted when a new step forms at the peak. Clearly the growth process involves a complex interaction between several steps, and local growth rates can vary quite dramatically. The large size of the steps indicates a bunching of atomic layer forming polyatomic steps.

This bunching phenomenon is well established, and a theoretical model was proposed by Frank [72] whereby the build up of molecular terraces occurs because of a "fence" of dislocations, which prevents the growth of a single layer. Here a "fence" is a boundary between two crystal structures at a small angle to each other [71].

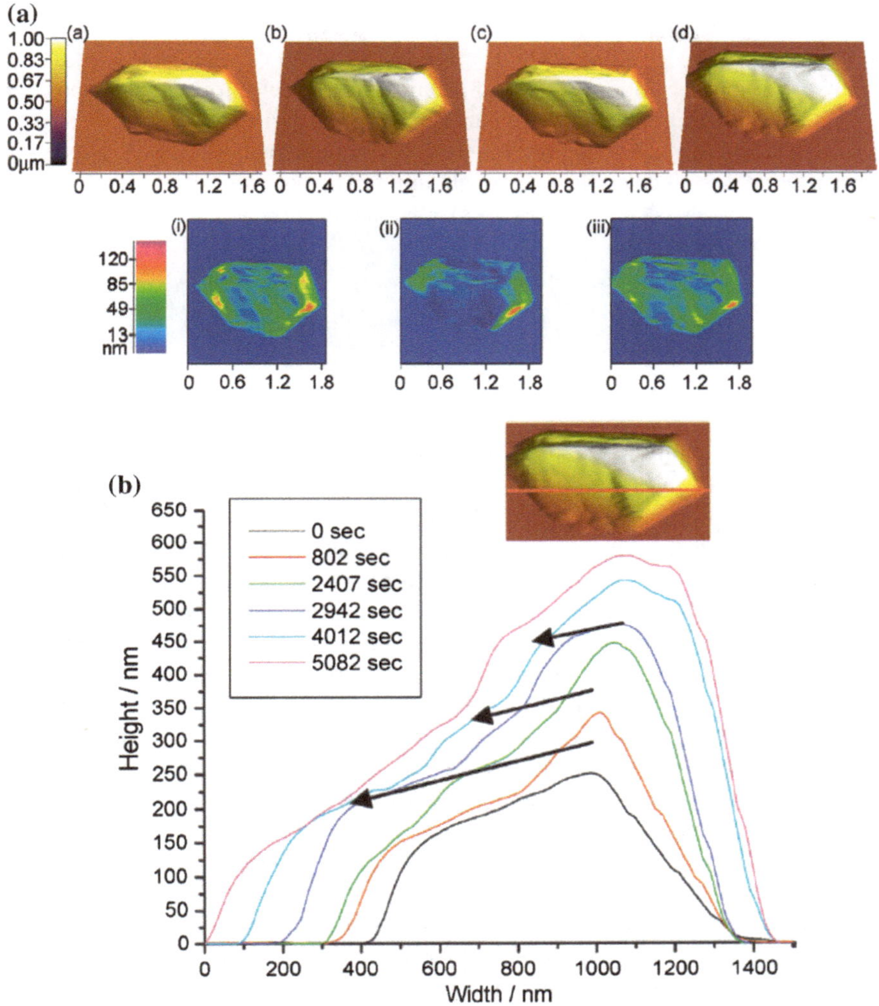

Fig. 4.26 **a** The sequential electrocrystallisation of an individual bismuth crystal at a potential of −0.32 V and at times of *a* 2937 s, *b* 3472 s, *c* 4007 s, and *d* 4542 s. Images underneath represent the local difference in height as calculated by MathCAD showing areas of growth, where *i* is the difference between (*a*) and (*b*), *ii* is the difference between (*b*) and (*c*), and *iii* is the difference between (*c*) and (*d*). The scale marker at the bottom of both the AFM and MathCAD images is in µm. **b** Line scans taken across the front of the bismuth crystal as indicated by the red line in the inset image (*i–iii blue* zero, *green* small, *yellow* medium, and *red* large height changes). *Arrows* on the graph indicate step-flow growth. Adapted from [71] with kind permission from Simon Bending and Laurie Peter, as well as with the permission of American Chemical Society, 2016

4.3.2 *Multiple 3D Nuclei and Coalescence for Film Growth*

For many electrodeposition systems of practical interest, electrochemical deposition occurs by 3D island growth, where nuclei firstly form and coalesce later during the growth to form a continuous film. In island growth, the nucleus density N is a key parameter because it determines the minimum film thickness for island coalescence, d_{crit}. In most cases, islands grow isotropically, and the critical film thickness can be estimated from:

$$d_{crit} = \frac{1}{2\sqrt{N}} \tag{3.9}$$

The critical film thickness is the thickness at which adjacent hemispherical islands with an average spacing of $N^{-1/2}$ begin to impinge on each other. Figure 4.27 shows the relationship between the critical thickness and number of nuclei per surface area. For instance, an island density of at least 10^{12} cm^{-2} is required to deposit a continuous 10-nm-thick film if island growth is isotropic (i.e., hemispherical islands). In thin film technology, to achieve island coalescence at small film thicknesses, high island densities are essential, as illustrated above. High island densities, however, usually lead to small grain size, which results in a decrease in the electron mean free path and hence an increase in electrical resistivity. This coupling between island density and grain size is a major limitation in materials synthesis. The control of island shape is a solution to this problem; the formation of large, quasi-2D islands resulting from anisotropic growth leads to a small critical thickness for coalescence but large grain size and hence low electrical resistivity. For example, if an island grows 10 times faster laterally than vertically, an island density of 10^{10} cm^{-2}, which is 100 times smaller than the case of isotropic growth, is sufficient to grow a continuous 10 nm thin film (Fig. 4.27). In addition, the resulting film has a 100 times larger grain size, which is likely to result in improved electrical conductivity. Electrochemical thin film deposition is usually implemented at large overpotentials to achieve high island density. However, deposition at high overpotentials usually results in isotropic island growth [31]. Here we demonstrate the feasibility of controlled growth by presenting a system where islands exhibit preferential lateral growth and eventually form a continuous film.

We here briefly explain an experimental electrodeposition work to demonstrate the effect of nucleation rate on the formation of 3D nickel nuclei and consequent coalescence to produce nickel thin film.

4.3.2.1 Electrodeposition of Nickel on Copper

Janjan et al. [73, 74] have studied the mechanism of nucleation and growth of nickel from a simple sulfate bath on polycrystalline copper (Cu) substrate, and

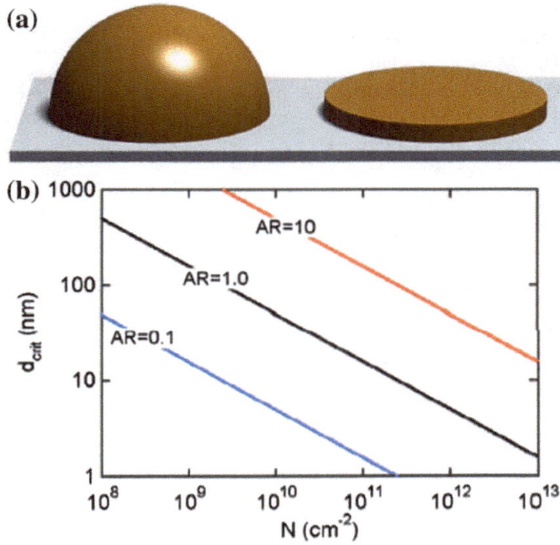

Fig. 4.27 **a** Illustration of hemispherical and disk-shaped islands resulting from isotropic growth and enhanced lateral growth, respectively. **b** Plot of critical thickness versus the island density for island coalescence for islands with aspect ratios of 0.1, 1.0, and 10. Reproduced from [31] with the permission from American Chemical Society, copyright (2016)

demonstrated 3D nucleation and growth mechanism from simple sulfate solution on the copper substrate surface with different concentrations. Different potential sweep rates from 20 mV/s up to 200 mV/s were employed. Figure 4.28a, b illustrate the cyclic voltammograms from different solutions on the copper substrate, respectively. It can be seen that nickel ions starts to be reduced from −0.7 V Ag/AgCl and reach a current peak at −1.4 V Ag/AgCl. Increasing scan rate appears to be associated with significant increase in peak currents exhibiting a diffusion controlled process. The presence of large solution resistance is evidenced from the CV shown in Fig. 4.28, where the oxidation peak is shifted to much more positive side. This can be because of the low supporting electrolyte concentration as compared to the analyte.

Electrochemical deposition was performed using standard chronoamperometry technique to study the nucleation and growth mechanism of nickel on copper. Deposition potentials were chosen according to the reduction peak appeared on the cyclic voltammograms. The deposition is conducted at constant potentials at potentiostatic mode during which current transients are recorded. Figure 4.29 shows current transients obtained at different deposition potentials is the two different solutions.

Typical potentiostatic current transient (I-t) curves (Fig. 4.29) demonstrate a diffusion control deposition behaviour with the characteristic peak corresponding to the nucleation-and-growth process and for longer times, merging into a common curve which is caused by a diffusion controlled process and described by the Cottrell equation (see Chap. 3). According to the I-t curves, each transient has one well-defined recognizable current maximum as a clear first peak followed by a

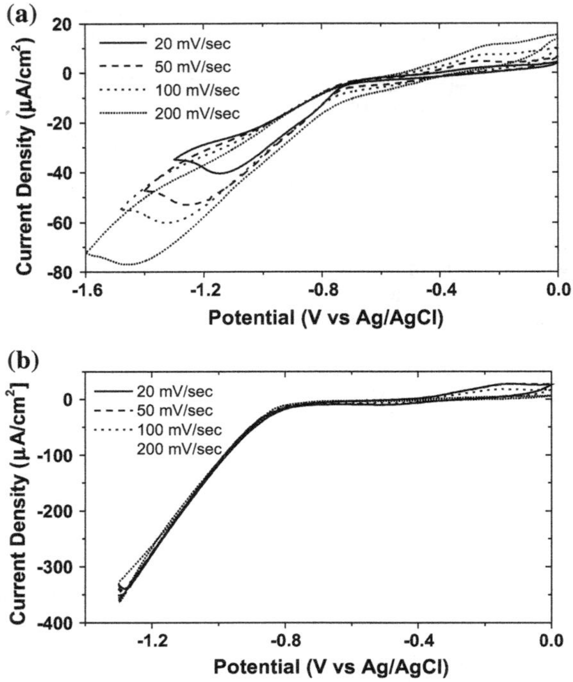

Fig. 4.28 Cyclic voltammograms taken from **a** 0.01 M nickel sulfate 0.05 M boric acid (solution 1) and **b** 0.1 M nickel sulfate 0.5 M boric acid (solution 2) on the copper substrate at different potential sweep speeds. Reproduced from [73] with the permission from Springer, copyright (2016)

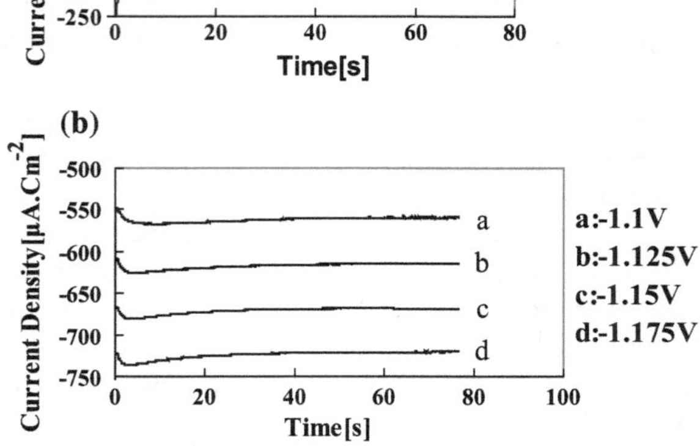

Fig. 4.29 Current transients recorded during electrodeposition of nickel on the copper substrate from **a** 0.01 M nickel sulfate and 0.05 M boric aside **b** 0.1 M nickel sulfate and 0.5 M boric acid. Reproduced from [73] with the permission from Springer

sharp fall and posterior growth. The I-t transients have a normal dependence to overpotentials, whereas the current density increases with increasing of overpotential. The peak corresponds to the nucleation of metallic sites on the surface which is followed by the reduction of current exhibiting three-dimensional (3D) growth. Increasing the peak current at higher overpotentials means that the number of sites (nucleation rate) increases owing a higher nucleation rate. Figure 4.30 shows obtained non-dimensional plots for electrodeposition of nickel on the copper substrate.

The experimental data are found to best describe the 3D instantaneous, diffusion-controlled nucleation and growth mechanism of nickel from the solutions. In some rare cases, increasing the overpotential shifts the intermediate curves to the lower theoretical curve corresponding to progressive nucleation. This transition indicates that the nucleation rate increases with overpotential.

Further to the achievements based on the non-dimensional curves, another criterion based on the variation of deposition current against $t^{1/2}$ and $t^{1/3}$ was used as mentioned is Fig. 4.31 for both solutions. The mechanism is instantaneous when I-$t^{1/2}$ plot has a linear behavior. Therefore, a linear plot of current against $t^{1/3}$ will describe the progressive nucleation and growth mechanism. It is seen that the behavior can show the instantaneous nucleation and growth.

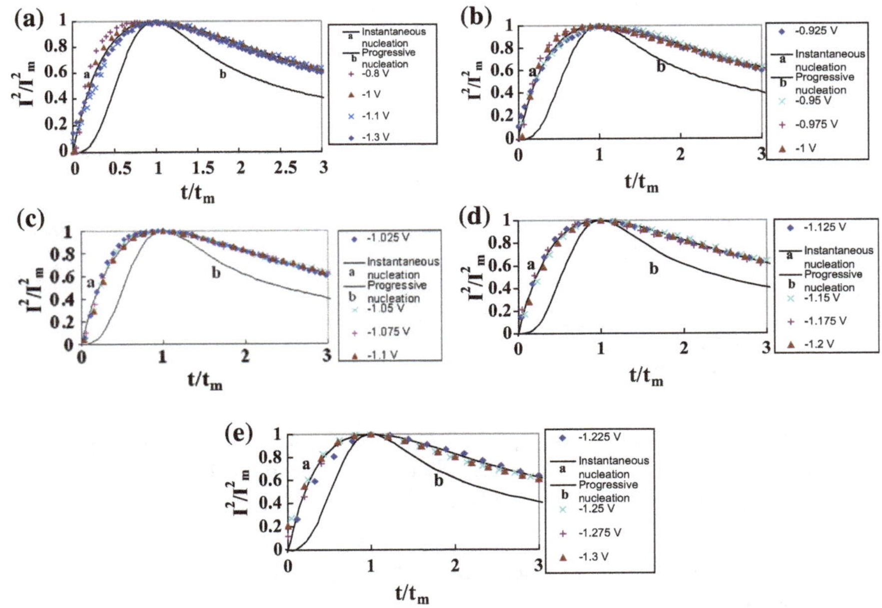

Fig. 4.30 Non-dimensional plots in form of (I^2/I_m^2) against (t/t_m) for of simple sulfate bath with different concentration and potential: **a** 0.01 M nickel sulfate 0.05 M boric acid. **b–e** 0. 1 M nickel sulfate 0.5 M boric acid. Reproduced from [73] with permission from Springer

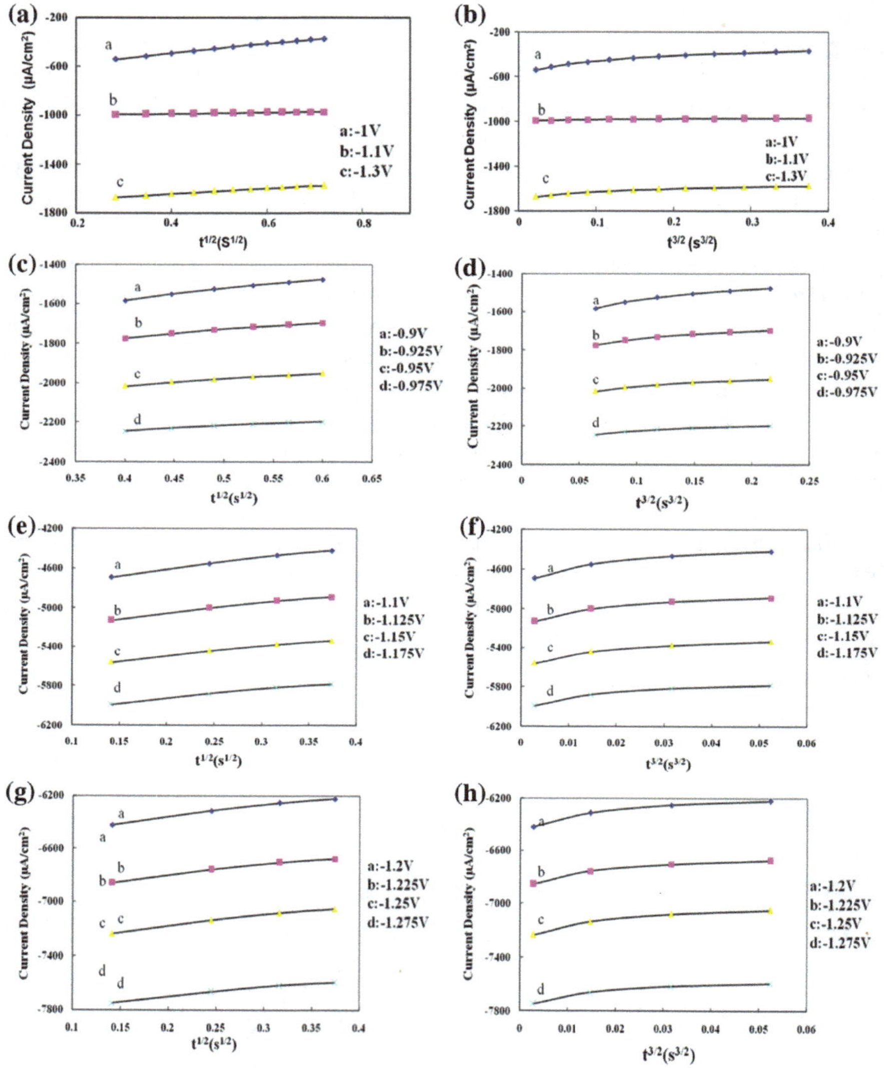

Fig. 4.31 Variation of current density versus $t^{1/2}$ and $t^{1/3}$ for simple sulfate bath with different concentration and potential: **a**, **b** 0.01 M nickel sulfate 0.05 M boric acid and **c–h** 0.1 M nickel sulfate 0.5 M boric acid. Reproduced from [73] with permission from Springer

4.3.2.2 Electrodeposition of Nickel on Silicon (3D Nucleation and Coalescence for Film Growth)

Nasirpouri and co-workers have carried out a systematic study on the electrode-postion of nickel on hydrofluoric acid (HF) treated [75, 76] H-terminated n-silicon

(111) wafers from simple sulfate solutions to manipulate various parameters including electrolyte composition and pH, presence and absence of chloride ions and overpotential towards the fabrication of continuous thin films using different characterisation methods. It is shown the electrodeposition mechanism significantly influence the magnetic properties and magnetisation reversal mode of nickel electrodeposit under different stages of electrodeposition.

Electrodeposition of Nickel from Dilute Sulfate Solution on Silicon

Electrodeposition of nickel was carried out from 0.01 M $NiSO_4{\cdot}6H_2O$ (mole per 1 l H_2O) and 0.05 M H_3BO_3 namely dilute solution Fig. 4.32a shows typical cyclic voltammograms taken from dilute solution on n-Si(111) at different potential sweep rates. Nickel reduction starts at a potential of −0.70 V reaching the reduction peak depending on the potential scan rate. Increasing scan rate appears to be associated with significant increase in the peak currents exhibiting a diffusion controlled process. According to Delahay equation (see Chap. 3), the variation of Ni deposition peak current against the square root of scan rate (v) is demonstrated in Fig. 4.32b. This is a criterion by which the diffusion-controlled process of the electrochemical reaction is approved. However, the distance between the onset of deposition and dissolution potentials of nickel on silicon proves the irreversible character of the electrochemical reaction in this solution. The presence of large solution resistance is evidenced from cyclic voltammograms shown in Fig. 4.32a, where the oxidation peak is shifted to more positive side. This can be due to the low supporting electrolyte concentration as compared to the analyte. Based on the observation of distinct reduction peaks of Ni^{2+}, it can be implied that the hydrogen evolution takes place negligibly in this solution.

The electrodeposition was then conducted under potentiostatic mode at constant potentials obtained from the cyclic voltammograms. Typical potentiostatic current transient (I-t) curves are shown in Fig. 4.1b. The variation of current against the deposition time implies that a typical diffusion control nucleation and subsequent growth takes place which has been described by Hills and Scharifker model. Despite of the absence an adequate amount of a supporting electrolyte, the electrodeposition of nickel on silicon was observed to be controlled by diffusion at this range of Ni^{2+} cations in the electrolyte.

The peak corresponds to the nucleation of metallic sites on the surface which is followed by the reduction of current exhibiting three-dimensional (3D) growth. The rise in current in each transient reflects the electroactive area increases because of either each independent nucleus grows in size and/or the number of nuclei increases. At this stage a diffusion controlled process directs the growth of deposit and then the current falls at final stage. But the dependence of the variation of current during electrodeposition upon the magnitude of the overpotential determines the mechanism of nucleation. Based on the diffusion controlled electrodeposition predictions of Scharifker-Hills, we constructed non-dimensional plots I^2/I_m^2

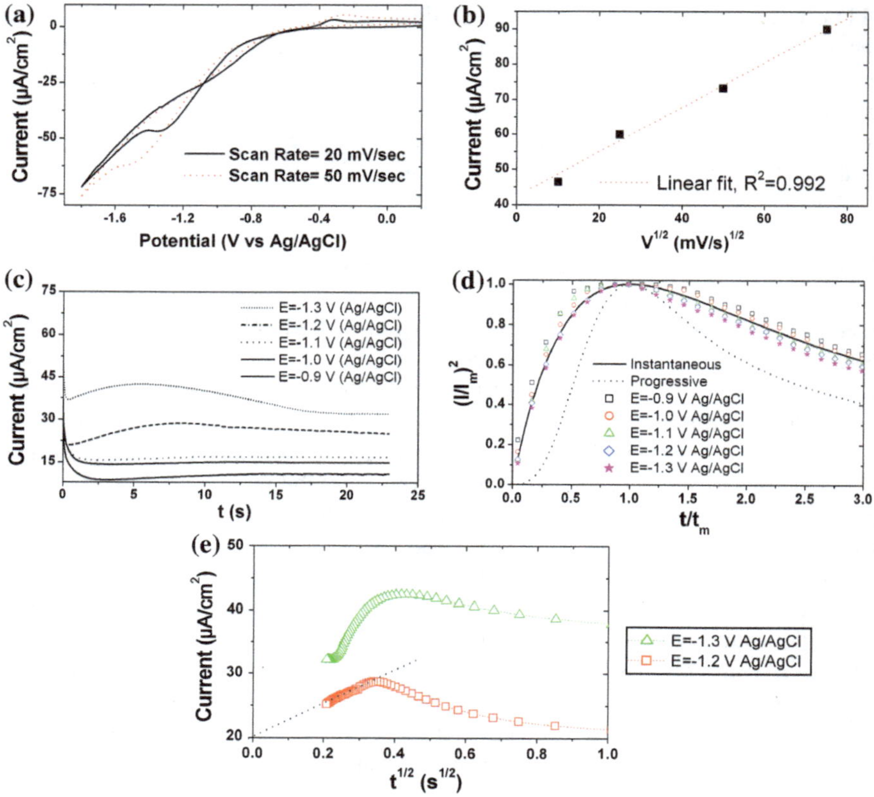

Fig. 4.32 **a** A typical cyclic voltammogram **b** variation of Ni deposition peak current against the square root of scan rate (V), **c** current transients recorded during the electrodeposition of nickel at different applied potentials and **d** non-dimensional plots calculated from the current transients for Ni electrodeposited from dilute sulfate solution containing 0.01 M $NiSO_4{\cdot}6H_2O$, 0.05 M H_3BO_3. Reproduced from [75] with permission from Elsevier, copyright 2016

versus t/t_m for our results. Non-dimensional plots are constructed from the I–t transients, shown in Fig. 4.32c, where the current I_m and the time t_m correspond to the maximum current peak. Typical plots together with theoretical curves derived from (3.55) and (3.56) for the instantaneous and progressive nucleation are shown in Fig. 4.32d for electrodeposition of Ni on n-Si(111) from dilute simple sulfate solution. The experimental data are described by the 3D instantaneous diffusion-controlled nucleation and growth mechanism of Ni from this solution. This should be notified that although the cyclic voltammetry reveals distinct reduction peaks for nickel, according to other studies which will be discussed later, perhaps, the only side reaction occurring during the electrodeposition of nickel may be the hydrogen reaction. Scharifker-Hills model descriptions of the nucleation and growth process anticipates that the metallic ions interact directly with the electronic states of the surface and it does not consider the possibility of an intercepting layer of hydrogen chemically reducing the ions during the electrocrystallisation. However, the controlling mechanism of diffusion still prevails even if a sheath of hydrogen covered the surface during the reduction of metallic ions. This was modeled for the electrodeposition process from aqueous electrolyte [77]. We only consider the Scharifker-Hill theory for the dilute solution, as the hydrogen evolution is not observable significantly according to the cyclic voltammograms shown in Fig. 4.32a.

The growth of nickel from dilute simple sulfate solution was examined using ex situ SEM, surprisingly confirm the predictions of the theory. Figure 4.33a, b illustrate the growth stages of the electrodeposition of nickel on n-Si(111). SEM images were taken from Ni electrodeposited on n-Si(111) for 10 s (Fig. 4.33a) and 20 s (Fig. 4.33b) at −1.1 V Ag/AgCl. Size distribution plots or histograms were constructed based on the SEM images which show a narrow distribution over the surface area on which the calculations were made. All the mature nuclei grown at different stages of electrodeposition have almost the same size. However, the longer the deposition time the bigger the diameter of nuclei grown. As is evident, the mechanism of nucleation and growth of nickel on n-type Si(111) is instantaneous.

The Effects of Nickel Content, Chloride Ions and PH

In Fig. 4.34, typical CVs taken at a potential scan rate of 20 mV/s in different solutions are shown. At the first instance, it is possible to investigate the effect of nickel content of electrolyte on the mechanism of nickel electrodeposition on Si substrate. For this purpose, a comparison between the CVs shown in Fig. 4.33a, b

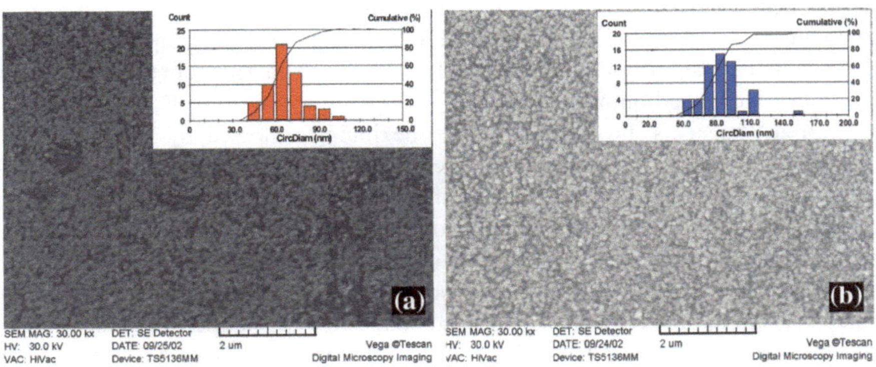

Fig. 4.33 Scanning electron micrographs and corresponding size distribution histograms taken from Nickel electrodeposited at −1.1 V Ag/AgCl from dilute sulfate solution containing 0.01 M $NiSO_4{\cdot}6H_2O$, 0.05 M H_3BO_3 for **a** 10 s and **b** 20 s. Mean particle size was calculated: 68(±13) nm for panel (**a**) and 87(±18) nm for panel (**b**). Reproduced from [75] with permission from Elsevier, copyright 2016

reveals solely the effect of nickel content on its deposition and dissolution on silicon. The observed peak in Fig. 4.33a, b, corresponds to two processes: (i) nickel electrodeposition and (ii) hydrogen evolution. Nickel reduction is indistinguishable from hydrogen evolution reaction. The onset of the reduction potential in solutions with higher Ni^{2+} concentration is slightly shifted towards more positive magnitudes which can be ascribed to the thermodynamically favorable conditions at higher concentration of electroactive species in the solution driven by the Nernst equation. This effect is remarkable when the increase of the nickel content is much higher when the composition of solution is changed to the Watts bath. Figure 4.34d, e illustrate the CVs recorded for the deposition of nickel on n-Si(111) from Watts solutions with different pHs.

The effect of addition of chloride ions to sulphate bath reveals particular changes on the reduction of nickel on silicon. In the absence of chloride ions, the reduction peak disappears with the addition of chloride ions in form of nickel chloride. By the way, the reduction potential of nickel on silicon is slightly shifted to more negative potentials, according to Fig. 4.34c. According to some earlier studies on the electrodeposition of nickel from sulfate and chloride solutions [51, 78], the negative shift in the onset of the electrodeposition potential of nickel after the addition of nickel chloride, can be attributed to the adsorption of Cl^- ions on n-type Si(111) surface with an infinitive binding energy [79]. It has also been pointed out that this adsorption affects the Ni^{2+} reduction process which is associated with the positive shift in onset of deposition potential with increasing Ni^{2+} concentration. Any increase in the nickel concentration indicates the occurrence of the reduction and dissolution peaks at higher currents as another outcome of the increase of electroactive species in the solution. When the deposition rate is further increased, the reduction peak on the CVs disappears due to the co-evolution of hydrogen [80] at more cathodic potentials.

To obtain precise information on the mechanism of nucleation and growth during electrodeposition of nickel from different solutions on silicon, we attempted to monitor the electrochemical deposition using ex situ SEM and AFM. At a distinct applied deposition potential of −1.1 V SCE, extracted from the CVs, nickel was electrodeposited on n-Si(111).

Current transients have a peak at the beginning followed by a dramatic fall leading to plateau. The current peak occurs at a definite critical time which corresponds to the nucleation process, which is measured on the current transients and reported in Table 4.2. This helps us exploit the effect of electrolyte composition on the rate of nucleation of nickel nuclei on the silicon substrate. Figure 4.34g, h illustrates SEM images of nickel electrodeposited from different solutions on the silicon substrate. Images were analyzed to obtain size-distribution of nuclei formed on the surface at the same deposition time for all solutions. The data of mean nickel nucleus size and density (number of nuclei per surface area) in different solutions are summarized in Table 4.2. The number of nuclei formed from different solutions reveals interesting results on the effect of chlorine ions and also pH on the electrodeposition of nickel from sulfate bath. Also, the size-distribution changes as the electrodeposition electrolyte changes.

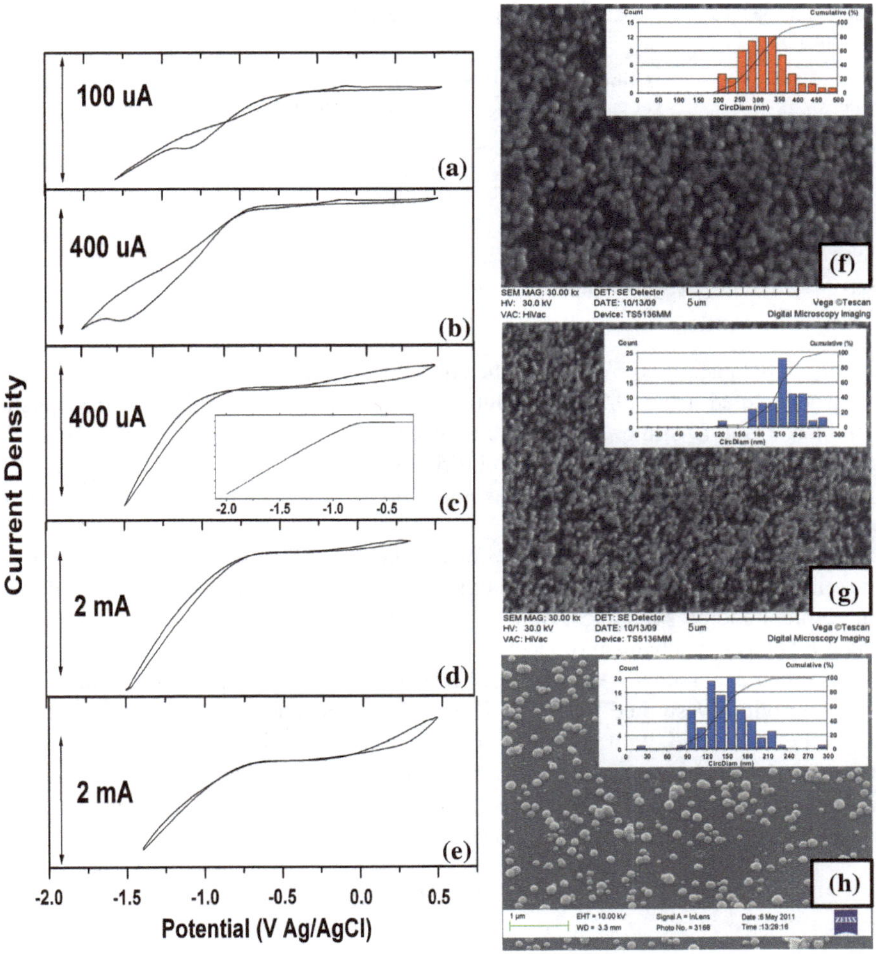

Fig. 4.34 *Left panel* Typical cyclic voltammograms taken from **a** dilute sulfate solution containing 0.01 M $NiSO_4{\cdot}6H_2O$ + 0.05 M H_3BO_3, **b** simple sulfate solution 0. 1 M $NiSO_4{\cdot}6H_2O$ + 0.5 M H_3BO_3, **c** simple sulfate with chloride 0.1 M $NiSO_4{\cdot}6H_2O$ + 0.05 $NiCl_2{\cdot}6H_2O$ + 0.5 M H_3BO_3, **d** watts bath 2.3 M $NiSO_4{\cdot}6H_2O$ + 0.6 $NiCl_2{\cdot}6H_2O$ + 0.5 M H_3BO_3 at pH = 3.2 and **e** watts bath 2.3 M $NiSO_4{\cdot}6H_2O$ + 0.6 $NiCl_2{\cdot}6H_2O$ + 0.5 M H_3BO_3 at pH = 2. *Right panel* SEM images and corresponding particle size distribution histograms taken from Ni electrodeposits at −1.1 V Ag/AgCl from **f** simple sulfate solution 0.1 M $NiSO_4{\cdot}6H_2O$ + 0.5 M H_3BO_3, **g** simple sulfate with chloride 0.1 M $NiSO_4{\cdot}6H_2O$ + 0.05 $NiCl_2{\cdot}6H_2O$ + 0.5 M H_3BO_3, and **h** watts bath 2.3 M $NiSO_4{\cdot}6H_2O$ + 0.6 $NiCl_2{\cdot}6H_2O$ + 0.5 M H_3BO_3 at pH = 2. Reproduced from [75] with permission from Elsveier, copyright 2016

As is evident from the data collected in Table 4.2, size of nuclei decreases with the addition of nickel content, chloride ions and also with the reduction of pH. Consequently, number of nuclei increases per surface area. However, an important

Table 4.2 Nucleation and growth parameters of the electrodeposition of nickel from different solutions on n-Si(111)

Electrolyte title	Critical time at current peak (s)	Mean diameter of nuclei (nm) (±standard deviation)	N ($1/cm^{-2}$) (on SEM)
Simple sulphate	3.8	318 (±59)	1.56×10^{10}
Simple sulphate with chloride	1.14	217 (±30)	2.56×10^{10}
Watt bath low pH	0.44	135 (±38)	1.51×10^{11}

Reproduced from [75] with permission from Elsevier

feature of data acquired by SEM is the change of size distribution of nickel granules electrodeposited from different solution. As is seen for the Ni electrodeposits obtained from a dilute solution, the mechanism of nucleation and growth is instantaneous. But with addition of chloride ions and more nickel content, the size distribution is widened, until it seems not to obey the instantaneous mechanism and will be discussed thereafter.

These effects not reported before for the mechanism of the electrodeposition of nickel on n-type silicon reveal markedly different nucleation and growth of nickel on semiconductor substrates. We interpret those effects in some respect by the change of basic mechanism of electrodeposition of nickel already understood on the other substrates such as vitreous carbon and paraffin impregnated graphite electrodes previously studied [81, 82]. The surface characteristics of those substrates could be compared with n-type silicon in terms of relatively weak interaction energy between ad-atoms and substrates. Electrodeposition of nickel from sulfate and chloride solutions proceeds simultaneously with the hydrogen discharge reaction, based on the proposed reactions:

$$\begin{aligned} &Ni + H^+ + e^- \rightarrow Ni - H_{ads.} \\ &Ni - H_{ads.} \rightarrow 2Ni + H_2 \\ &Ni - H_{ads.} + H^+ + e^- \rightarrow Ni + H_2 \end{aligned} \tag{4.10}$$

In sulfate electrolytes the hydrogen evolution reaction occurs prior to nickel deposition indicated by Delahay equation used to relate a bigger portion of the consumption of current by the proton (H^+) than by nickel (Ni^{2+}) reduction. When the chloride solution is used, it has been proved that the deposition process in this type of electrolyte starts with nickel reduction, because chloride anions are preferentially adsorbed on the surface in a competition with protons. Therefore, in our case, when simple sulfate solution is employed nickel reduction is accompanied by adsorption of protons. Therefore, it covers the nucleation sites on n-Si(111) substrate and to some extent inhibits the nucleation of nickel deposit. When chloride is added to the sulfate solution giving rise solution 3, the mechanism of chloride adsorption likely becomes activated partly enhancing the nucleation of nickel by inhibiting hydrogen evolution.

In Watts bath, the situation is rather complicated. There have been principle studies on the mechanism of the electrodeposition of nickel from Watts bath on vitreous carbon electrodes [83–87] which in general propose the following reactions:

$$\begin{aligned} & Ni^{2+} + X^- \leftrightarrow NiX^+ \\ & NiX^+ + e^- \leftrightarrow NiX_{ads.} \\ & NiX_{ads.} + e^- \leftrightarrow Ni + X^- \end{aligned} \tag{4.11}$$

The anion X^- has been variously assumed to be OH^- or Cl^-. However, the morphology and mechanism of nickel electrodeposited on silicon from Watts has not been studied. However, the effects of additives in electrolyte and pH have been reported on the mechanism of reduction Ni^{2+} ions and protons [51] and also in particular on the crystalline microstructure of Ni electrodeposited from Watts bath [88]. The significant increase in the nucleation rate of Ni electrodeposited form Watts bath is in principle due to the increase of Ni^{2+} ion concentration in Watts solutions. This is in a good agreement with the model reported by Fleischmann and coworkers [81, 85]. It has been demonstrated that the nucleation constant (A) in form of:

$$A = A_0 exp\left(\frac{-A_k}{kT}\right), \tag{4.11}$$

in which A_k is the work required for the formation of the critical cluster including n_c atoms. In the model, the nucleation constant is inversely proportional to the number atoms involved in the formation of the critical cluster and, thus, nuclei size. The nucleation constant (A) for the electrodeposition of nickel from Watts bath on vitreous carbon substrate has been shown [81, 85] to be decreased by increasing Ni^{2+} ion concentration Watts bath. Therefore, the reduction of size of nickel nuclei exhibiting higher nucleation rate appears to be reasonable which is observed in our experiments.

On the other hand, the nucleation constant has been demonstrated to be independent of pH and Cl^- concentration in Watts bath on carbon substrates. But we still see an increase in the number of nuclei formed on silicon by the reduction of pH from 3.2 to 2. At lower pH, hydrogen evolution takes place on the silicon substrate. In the cyclic voltammograms shown in Fig. 4.34d, e separate peaks for nickel deposition and H_2 evolution were not observed due to probably a significant barrier to the nucleation of nickel on silicon. The contribution of hydrogen evolution to the current passing through the cell is higher when the deposition potential approaches more negative values. Ni electrodeposition from sulfate supporting electrolytes at pH 2 and 4 has been investigated and a strong influence of H_2 evolution and its effect on the morphology of Ni electrodeposits have been observed in this electrolyte at conditions indicating excess of H_3O^+ over Ni^{2+}, e.g., at pH 2. Nasirpouri and co-workers have shown [45] that a considerable hydrogen evolution makes an appreciable contribution to the morphology change of nickel nanostructures from nanowire to nanoparticle electrodeposited on highly oriented pyrolytic graphite. Our results here clearly indicate the presence of a significant hydrogen evolution in Watts bath with a low pH = 2. Figure 4.34h illustrates a series of SEM images of nickel electrodeposited on n-Si(111) from Watts bath at pH = 2. From the SEM image we observe the accumulation of fine distinct islands of about 200 nm at the beginning of electrodeposition on silicon. In the course of markedly reduction of hydrogen, we assume that H^+ ions are rapidly adsorbed on silicon and under enough overpotential Ni starts to reduce with smaller critical size on the surface.

Microstructure of Electrodeposited Nickel Films on n-Si(111): Transition from Particle to Continuous Film

Nasirpouri et al. have found that the nucleation rate and therefore initial size of nuclei increase with increasing nickel content, lowering the pH and addition of chloride into solution. The effect of electrodeposition conditions was examined on the microstructure of nickel films electrodeposited on n-Si(111). Figure 4.35a compares X-ray diffractograms obtained from a typical nickel film electrodeposited on n-type silicon (111) with a nominal thickness of 200 nm.

As is evident, the main Bragg diffraction peak appears along (111) crystallographic direction for three different nickel films electrodeposited from three types of solutions. However, there are two other Bragg diffractions corresponding to Ni crystalline structure occurring along (200) and (220) orientations. These two peaks are more evident for the nickel films electrodeposited from simple sulfate solution without chloride, while intensity of the two peaks decreases significantly for the

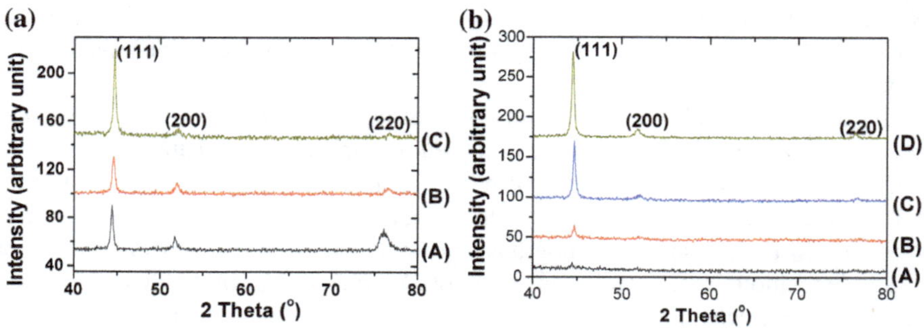

Fig. 4.35 X-ray diffractograms obtained from **a** nickel electrodeposited at −1.1 V Ag/AgCl with a nominal thickness of 200 nm from (*A*) simple sulfate solution 0.1 M $NiSO_4{\cdot}6H_2O$ + 0.5 M H_3BO_3 and (*B*) simple sulfate with chloride 0.1 M $NiSO_4{\cdot}6H_2O$ + 0.05 $NiCl_2{\cdot}6H_2O$ + 0.5 M H_3BO_3 and (*C*) watts bath 2.3 M $NiSO_4{\cdot}6H_2O$ + 0.6 $NiCl_2{\cdot}6H_2O$ + 0.5 M H_3BO_3 at pH = 2, **b** nickel electrodeposited from Watts bath 2.3 M $NiSO_4{\cdot}6H_2O$ + 0.6 $NiCl_2{\cdot}6H_2O$ + 0.5 M H_3BO_3 at pH = 2 with different nominal thickness of (*A*) 25 nm, (*B*) 50 nm, (*C*) 200 nm and (*D*) 500 nm. Reproduced from [75] with permission from Elsevier, copyright 2016

films electrodeposited from simple sulfate solution containing chloride and Watts bath at pH = 2. Furthermore, intensity ratio of (111) to (200) and (220) diffractions is higher for the films electrodeposited from the Watts bath. It can be understood that, in general, the structure of all types of nickel films electrodeposited is polycrystalline with no particular growth orientation. However, comparing the nickel films electrodeposited from the three different solutions implies that the films obtained from the Watts bath intends to grow more readily along (111) direction rather than the two other orientations. This is more likely consistent with our finding on the reduction of the critical size of stable nuclei on silicon exploiting the substructure of the substrate in the film.

Figure 4.35b indicates X-ray diffractograms obtained for nickel films electrodeposited from Watts bath at a pH of 2 at different nominal thicknesses. When the nominal thickness is less than 50 nm, (111) peak is the only Bragg diffraction which takes places. But, (200) peak and, after that, (220) are the two other diffractions appearing when the nominal thickness of the films exceeds 100 nm. Further growth of nickel films from the Watts bath reveals the formation of polycrystalline structure of nickel.

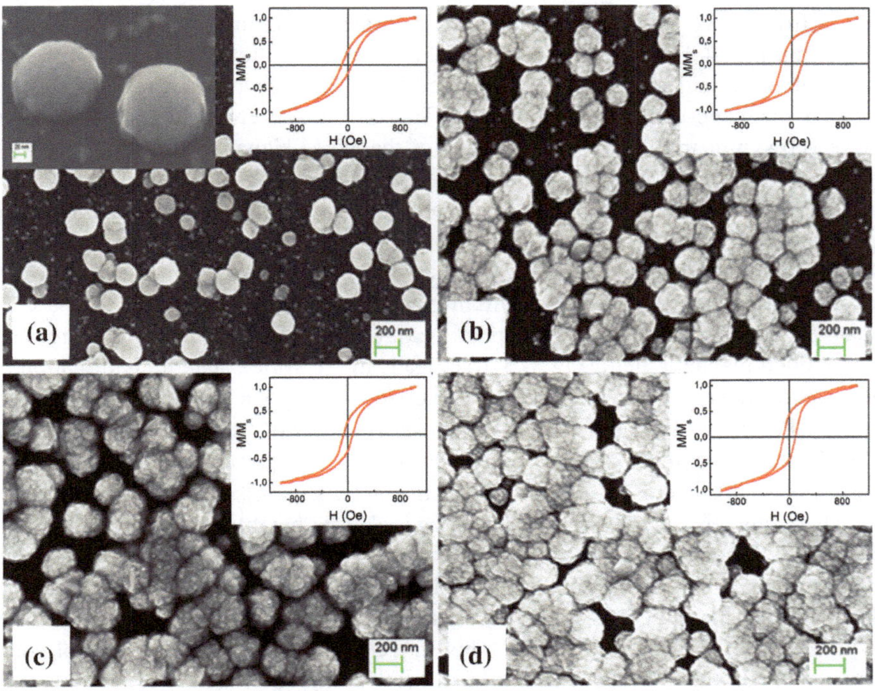

Fig. 4.36 SEM images of Ni granular films with nominal thickness: **a** 8 nm, **b** 50 nm, **c** 100 nm, **d** 1000 nm. *Inset* in **a** shows the side view of nanogranules. Inserted M-H curves demonstrate magnetization reversal in the corresponding films. Reproduced from [89] with the permission of Elsevier, copyright (2016)

In order to understand this growth behavior, we have examined the sequence of the electrodeposition of nickel from the Watts bath at pH = 2. Figure 4.36 illustrates SEM micrographs taken during different stages of the electrodeposition of nickel on n-Si(111) characterized by the nominal thickness of the film. It interestingly shows that the electrodeposition of nickel starts with the formation of individual Ni granules on silicon at the initial stages of growth, as shown in Fig. 4.36a for 8 nm nickel electrodeposited on Si. As the electrodeposition time increases, the individual Ni granules become bigger in diameter and tend to coalesce in the form of a continuous film (Fig. 4.36b–d). This behavior is further studied by investigating the magnetic properties [75, 89].

Island Growth for Tuning Magnetisation Mode: Transition from Particle to Continuous Film

Samardak et al. [89–91] have studied in two distinct magnetic structures obtained by electrodeposition on n-type Si(111) surface from a modified Watts bath at a low pH of 2 exhibiting the formation of two distinct states (i) individual and (ii) agglomerated granules. The AFM and magnetic force microscopy (MFM) images taken from nickel individual granules with different nominal thickness are illustrated in Fig. 4.37. As seen for the film with a nominal thickness of 8 nm, stand-alone granules or granules separated by a significant gap (Fig. 4.37a) have bipolar magnetic contrast at saturation (Fig. 4.37b). At zero field (Fig. 4.37c) we observed symmetric contrast distribution of the MFM signal with distinct black or white areas (the color depends on the vortex polarity). It is assumed that vortex states exist in individual granules. The proof of this is the magnetic signal amplitude profile shown in Fig. 4.37f. For the bipolar contrast measured at the tip height of 80 nm the amplitude is about 25 pA, meanwhile at zero field even at the tip height of 40 nm it is less than 15 pA. This implies the out-of-plane vortex with a core magnetized perpendicular to the applied field. Changing the polarity and increasing the applied magnetic field (Fig. 4.37d, e) leads to the vortex state transformation to the saturated single domain state characterized by bipolar contrast in the MFM image. Provided that the density of granules (number per surface area) increases, they become in contact with each other, forming the agglomerates and then a solid film. For such systems the multi-domain magnetic structure is more preferable.

This argument is proved by MFM image taken at height 100 nm and shown in Fig. 4.38. As one can see, there is a stripe domain structure at zero field, Fig. 4.38c. The domain walls spread on a few granules and the stripe period is different. Formation of the same domain configuration was shown in the thick magnetic films with weak out-of-plane anisotropy. This suggestion approves the increase of coercivity and decrease of remanence for the films with thicknesses of 500 and 1000 nm. If the applied magnetic field increases, the domain walls disappear and the MFM probe senses magnetostatic perturbation of individual granules or agglomerates on the sample surface, Fig. 4.38b, d. Micromagnetic simulation proved the scenario given based on the experimental results explained. Micromagnetic simulation results

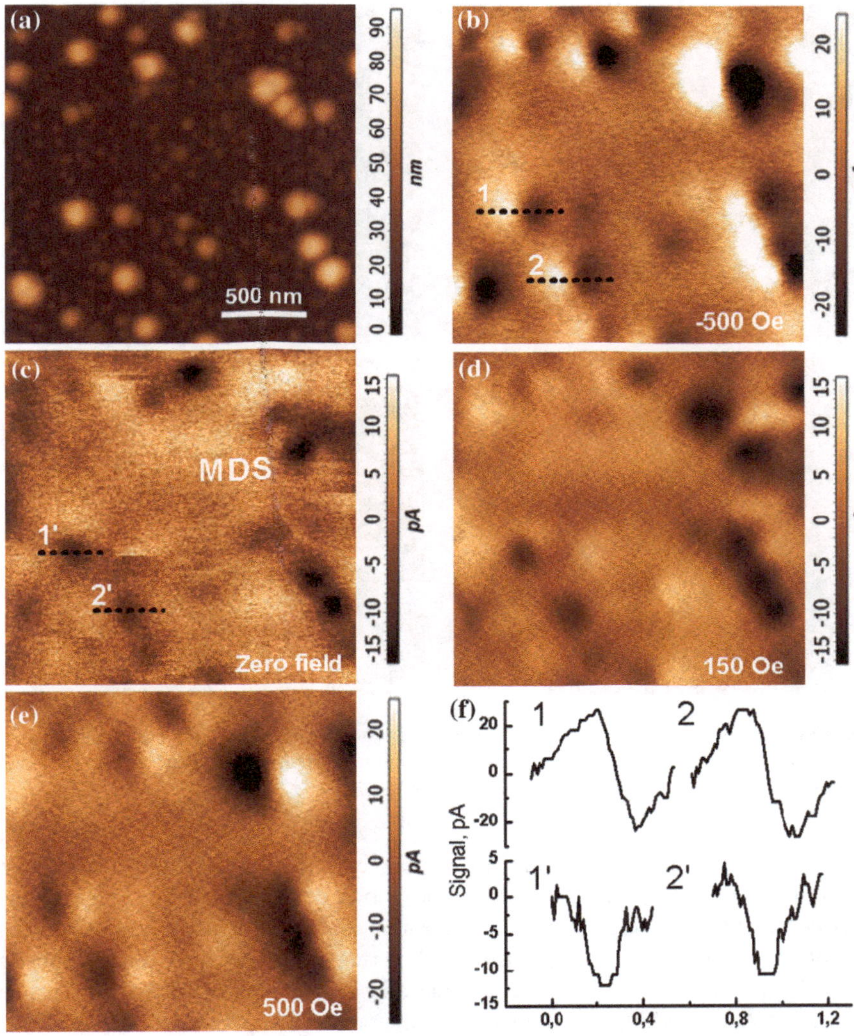

Fig. 4.37 Surface topography (AFM image) (**a**) and magnetic structure (MFM images) at zero field (**c**), small field (**d**) and saturation (**b**, **e**) states of nickel electrodeposits with nominal thickness of 8 nm. **f** Magnetic signal amplitude from granules at saturation (marks *1* and *2*) and at zero field (marks *1′* and *2′*). Reproduced from [89] with permission from Elsevier, copyright 2016

are shown in Fig. 4.39 along with the calculated magnetisation loops for three cases for the nanogranular Ni film including of: (i) a stand-alone granule, (ii) a small array of granules with different sizes, (iii) an agglomerate. Granules were represented as hemispheres in accordance with our findings.

For the stand-alone granule, micromagnetic simulation results imply that a vortex state is predominant (Fig. 4.39a). As seen in the three dimensional image,

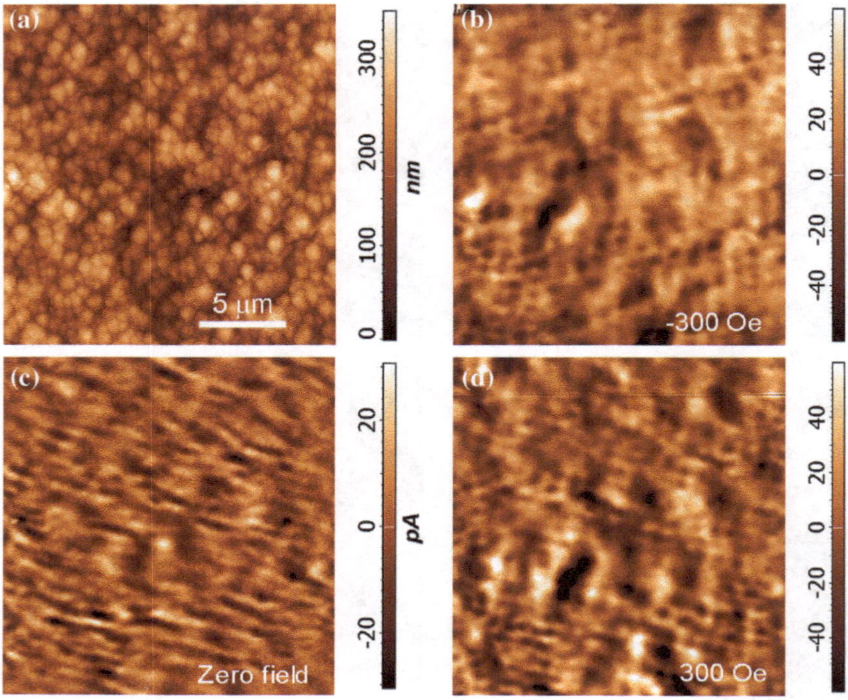

Fig. 4.38 Surface topography (**a**) and magnetic structure at zero field (**c**) and saturation (**b**, **d**) states of nickel electrodeposits with a nominal thickness of 500 nm. Reproduced from [89] with permission of Elsevier, copyright 2016

spins lie in the plane and form the closed magnetic flux. At the top of the granule the out-of-plane spins corresponding to the vortex core are observed. The typical M-H curve characterizing the vortex nucleation and annihilation is represented in the bottom image of Fig. 4.39a. MFM findings were supported by micromagnetic simulations compared for a single granule and a planar nanodisk with the same diameter 160 nm, which correspond to real granule diameter. It was found that z-component (out-of-plane) of vortex magnetization (M_z) in the granule in three times larger, than value of M_z for the planar nanodisk. Diameter of the vortex core in the granule was much bigger than in the nanodisk. This fact explains why the MFM contrast induced by vortex in granules is much stronger than in the case of planar nanodisks. Furthermore, a small array of granules with size distribution and inter-granular spacing similar to the real system was considered for micromagnetic simulations and the results are shown in Fig. 4.39b. As a result of micromagnetic simulation the open M-H curve was obtained as shown in the bottom image of Fig. 4.39b. This curve matches the experimental hysteresis loop measured from whole sample by VSM. We believe that simulation of an array possessing the large number of granules with the real size distribution finally can lead to full matching

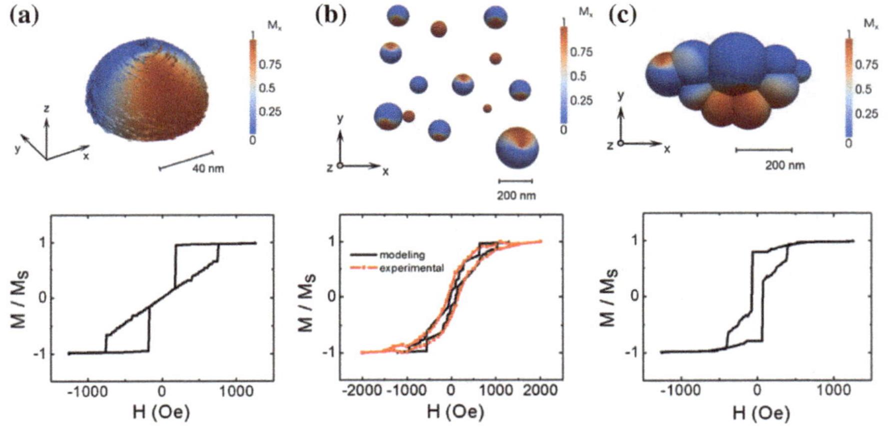

Fig. 4.39 Micromagnetic simulation results. **a** Single hemisphere in the out-of-plane vortex state. **b** The ensemble of nanogranules in the vortex and single domain states. **c** The agglomerate of granules in the flux-closure state with two vortices. *Insets* consist of simulated (*black*) and experimental (*red*) M-H curves for nickel electrodeposits. Reproduced from [89] with permission of Elsevier

between experimental and simulated M-H curves. For the third case when the agglomerate consists of a few closely-spaced granules, we observed the flux-closure magnetic structure with two out-of-plane vortices (Fig. 4.39c). Magnetic structure at zero field has blue (dark) and red (bright) contrasts corresponding to opposite orientation of magnetization in granules. Simulated M-H curve has several characteristic features related to existed vortex states. If to come from positive saturation, the magnetization decreases abruptly at a negative critical field and then decreases less rapidly up to negative saturation. As is confirmed by MFM images, the first abrupt switch corresponds to a transition from a single domain state into a flux-closure state with further formation of vortices. Following simulation of large agglomerates consisting of many granules confirms the formation of the multi-domain state in accordance with our MFM study.

4.3.2.3 Electrodeposition of Nickel on HOPG

Nasirpouri and coworkers have studied [45, 46] the formation of nickel film on HOPG. Cyclic voltammograms taken from a dilute electrolyte and the Watts bath, show that nickel reduction occurs at more negative potentials on the first potential sweep. However, nickel start to deposit on HOPG at less negative potentials with lower nickel content. A typical cyclic voltammogram is shown in Fig. 4.40 for the deposition and dissolution of nickel from the Watts bath. The semi-reversible behavior of the electrodeposition of nickel indicates it can be deposited more easily

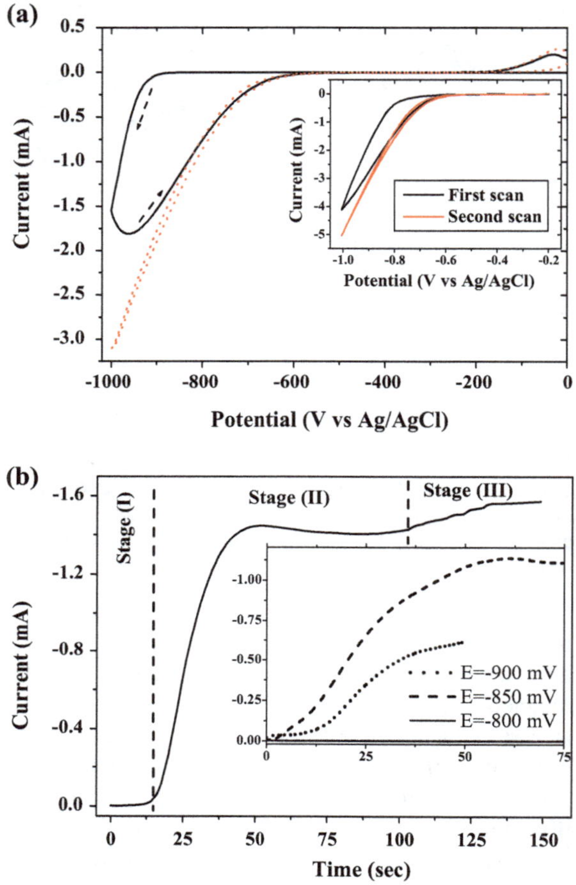

Fig. 4.40 A typical cyclic voltammogram taken in Watts bath on a freshly cleaved HOPG at scan rate of 20 mV/s. The *inset to panel* **a** shows a cyclic voltammogram taken in the same solution on silver electrodeposited HOPG demonstrating a more positive reduction potential. **b** Current transient recorded during one-step potentiostatic electrodeposition of nickel at −0.850 mV versus Ag/AgCl from a Watts bath. *Inset to panel* **b** illustrates the current transient for electrodeposition of nickel at −0.850 versus Ag/AgCl from a dilute sulphate solution. Reproduced from [45] with permission from Elsevier

on the second potential sweep due to the breakdown of the nucleation barrier on HOPG.

Current transients recorded during electrodeposition of nickel from dilute and Watts solutions (Fig. 4.40b) clearly show that a three-stage nucleation and growth mechanism takes place, arbitrarily divided on the graphs. The general mechanism of the electrodeposition of nickel from sulphate solutions on vitreous carbon substrates shows the typical behavior described earlier by Abyaneh et al. [83, 84] including: (i) the deposition of a monolayer of nickel via two-dimensional nucleation and crystal growth under special conditions accompanied by hydrogen evolution on both the edges and tops of the growth centers, (ii) three dimensional nucleation and crystal growth on top of the first monolayer and (iii) the overlap of growth centers initiated in the last parts of stage ii followed by outward growth into the solution, which can be associated with the death of the initial growth centers and the rebirth of new growth centers by further nucleation.

Figure 4.41 illustrates SEM images of the progress of the three stages of electrodeposition of nickel on HOPG at infinite deposition times solely to recognize the three-stages. The first stage described above can be negligible for the electrodeposition of metallic centers on HOPG as there is no evidence of any rise in current at the beginning of the current transient. This may be because of the very low surface energy of HOPG and the low bonding energy between the metal atoms and the substrate. After a certain time, the current starts to increase implying that Ni particles are being readily formed, as shown on Fig. 4.41a. However, nucleation occurs preferentially at step edges on the graphite surface. In the case of metal atoms residing on HOPG surfaces, step edge selectivity is promoted due to its ability to catalyze electron transfer to metal ions in solution [50, 92]. Previously, it has been demonstrated that the threshold required for nucleation at steps is smaller than the nucleation on terraces [93]. As illustrated in Fig. 4.41b, c, when the deposition process continues for long enough, the coalescence of nickel centers at step edges and also terraces occurs (i.e., stage (ii)) until the centers start to overlap (stage iii). At this point, overgrown nickel is found to form. Figure 4.41d shows the overlapped nickel centers indicating the destructive effects of hydrogen evolution on grown centers. The simultaneous H_2 reduction is evident from the dark circles sporadically created by H_2 bubbles growing on the electrode surface. The Ni

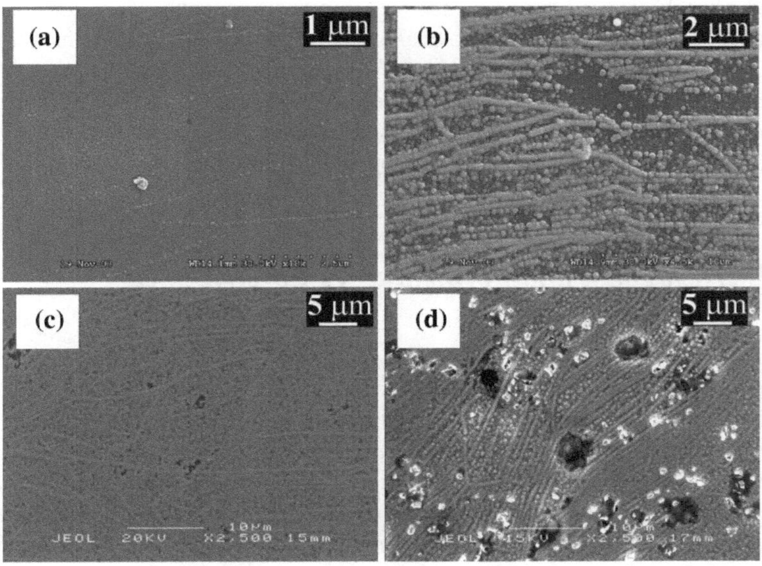

Fig. 4.41 Scanning electron micrographs taken of nickel electrodeposited from a Watts bath (solution 2) at a potential of −0.850 V versus Ag/AgCl for **a** 5 s, **b** 20 s, **c** 75 s and **d** 150 s, representing the three stages of electrodeposition of nickel on HOPG. Reproduced from [45] with permission from Elsevier

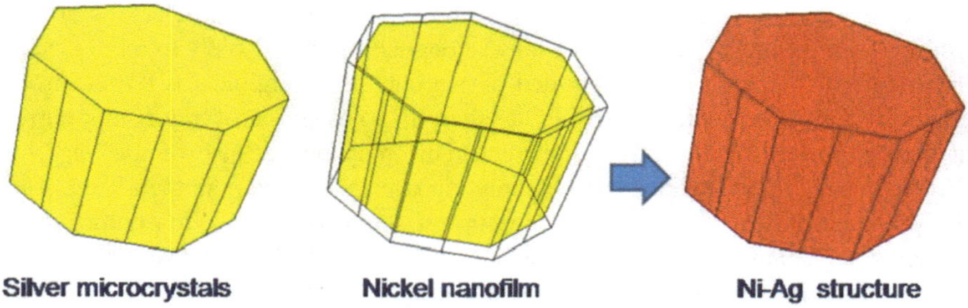

Fig. 4.42 A schematic presentation of the dual-bath pulse electrodeposition method to produce core-shell mesocrystal

deposit can also be observed under the evolving bubbles, which indicates catalytic reduction of H_2 by deposited Ni [51].

4.3.3 3D Faceted Core-Shell Mesocrystals

Direct electrodeposition of nickel from aqueous solutions does not form highly faceted crystals, as we have also observed in our experiments. Therefore, we have used dual-bath electrodeposition to coat a thin magnetic shell layer of nickel on top of a highly faceted mesocrystal 'template' like silver. It has been shown that Ag can be deposited in the form of highly faceted hexagonal or truncated hexagonal crystals. Using the same method, highly faceted silver hexagonal mesocrystals with different sizes and shapes were electrodeposited from silver nitrate solution.

4.3.4 Hexagonal and Polyhedral Ag Core-Ni Shell Mesocrystals

Nasirpouri et al. have reported the electrodeposition of hexagonal and polyhedral core/shell mesocrystals. Figure 4.42 illustrates a schematic picture of the dual-pulse method to produce this type of core-shell mesocrystal. Two types of electrodeposited Ag mesocrystals studied in this section including (a) hexagonal Ag platelets electrodeposited at −60 mV in a solution of 0.6 M $AgNO_3$ for 5 s and (b) truncated Ag polyhedra with sixfold rotational symmetry electrodeposited at −70 mV in a solution of 0.1 M $AgNO_3$ for 20 s (Fig. 4.43). A Ni shell film with a nominal thickness of ~100 nm was then electrodeposited on silver mesocrystals from a Watts bath at a potential of −800 mV versus Ag/AgCl reference electrode. This value is estimated by counting the pre-electrodeposited silver metallic centers on HOPG, calculating the total surface of silver mesocrystals and applying the

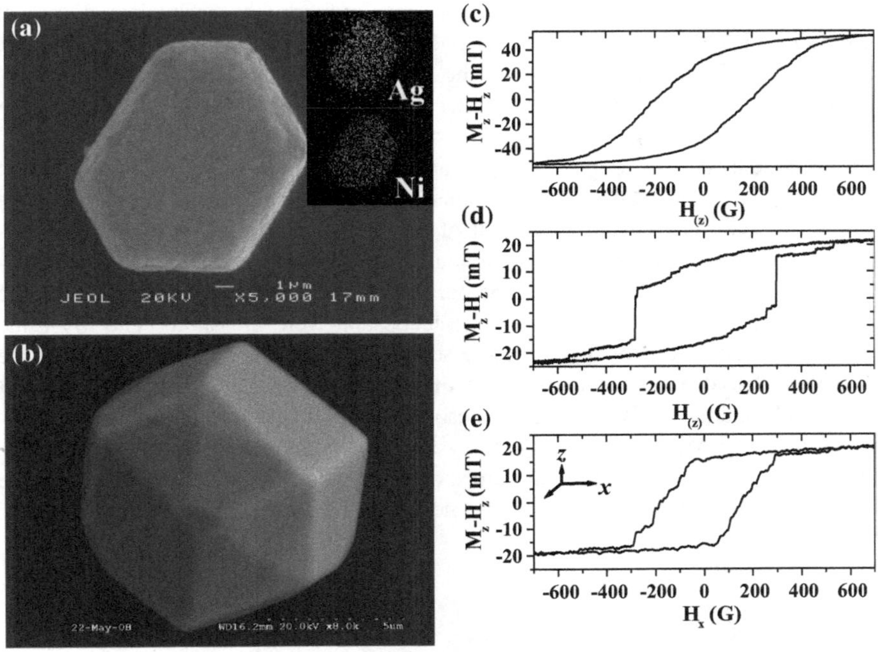

Fig. 4.43 *Left column* SEM images of **a** a hexagonal platelet electrodeposited at −60 mV in a solution of 0.6 M $AgNO_3$ for 5 s and **b** a truncated sixfold symmetric Ag polygon electrodeposited at −70 mV in a solution of 0.1 M $AgNO_3$ for 20 s. Ag cores have subsequently been plated with Ni in a Watts bath to create Ag-Ni core-shell mesocrystals. *Insets to panel* **a** show Ag and Ni elemental surface maps obtained from EDX analysis. *Right column* Hysteresis loops measured using micro-Hall probe magnetometry of individual nickel-silver mesostructures. Panel **c** is for the mesostructure shown in (**a**). Panels (**d**, **e**) are for the mesostructure shown in panel (**b**), when the external magnetic field is applied along the base (x-direction) and apex (z-direction), respectively. Reproduced from [45] with permission from Elsevier

Faraday's law. On the basis of this calculation there is likely to be a substantial error bar on the nominal thickness value. As was shown in the inset to Fig. 4.41, the cyclic voltammograms taken from the Watts bath on freshly cleaved HOPG and silver electrodeposited HOPG, showed a considerable deviation from the equilibrium potential of reduction of nickel ions to solid nickel. When nickel is electrodeposited on the bare HOPG, it starts to reduce at potentials more negative than −840 mV versus Ag/AgCl reference electrode. In the presence of silver metallic centers on HOPG, the reduction of nickel is facilitated as it starts to grow at potentials more negative than −670 mV versus Ag/AgCl. At this potential, the nucleation rate of nickel on HOPG is considerably slower and the nickel shell is solely electrodeposited on silver centers. The inset to Fig. 4.43a demonstrates the surface elemental distribution of nickel and silver, taken using the EDX analyzer, indicating that nickel is only detected on top of the silver mesocrystal. Surface roughening of the polycrystalline nickel shell is clearly seen, either as a result of the well established three-dimensional nucleation and growth mechanism of nickel electrodeposition [A. Saraby-Reintjes and M. Fleischmann, Electrochimica Acta, 29 (1984) 557–566.] or due to the presence of hydrogen evolution [W Schwarzacher J. Phys.: Condens. Matter 16 (2004) R859] on the crystalline Ag core.

Magnetic hysteresis loops of individual Ni-Ag mesostructures, with the different shapes shown in Fig. 4.43a, b, were measured and reproduced in Fig. 4.43c–e. In Fig. 4.43c, the magnetization reversal of the hexagonal platelet structure exhibits regular coherent rotation of ferromagnetic moments as expected for an approximately thin film sample with in-plane anisotropy and no obvious facets. The same was true for all mesoscrystals with this shape that were measured. The M_z value recorded at saturation for this structure seems to be greater in magnitude than the truncated sixfold polyhedron shown in Fig. 4.43b, c. We have examined different hexagonal platelet Ag-Ni core-shell mesocrystals and find that the measured saturation values of M_z do not seem to be systematically dependent on the crystal size. We assume then that the different values of M_z can be ascribed to deviations of the actual thickness of Ni shells from the nominal estimated values. Figure 4.44b, c

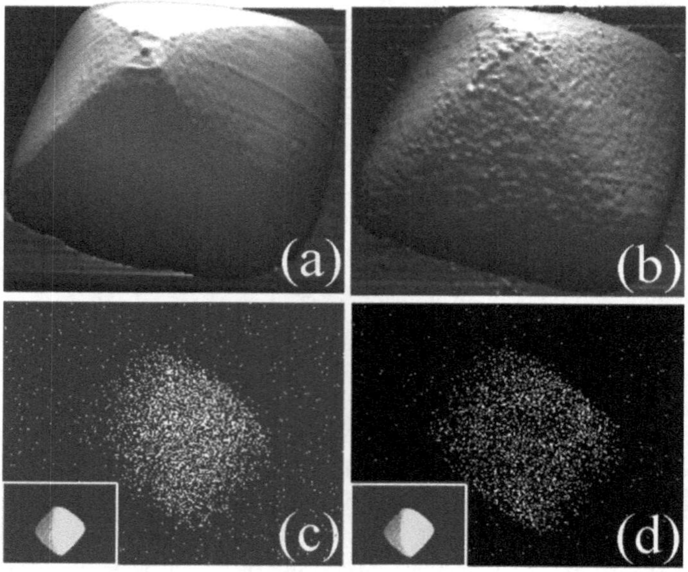

Fig. 4.44 Electrodeposition steps used for the fabrication of the Ni shell film on a three dimensional highly faceted Ag mesocrystal. **a** Three dimensional AFM image of a typical Ag pyramidal mesocrystal with the height from the base to the apex of 5.2 and 10 μm base sidelength, **b** Ag mesocrystal after electrodeposition of a 100 nm Ni shell on *top*. Energy dispersed X-ray elemental maps of silver (**c**) and nickel (**d**) show that Ni is solely electrodeposited on *top* of Ag centers, grown primarily on HOPG. *Insets to panels* **c**, **d** are SEM images of the actual Ag-Ni mesocrystals on which EDX analysis was performed. Reproduced from [94, 95] with permission from American Institute of Physics (AIP)

show magnetization curves for the truncated sixfold polyhedron measured for two different applied field directions. For the field applied along the z-direction we observe sharp steps in magnetisation reversal steps that appear to occur at characteristic values of applied field. We link these steps to the abrupt reversal of the faceted vertical sidewalls of the structure which have in-plane magnetic anisotropy. A comparison of the magnetic behaviour to these two different-shaped structures clearly shows how magnetisation reversal can be tuned via shape control. The mechanism of magnetization of such highly faceted structures is not yet completely understood, but micromagnetic simulations carried out on simpler scaled-down pyramidal-shaped Ni-Ag structures exhibit the formation of several stable and metastable states among which magnetic vortex states are prominent in larger structures [94]. In this case sweeping the magnetic field makes it possible to switch the magnetic state abruptly between two vortex states, leading to kinks in the magnetisation curves.

4.3.5 *Truncated Icosahedral and Pyramidal Ag Core-Ni Shell Mesocrystals*

A two-step dual-bath potentiostatic method similar to the previous section was used by Nasirpouri et al. [95–97] to electrodeposit truncated icosahedral and pyramidal silver-nickel core-shell structures on a HOPG working electrode at ambient temperature. A Ni shell with a nominal thickness of 100 nm was subsequently electrodeposited on silver mesocrystals from a Watts bath at a potential of −800 mV versus Ag/AgCl reference electrode. Cyclic voltammograms, taken in the Watts bath on bare HOPG and with silver electrodeposited on HOPG, showed a considerable difference in the equilibrium reduction potential of nickel ions to solid nickel. When nickel is electrodeposited on the bare HOPG, it starts to reduce at potentials more negative than −840 mV versus Ag/AgCl reference electrode. In the presence of silver metallic centers on HOPG, the reduction of nickel is facilitated as it starts to grow at potentials more negative than −670 mV versus Ag/AgCl. At this potential, the nickel shell is solely electrodeposited on the silver centers. Ex situ atomic force microscopy images taken from a pyramidal silver mesocrystal before and after electrodeposition of nickel are shown in Fig. 4.44a, b.

Surface roughening of the polycrystalline nickel shell is clearly seen as a result of the well-established three-dimensional nucleation and growth mechanism of electrodeposition of nickel [101] or due to the presence of hydrogen evolution [Walther] on the crystalline Ag core. The distribution of nickel and silver as evidenced by the EDX analyzer also proves that nickel is only deposited on top of the silver mesocrystals, as shown in Fig. 4.44c, d for a pyramidal Ag-Ni mesostructure.

Magnetic hysteresis loops of individual Ag-Ni mesostructures with different shapes such as truncated icosahedral and pyramidal structures, typically with a 100 nm thick nickel shell on a Ag microcrystal of height 3–5 μm from base to apex

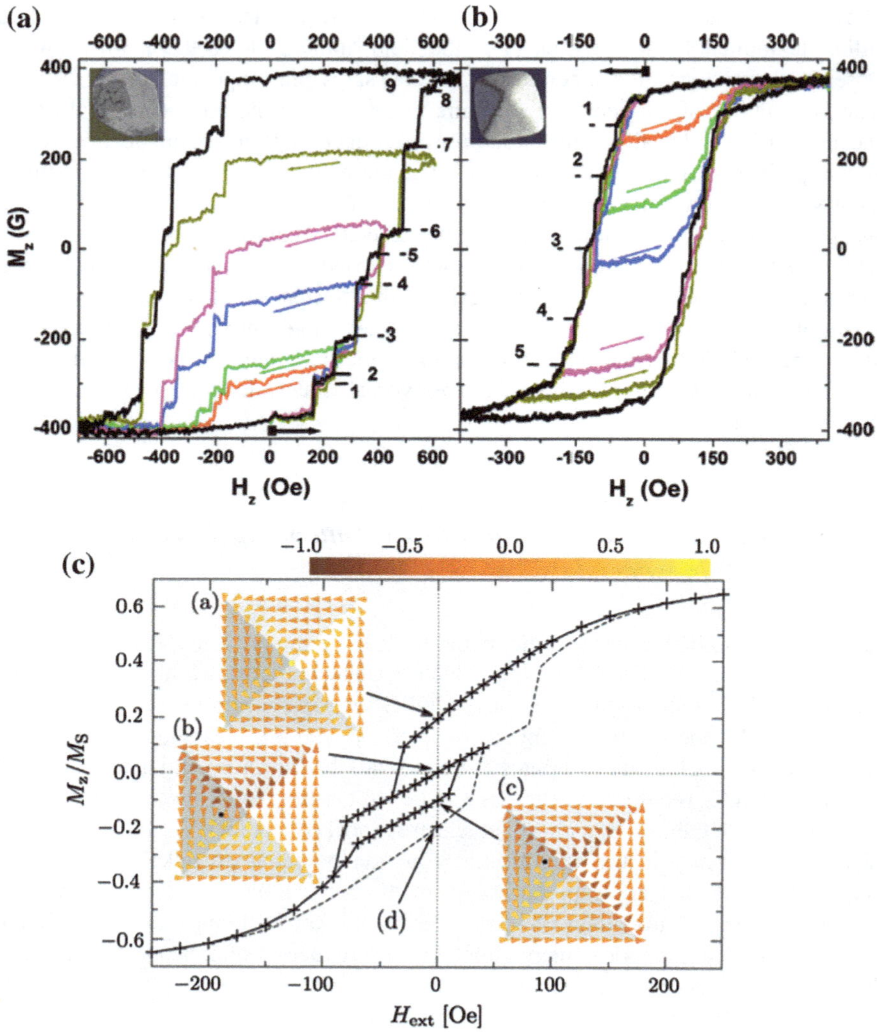

Fig. 4.45 Hysteresis loops of **a** a truncated icosahedral and **b** pyramidal silver/nickel core-shell structure (shown in *insets*) measured in a uniform external magnetic field applied perpendicular to the base of the element at 5 K. Different curves correspond to back-tracking of different magnetic states found as magnetization steps during the magnetization reversal. Not all stable states are shown, to avoid overlap of the curves. The magnetization loops obtained by micromagnetic simulations, for a pyramidal Nickel shell with a base side length of 500 nm, a height of 250 nm, and thickness of 5 nm are shown. Reproduced from [94, 95] with the permission of AIP publishing

and 6–10 μm base side length were discussed. Figure 4.45a shows the magnetization curves obtained for one particular pyramid (a) and a truncated icosahedron (b) at the same applied field sweep rate. Surprisingly, the magnetization reversal occurs in a *stepwise fashion* in both hysteresis loops. Each step corresponds to a different, stable magnetic state, as verified by performing minor loops (back-tracking) for several magnetization steps. Depending on the magnetic history, we find *seven* different stable states at zero magnetic field for the pyramidal structure, and *ten* for the more complex icosahedral shaped magnetic element. It was demonstrated not only existence of different magnetic states for the 3D mesostructures, but also the controllable, sharp switching between different magnetization states. To verify the above scenario the finite-element, micromagnetic solver Nmag was used in order to simulate the magnetization process of pyramidal shell structures. Figure 4.45c shows the obtained simulation results for the reversal behavior of a Ni shell of the same shape as the grown structure of Fig. 4.44 but downscaled by a factor of 20. Metastable states are found to be responsible for the magnetisation reversal of the core-shell mesocrystals.

References

1. M. Ohring, *The Materials Science of Thin Films* (Academic Press, San Diego, California, 1992)
2. E. Budevski, G. Staikov, W.J. Lorenz, *Electrochemical Phase Formation and Growth* (VCH Verlagsgesellschaft mbH, Weinheim, Germany, 1996)
3. J.W. Schultze, F.D. Koppitz, Electrochim. Acta **21**, 327 (1976)
4. E. Budevski, V. Bostanov, G. Staikov, Ann. Rev. Muter. Sci. **10**, 85 (1980)
5. D.M. Kolb, in *Advances in Electrochemistry and Electrochemical Engineering*, vol. 11, ed. by H Gerisher, C.W. Tobias (Wiley-Interscience, New York, BOOK staikov 1978)
6. F. Nasirpouri, Ionics **17**(4), 331–337 (2011)
7. Reproduced from reference [6] with the permission of Springer
8. F. Nasirpouri, Prot. Met. Phys. Chem. Surf. **47**(4), 534–539 (2011)
9. Reproduced from reference [8] with the permission of Springer
10. B.J. Hinch, C. Koziol, J.P. Toennies, G. Zhang, Vacuum **42**, 309 (1991)
11. T.P. Moffat, J. Phys. Chem. B **102**, 10020 (1998)
12. H. Siegenthaler, K. Juttner, J. Electroanal. Chem. **163**, 327 (1984)
13. J.R. Vilche, K. Juttner, Electrochim. Acta **2**, 1567 (1987)
14. G.M. Brisard, E. Zenati, H.A. Gasteiger et al., Langmuir **13**, 2390 (1997)
15. G.M. Brisard, E. Zenati, H.A. Gasteiger et al., Langmuir **11**, 2221 (1995)
16. R. Vasilic, N. Vasiljevic, N. Dimitrov, J. Electroanaly. Chem. **580**, 203 (2005)
17. M.J. Henderson, E. Bitziou, A.R. Hillman et al., J. Electrochem. Soc. **148**, E105 (2001)
18. G.R. Stafford, U. Bertocci, J. Phys. Chem. C **111**, 17580 (2007)
19. G.Y. Wu, S.E. Bae, A.A. Gewirth et al., Surf. Sci. **60**, 1886 (2007)
20. A.W. Bott, Curr. Sep. **18**, 79 (1999)
21. M.E. Martins, R.C. Salvarezza, A.J. Arivia, Electrochim. Acta **41**, 244 (1996)
22. R. Schumacher, Angew. Chem. Int. Ed. Engl. **29**, 239 (1990)
23. S.J. Martin, V.E. Granataff, G.C. Frye, Anal. Chem. **63**, 2272 (1991)
24. C. Chagnard, P. Gilbert, A. Neal Watldns et al., Sensors and Actuators B **32**, 129 (1996)
25. G. Herzog, D.W.M. Arrigan, Electroanalysis **17**, 1816 (2005)
26. X. Zeng, S. Bruckenstein, J. Electrochem. Soc. **146**, 2555 (1999)
27. J. Gray, W. Schwarzacher, X. Zhu, Mat. Res. Soc. Symp. Proc. **781E**, Z10.1 (2001)
28. G.Y. Wu, W. Schwarzacher, J. Electroanal. Chem. **629**, 164 (2009)

29. J. Nutariya, J. Velleuer, W. Schwarzacher, N. Vasiljevic, Meet. Abstr. Electrochem. Soc. **100**, 1234 (2010)
30. A.J. Bard, L.R. Faulkner, Electrochemical Methods. Fundamentals and Applications, 2nd edn. (Wiley, New York, 2001)
31. L. Guo, P.C. Searson, Electrochim. Acta **55**, 4086–4091 (2010)
32. L. Guo, A. Radisic, P.C. Searson, J. Phys. Chem. B **109**, 24008 (2005)
33. L. Guo, P.C. Searson, Electrochem. Solid State Lett. **10**, D76 (2007)
34. S. Strbac, O.M. Magnussen, R.J. Behm, Phys. Rev. Lett. **83**, 3246 (1999)
35. S. Strbac, F. Maroun, O.M. Magnussen, R. Behm, J. Electroanal. Chem. **500**, 479 (2001)
36. F.A. Moller, O.M. Magnussen, R.J. Behm, Phys. Rev. Lett. **77**, 3165 (1996)
37. F.A. Moller, O.M. Magnussen, Behm. R. J. Phys. Rev. Lett. **77**, 5249 (1996)
38. D.M. Kolb, G.E. Engelmann, J.C. Ziegler, Angew. Chem. Int. Ed. **39**, 1123 (2000)
39. G.E. Engelman, J.C. Ziegler, D.M. Kolb, J. Electrochem. Soc. **145**, 2970 (1998)
40. D.M. Kolb, R. Ullmann, J.C. Ziegler, Electrochim. Acta **43**, 2751 (1998)
41. D.M. Kolb, R. Ullmann, T. Will, Science **275**, 1097 (1997)
42. R. Stiger, B. Craft, R.M. Penner, Langmuir **15**, 790 (1999)
43. G. Oskam, J.G. Long, A. Natarajan, P.C. Searson, J. Phys. D Appl. Phys. **31**, 1927 (1998)
44. Reginald M. Penner, J. Phys. Chem. B **106**, 3339–3353 (2002)
45. F. Nasirpouri, S.J. Bending, L.M. Peter, H. Fangohr, Thin Solid Films **519**(23), 8320–8325 (2011)
46. Reproduced and Adapted from reference [45] with the permission of Elsevier
47. H. Liu, F. Favier, K. Ng, M.P. Zach, R.M. Penner, Electrochim. Acta **47**, 671 (2001)
48. E.C. Walter, B.J. Murray, F. Favier, G. Kaltenpoth, M. Grunze, R.M. Penner, J. Phys. Chem. B **106**, 11408 (2002)
49. P. Zach, R.M. Penner, Adv. Mater. **12**, 878 (2000)
50. M.P. Zach, K. Inazu, J.C. Hemminger, R.M. Penner, Chem. Mater. **14**, 3206 (2002)
51. M. Streckova, R. Orinakova, R. Rozik, L. Trnkova, M. Galova, Helv. Chim. Acta **89**, 622 (2006)
52. A.M. El-Sherik, U. Erb, J. Page, Surf. Coat. Technol. **88**, 70 (2006)
53. A. Milchev, Russ. J. Electrochem. **44**, 619–645 (2008)
54. Y. Yu, Q. Zhang, B. Liu, J.Y. Lee, J. Am. Chem. Soc. **132**, 18258–18265 (2010)
55. G. Hills, A. Kavepour, B. Scharifker, Electrochim. Acta **28**, 891 (1983)
56. A. Milchev, *Electrocrystallization: Fundamentals of Nucleation and Growth* (Kluwer Academic Publishers, Dordrecht, 2002)
57. F. Nasirpouri, S.J. Bending, L.M. Peter, Galvanotechnik **104**, 1330–1339 (2013)
58. G. Gunawardena, G. Hills, I. Montenegro, J Electroanal. Chem. **138**, 241–254 (1982)
59. U. Schmidt, M. Donten, J. Osteryoung, J. Electrochem. Soc. **144**, 2013–2021 (1997)
60. W. Simons, D. Gonnissen, A. Habin, J Electroanal. Chem. **433**, 141–151 (1997)
61. A. Milchev, Electrochim. Acta **28**, 947–953 (1983)
62. M. Jose-Yacaman, M. Avalos-Borja, Catal Rev Sci Eng **34**, 55 (1992)
63. Radmilovic et al, J. Solid State Electrochem. **2**, 162–169 (1998)
64. E. Michailova, A. Milchev, J. Appl. Electrochem. **18**, 614 (1988)
65. Ko Wen-Yin, Chen Wei-Hung, Tzeng Shien-Der, Gwo Shangjr, Lin Kuan-Jiuh, Chem. Mater. **18**(26), 6097–6099 (2006)
66. Ko Wen-Yin, Chen Wei-Hung, Cheng Ching-Yuan, Lin Kaun-Jiuh, Nanoscale Res. Lett. **4**, 1481–1485 (2009)
67. Z.-L. Xiao, C.Y. Han, W.-K. Kwok, H.-H. Wang, U. Welp, J. Wang, G.W. Crabtree, J. Am. Chem. Soc. **126**, 2316–2317 (2004)
68. L.D. Marks, Philos. Mag. A **49**, 81–83 (1984)
69. L.D. Marks, Rep. Prog. Phys. **57**, 603–649 (1994)
70. A. Muller, S.E.C. Dale, M.A. Engbarth, S.J. Bending, L.M. Peter, CrystEngComm **12**, 2135–2138 (2010)
71. E.C. Dale Sara, J. Bending Simon, M. Peter Laurence, Langmuir **25**(19), 11228–11231 (2009)

72. F.C. Frank, Discuss. Faraday Soc. **5**, 48 (1949)
73. S.M. Janjan, F. Nasirpouri, M.G. Hosseini, Russ. J. Electrochem. **47**(7), 787–792 (2011)
74. Copyrigth (2016), obtained from Springer for full article
75. F. Nasirpouri, S.M. Janjan, S.M. Peighambari, M.G. Hosseini, A. Akbari, J. Electroanal. Chem. **690**, 136–143 (2013)
76. Copyright (2016) with the permission of Elsevier
77. J.E.T. Anderson, ECS Trans. **16**, 46 (2009)
78. M. Šupicová, R. Rozik, Trnkova L.R. Orinakova, M. Galova, J. Solid State Electrochem. **10**, 61 (2006)
79. S. Rivillon, Y.J. Chabal, L.J. Webb et al., J. Vac. Sci. Technol. A **23**, 1100 (2005)
80. Y.H. Ogata, K. Kobayashi, M. Motoyama, Curr. Opin. Solid State Mater. Sci. **10**, 163–172 (2006)
81. A. Saraby-Reintjes, M. Fleischmann, Electrochim. Acta **29**, 557–566 (1984)
82. R. Orinakova, M. Streckova, L. Trnkova, R. Rozik, M. Galova, J. Electroanal. Chem. **594**, 152–159 (2006)
83. M.Y. Abyaneh, M. Fleischmann, J. Electroanal. Chem. **119**, 187 (1981)
84. M.Y. Abyaneh, M. Fleischmann, J. Electroanal. Chem. **119**, 197 (1981)
85. M. Fleischmann, A. Saraby-Reintjes, Electrochim. Acta **29**, 69 (1984)
86. J. Amblard, M. Froment, G. Maurin, N. Spyrells, E. Trevisan-Souteyrand, Electrochim. Acta **28**, 909 (1983)
87. M. Motoyama, Y. Fukunaka, T. Sakka, Y.H. Ogata, J. Electrochem. Soc. **153**, C502 (2006)
88. A. Samardak, E. Sukovatitsina, A. Ognev, M. Stebliy, A. Davydenko, L. Chebotkevich, Y.K. Kim, F. Nasirpouri, S.-M. Janjan, F. Nasirpouri, J. Magn. Magn. Mater. **371**, 149–156
89. Copyright (2016),with the permission of Elsevier
90. A. Ognev, A. Samardak, E. Sukovatitsina, M. Stebliy, A. Davydenko, L. Chebotkevich, Y.K. Kim, F. Nasirpouri, S.-M. Janjan, F. Nasirpouri, J. Phys: Conf. Ser. **345**(1), 012010 (2012)
91. R.J. Bowling, R.L. McCreery, C.M. Pharr, R.C. Engstrom, Anal. Chem. **61**, 2763 (1989)
92. R.S. Robinson, K. Sternitzke, M.T. McDermott, R.L. McCreery, J. Electrochem. Soc. **138**, 2412 (1991)
93. A. Knittel, M. Franchin, T. Fischbacher, F. Nasirpouri, S. Bending, H. Fangohr, New J. Phys. **12**, 113048 (2010)
94. F. Nasirpouri, M.A. Engbarth, S.J. Bending, L.M. Peter, A. Knittel, H. Fangohr, M. Milosevic, Appl. Phys. Lett. **98**(22), 222506 (2011)
95. Reprint copyright (2016) with the permission of American Institute of Physics (AIP publishing)
96. A. Knittel, M. Franchin, F. Nasirpouri, S.J. Bending, H. Fangohr, J. Appl. Phys. **111**(7), 07D127 (2012)
97. S. Morin, A. Lachenwitzer, O.M. Magnussen, R.J. Behm, Phys. Rev. Lett. **83**, 5066 (1999)

Chapter 5
Template Electrodeposition of Nanowires Arrays

Abstract Template electrodeposition is known as an outstanding method of fabrication of nanowires. In this chapter we start with an introduction to template fabrication and properties. Anodic aluminium oxide (AAO) and track-etched (TE) membranes are the most common templates which will be discussed in this chapter. The electrodeposition of nanowire arrays can be carried out for wide variety of materials from metals and alloys to compounds and nanostructures. Electrochemical deposition mechanisms will be discussed using the electrochemical relations. Several techniques of templates electrodeposition including alternating current, direct current and pulse current and related topics will be explained. Recent progress of electrodeposited magnetic nanowires including metal and alloy, multilayered, core-shell and diameter modulated nanowires are presented.

5.1 Introduction

Template known as a particular form of nanoporous materials provides limited spaces for nucleation and growth of nanomaterials. Template-based synthesizing methods have been paid much attention owing to their capability of producing high aspect ratio nanostructures with considerably high areal density. There are different approaches towards filling the pores of a template which finally leads to formation of one-dimensional nanostructures such as evaporation, melt injection and electrodeposition. The electrodeposition is the most convenient method of template based synthesizing of nanowires with controlled process and properties. Electrodeposition is performed with a reduction reaction of desirable materials taking place at interface of electrolyte/pore bottom until the pores are completely filled. In this chapter we start with an introduction to template fabrication and properties. The electrodeposition of nanowires arrays can be carried out for wide variety of materials from metals and alloys to compounds and semiconductors. Electrochemical deposition techniques and related topics will be explained. Recent progress of electrodeposited nanowires is presented.

F. Nasirpouri, *Electrodeposition of Nanostructured Materials*,
Springer Series in Surface Sciences 62, DOI 10.1007/978-3-319-44920-3_5

5.2 Template

A template is typically defined as an insulating host material composed of tubular long pores with diameters in nanometer range. This is an elegant replacement for common lithographical soft patterns which could be fabricated for device production. Templates have found advantages over their lithographically made one in durability of use, versatility in various fabrication and synthesizing methods, and capability of production of high aspect nanostructures ratio up to 10^3, therefore, templates may be used for fabricating of nanodot, nanoparticles, nanotubes, and nanowires. The topic of this chapter focuses on the electrodeposition of nanowires; hence, we further talk about those templates which are specifically more useful for the template electrodeposition of nanowires.

Templates can be characterised by their fabrication method, material, areal pore density (cm^{-1}), aspect ratio, pore diameter and its size distribution, pore arrangement, and service conditions. The templates are generally classified into two major categories: (1) track-etched templates, (2) anodic oxide templates.

5.2.1 Track-Etched Templates

Nuclear track filters (NTF) or so-called track-etched (TE) membranes are the most commonly used template for electrodeposition. The fabrication process of a track-etched template involves two major stages. First, an appropriate material such as a polymer or mica is subjected to the irradiation of heavy charged particles like Ar^{+9} from a nuclear radiation resource accelerated by a strong electrical field up to a few mega-electron volts (MeV/amu). The polymer or mica must be thin enough like a thin foil or film to allow the particles pass through. This process so-called Coulomb explosion phenomenon leaves several tracks of radiation damage in the film via. Second stage involves selective etching of tracked foil in a suitable reagent, such as hydrofluoric acid or NaOH. This stage *develops* and *fixes* the tracks, which results in fine cylindrical pores in the membrane along the tracks, while leaving the rest of the material essentially unchanged. Various materials have been used to produce track-etched templates such as natural mica, organic foils and polymers but for template electrodeposition purposes, polycarbonate (PCTE) and polyester (PETE) have been most frequently used. Figure 5.1 shows surface topography of commercially available PETE membrane from the top surface.

Since the irradiation of the heavy particles is the origin of the formation of pore, they exhibit random distribution with a rather uniform diameter along their entire length. Nominal pore sizes ranging from 10 nm up to micrometer in diameter are common for track-etched templates. Furthermore, the areal pore density ranges 10^7–10^9/cm^2.

Many process parameters control the shape and morphology of tracks in this type of templates which may include, the nature of the material, the accelerating

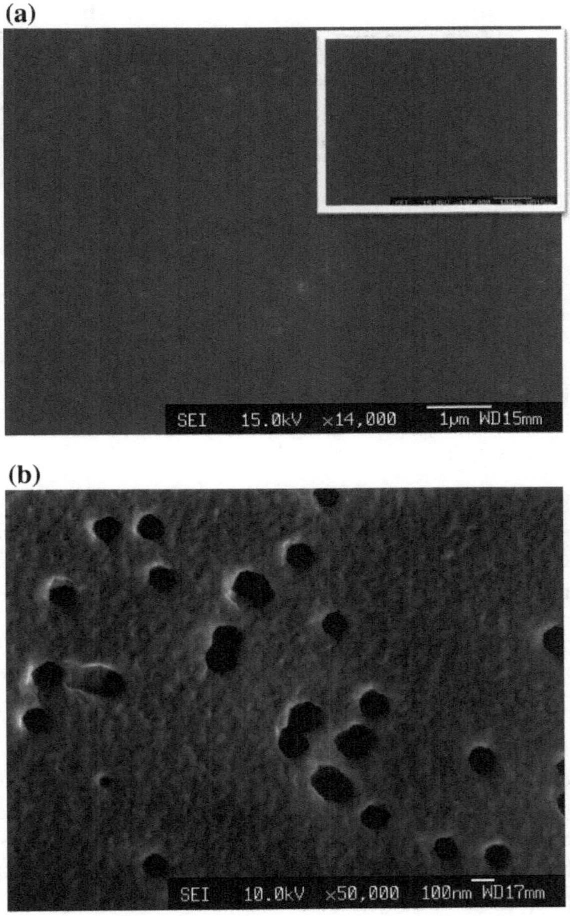

Fig. 5.1 Typical SEM images taken from *top* surface of a commercially available (Osmonics) **a** polycarbonate template with a nominal pore diameter 30 nm, and **b** polyester template with a nominal pore diameter 100 nm

voltage, mass or momentum of incident heavy particles, the ion beam energy and deposition rate, pre- and post-irradiation storage and environment, and the etching conditions, etc. [1, 2]. The morphology of tubes formed in this process is of most importance for the template electrodeposition of materials. Particular issues may be associated with the track-etched templates such as unsmooth internal surface and variation of diameter along the tubular pores. Figure 5.2 depicts two common issues associated with the morphology and diameter of commercially track etched polycarbonate (PCTE) templates. In Fig. 5.2a it is clear that the metal nanowire electrodeposited exhibits a diameter bigger in inner section than in the end. This means that the corresponding pore diameter at the surface of the template is smaller than at inside. This is known as *toothpick-shaped* nanowire obtained from commercially available polycarbonate template [3]. Another phenomenon is the surface roughness of internal pores in the commercially available track-etched templates.

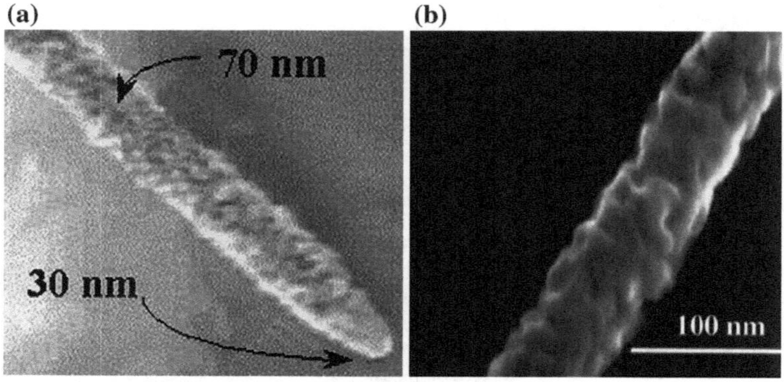

Fig. 5.2 SEM images of a metal nanowire electrodeposited in commercially available PCTE template with **a** high wall roughness, and **b** tooth-pick shaped nanowires effect. Reproduced from [3, 4] with the permission of Elsevier, copyright 2016

Figure 5.2b displays a SEM image of a metal nanowires electrodeposited in commercially available PCTE which exhibit high wall roughness [4]. It is worth noting that the track-etched process has been modified to eliminate such defects.

5.2.2 Anodic Aluminum Oxide (AAO) Templates

Anodic aluminium oxide (AAO) templates have attracted much more interests recently due to its self-organizing nanoporous structure, introduced by Masuda et al. [5]. This kind of self-organization of nanoporous anodic aluminum oxide is based on naturally occurring long-range ordering. Fabrication method of AAO templates, *multi step anodic oxidation or anodization*, is very flexible which makes it possible to provide pores with diameters less than 10 nm up to 200 nm and 10^9–$10^{11}/cm^2$ pore density. Moreover, thickness of template, determining lengths of nanowires, only depends upon oxidation time and leads to formation of 1–50 μm thick oxide layers. In the anodization process, an electrical circuit is established between a cathode and a film of aluminum which serves as the anode. Then, anodic oxidation or anodization of the film occurs in accordance to the following reaction [6]:

$$2Al + 3H_2O = Al_2O_3 + 3H_2, \Delta G^\circ = -864.6\,kJ, \quad (5.1)$$

where ΔG° is the standard Gibbs free energy change.

Over the last decade, these highly ordered nanoporous films have been used as templates for nanofabricating metal and semiconductor nanostructures in magnetic and optoelectronic applications. The typical structure of the film consists of a thin non-porous barrier oxide layer which lies adjacent to the Al substrate and a porous layer on the top. The top layer contains an array of hexagonally close-packed cylindrical cells at the center of which a straight hole is located.

5.2.2.1 Current-Transients Achieved by Multi-step Anodic Oxidation of Aluminum

Nasirpouri and co-workers have studied the mechanism of ordering and surface morphology and structure of AAO templates using two [7, 8] and three step anodisation processes [9, 10]. On the mechanism of self-ordering in AAO nano-porous materials, so far O'Sullivan and Wood presented a model which was based on an electric field distribution at the pore tip. This model is able to give microscopic explanations for the dependence of, e.g., pore diameters and pore distances on applied voltage or electrolyte composition, but cannot easily explain the self-ordering behavior. The self-organized arrangement of neighboring pores in hexagonal arrays can be explained by any repulsive interaction between the pores [11]. A possible origin of these forces between neighboring pores is the mechanical stress, which is associated with the expansion during oxide formation interface and leads to form curved shape metal/oxide interface. It is claimed that the pores are formed during electropolishing and/or anodizing on the aluminium surface and can become hexagonally ordered at certain voltages and times of the initial electropolishing [12] or by long-term anodization and reanodization or also by a dynamic process depending on the mobility of ions within the barrier oxide and of Al atoms within the metal [13].

The self-ordered anodic aluminium oxide (AAO) membranes have been generally characterized by removing the aluminum substrate and the structure has been observed from the bottom view of grown layer, in which the pore ordering can be achieved easily than the top surface of grown films which is more important in applications [14–17]. However, Nasirpouri et al. have studied the topological properties of AAO templates as is discussed here.

The effects of multi step anodic oxidation on the ordering of pores in AAO templates were studied using potentiostatic anodisation current transients, morphological examinations using AFM and SEM, and surface roughness scaling theorem (see Chap. 3). Figure 5.3 illustrates the current transients recorded during two and the three step anodic oxidation processes.

Figure 5.3a, c illustrate the variation of current recorded during a potentiostatic anodisation at a constant voltage of 40 V in 0.3 M oxalic acid at 0 °C in the first step. As shown, there exists a region associated with almost uniform and constant values of current without any possible expected drop in the graphs. Any drop in current transients at the first stages of anodic oxidation process may occur due to the formation of a compact and adhere oxide layer namely *barrier layer*, Fig. 5.3f. The observed smooth variation of current may be explained by the formation of an oxide layer on aluminium, naturally produced in the atmosphere on the samples during the rest time in the lab after electropolishing. The pre-formed oxide film may act as a barrier layer and further anodic oxidation develops the porous layer of AAO

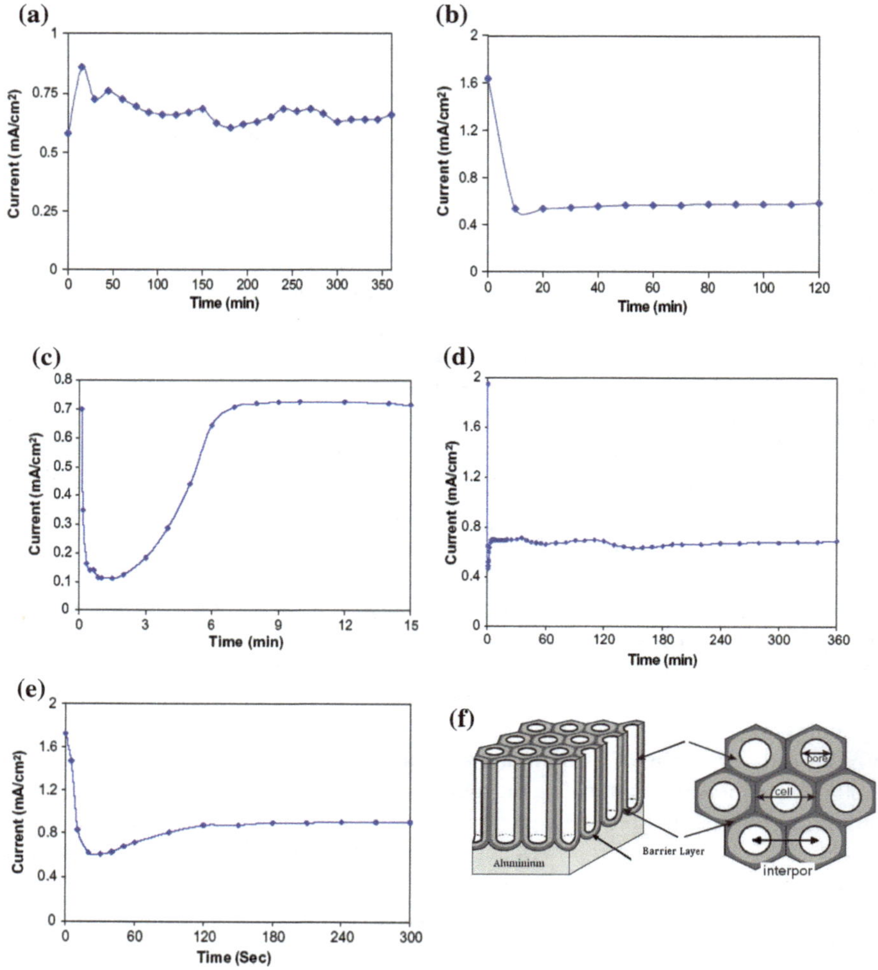

Fig. 5.3 Typical current transients recorded during multistep anodising of high purity Al for **a**, **b** two-step, and **c–e** three step anodisation. **a** First step for 6 h and **b** second step for 2 h, **c** first step for 15 min, **d** second step for 6 h and **e** third step for 5 min. Al substrates were surface prepared under a sequence of chemical surface treatments in KOH 200 g/l, Na_2CO_3 50 g/l and HNO_3 50% (in vol.) solutions. Then, Al foils were eloctropolished in a mixture of $HClO_4$ and ethanol (1:4 in vol.) at 15 V_{dc} below 5 °C. Anodization was conducted under constant potential of 40 V_{dc} in 0.3 M oxalic acid electrolyte. The temperature of the electrolyte was maintained at 0 ± 2 °C during anodization using a cooling system. Oxide film formed in previous steps was removed by wet chemical dissolution in a mixture of 0.2 M chromic acid and 0.4 M phosphoric acid at 60 °C for an appropriate time depending on the anodizing time. **f** Schematic picture of an ideally grown AAO template. Adapted from [7–10] with permission from Elsevier, copyright (2016)

films. The current transient curves show plateaus regions in a range of 0.6–0.9 mA/cm^2 during first step which may be an evidence of the growth of porous film on the substrate. The current transients recorded for the second steps shown in Fig. 5.3b, d and the third (Fig. 5.3c) steps just after dissolving the oxide layer show a drastic drop in the beginning of these steps. Since the second and the third steps were conducted immediately after dissolving the grown oxide layer at the first step, the sudden drop followed by a raise of current up to the same value of the first step (0.6–0.9 mA/cm^2), can be attributed to the formation of barrier oxide layer on oxide-free surface of substrate [7, 9].

5.2.2.2 Self-organization Mechanism of Nanopores

Nasirpouri and co-workers investigated the topographical properties of AAO films, which contribute into ordering the nanopore domains. AFM was used as a useful method to control the nanopore topological characteristics of anodic aluminum oxide. AFM allows the assessment of the ordering in nanopore domains, pore density and the external shape of AAO films. However, It must be taken into account that the internal diameter and shape of nanopores, which are defined as through pore arrangement, cannot be evaluated by this method. This structural characteristic of AAO membranes is easily characterized by scanning electron microscopy across the cleavage surfaces of films.

Figure 5.4 illustrates AFM images of a typical growth sequence employed to produced highly ordered AAO templates. Figure 5.4a shows Al surface electropolished in $HClO_4$/ETOH solution. After electropolishing, Al has an almost flat surface, exhibiting small etch pits and bumps, which could be seeds for pore nucleation. Consequently, the electropolished aluminium was anodized for first time. It has been shown that the pits or pores nucleate on the natural barrier layer or in the bottom of porous layer during the initial stages of anodization. However, it is assumed that the pits formed in the electropolishing contributing in the nucleation of pores on the aluminium in the order of 10^{10}–10^{12}.

Figure 5.4b displays the surface morphology of anodized aluminium for 6 h in 0.3 M oxalic acid at 0 °C, obtained by contact mode-AFM. As it can be seen, pores are randomly distributed in AAO film with a broad size distribution. During the first step anodization process, the pores nucleate on the electropolished surface at almost random positions, i.e., lattice imperfections or pits formed by electropolishing. As the anodization time is increasing the pores merge and form the curved metal/oxide interface due to stress inducing by volume expansion. Because of the random nucleation positions of initial pores, the hexagonally ordering of pores is just achieved in the first stages of anodizing at the bottom of porous oxide layer and cannot grow up to the thick anodized layers.

Interesting feature appeared after removing the first anodizing oxide layer. Curved shape hemispherical interface remains on the aluminum substrate. This structure is shown in Fig. 5.4c, which demonstrates AFM images of aluminum surface after removing the oxide layer. The uniform hemispherical shape of barrier

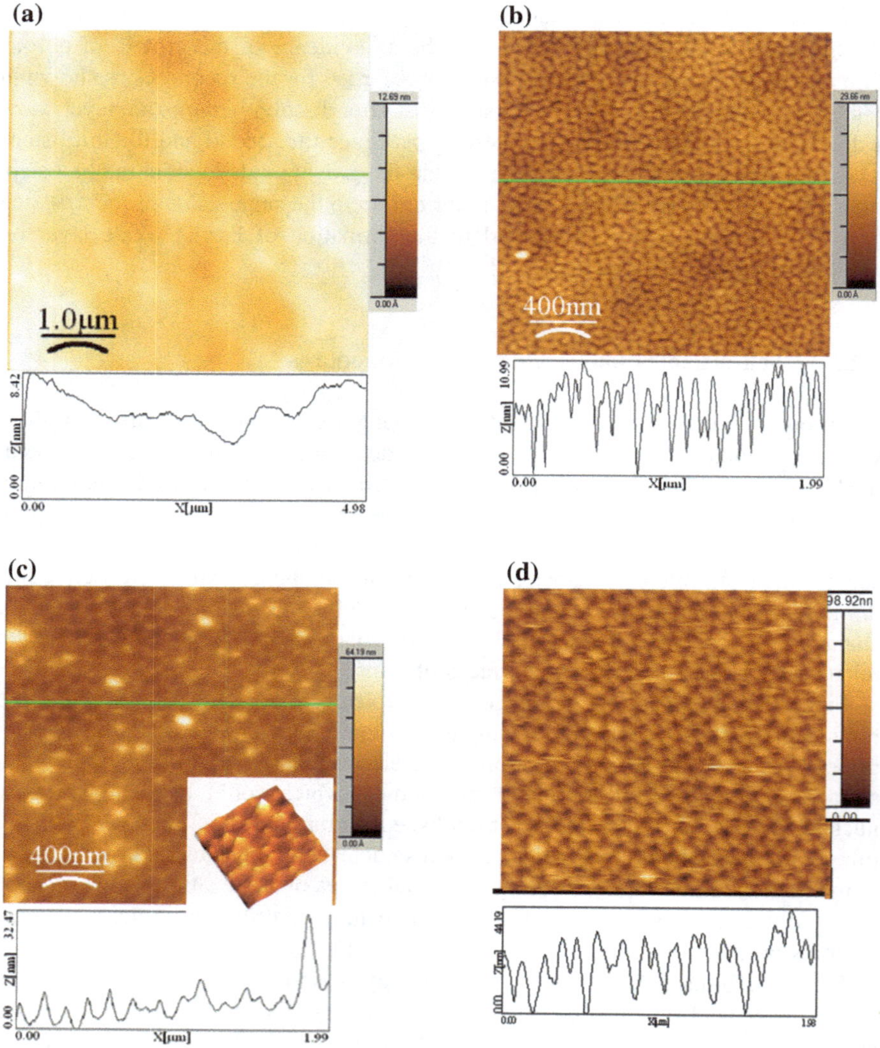

Fig. 5.4 AFM images taken during a typical growth sequence of highly ordered AAO template **a** electropolishing in a mixture of $HClO_4$ and ethanol (1:4 in vol.) at 15 V_{dc} below 5 °C. **b** First step anodic oxidation for 6 h at a constant potential of 40 V_{dc} in 0.3 M oxalic acid electrolyte at 0 ± 2 °C, **c** removal of first step oxide in during anodization by wet chemical dissolution in a mixture of 0.2 M chromic acid and 0.4 M phosphoric acid at 60 °C for 35 min., and **d** two-step anodic oxidation for 1 h. From [7] with the permission of Elsevier

layer covers the substrate surface. These hemispheres act as nucleation centers for second step anodic oxidation. Figure 5.4d shows that two-step anodised sample reveals highly uniform pores with hexagonally arranged lattice.

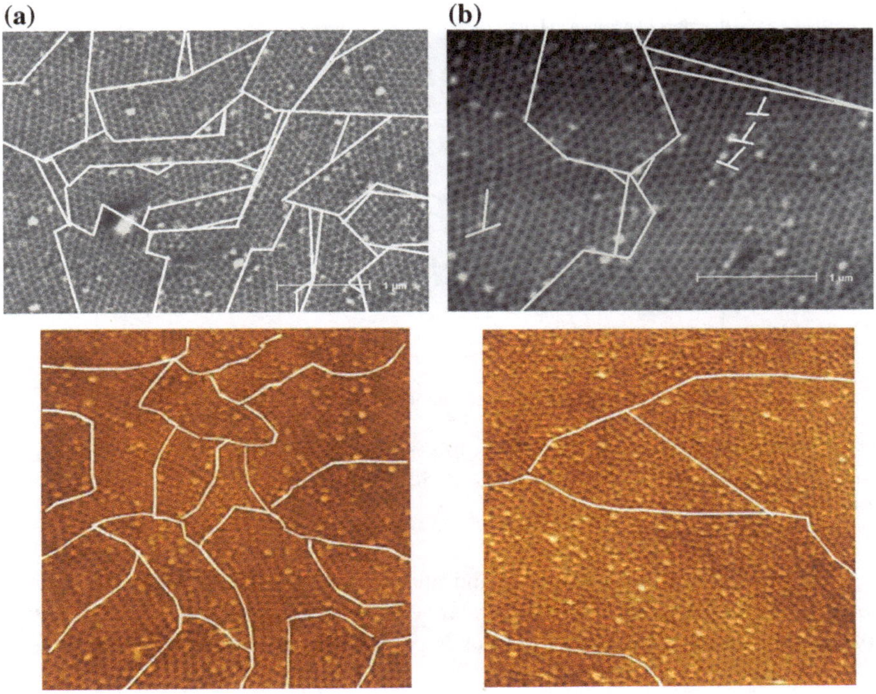

Fig. 5.5 SEM and AFM micrographs of identified ordered domains on *top* surfaces of AAO films obtained in oxalic acid after: **a** 4 h first step and 45 min second anodization step; **b** 15 h first step and 45 min second anodization step. Adapted from [7] with permission from Elsevier

Anodizing the sample for second time develops the pore growth exactly on the concave pattern created during the first step anodization. As the first step anodizing duration increases the hexagonally ordered pore areas, domains, in the bottom of porous oxide layer occur in the larger surfaces. In fact, nanopores exactly grow upon the relevant hemispheres and form direct pillars, which can be detected on top surfaces of AAO films. Figure 5.5 demonstrates a typical topography of AAO films, anodized in the different first step anodizing time. The pore alignments are different in the domains and can be found out by domain boundaries, along which the pores gradually merge. Moreover, some other kinds of defects such as point defects and misfit dislocations can be seen in the topological studies. The misfit dislocation of the pores line interrupts the periodic arrangement of the pores.

As another important result, the ordered domains size is a function of time (Fig. 5.6). Two kinds of data fitting methods have been used to our experimental data. The linear method gives a function as: $\boldsymbol{D} = 0.55\,\boldsymbol{t}$, where D is the domain size in square micrometer and t is the first anodizing step duration in hr. Li and co-workers [13] have shown a linear behavior in their investigations. We have also fit our data in the parabolic form and we obtained: $\boldsymbol{D} = 0.52\,\boldsymbol{t}^{0.7}$. This is very similar

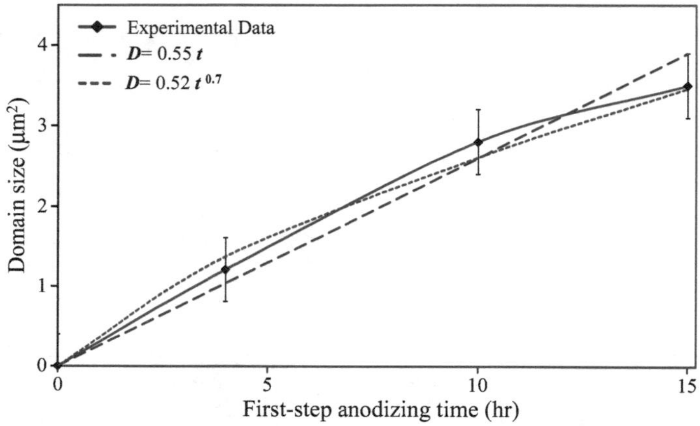

Fig. 5.6 The ordered domain growth versus first-step anodizing time. Data represent the average domain size based on the identified areas of SEM micrographs. Reproduced from [7] with permission from Elsevier

to grain growth behavior in metals and alloys. For metals and alloys, the driving force of grain growth is the grain boundary energy per unit area. For grain growth at a fixed temperature, the average radius R of the grain is a function of the time t:

$$R = B^n \tag{5.1}$$

in which B is a temperature-dependent parameter and n is about 0.4–0.5 [14]. As a result, we considered a quasi-linear growth behavior of ordered domain size on the time.

Consequently, the effect of second step anodizing time has been investigated. In Fig. 5.7, the fracture surfaces of highly ordered AAO films, obtained in different reanodizing time, elucidate the same through pore arrangement. Thus, as the pore ordering takes place, increasing the second step anodizing time does not affect the achieved arrangement. The thicknesses of AAO films were measured approximately 1 and 6 μm after 45 min and 6 h anodizing in second step, respectively [7, 8].

5.2.2.3 Enhancement of Ordering Degree of AAO Templates by Three-Step Anodic Oxidation

Once anodic oxidation is conducted for one more step, the third step, a different degree of ordering can be achieved. The results achieved by Nasirpouri et al. [9, 10] shown in Fig. 5.8 correspond to AAO film anodized three times with a shorter and an equivalent anodization time at different first step times. As is evident, there is a considerable change of pore size distribution and also ordered domain size. The histograms show reasonable enhancement of size distribution. Not only histograms

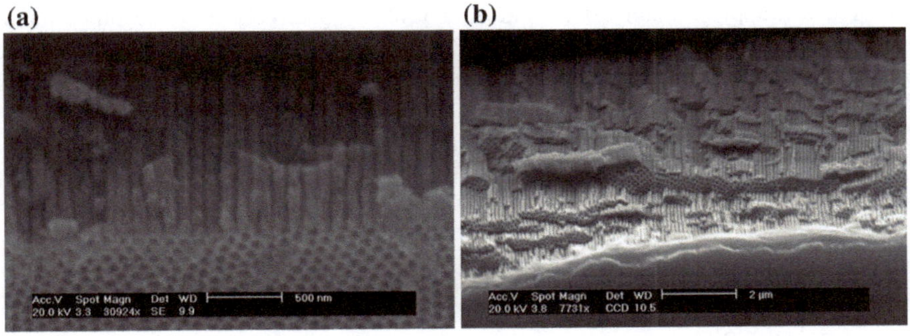

Fig. 5.7 Cross-sectionla view of highly ordered AAO templates grown by two-step anodisation in oxalic acid with 40 nm pore diameter after 6 h first step anodization and **a** 45 min and **b** 6 h in second step anodization. Reproduced from [7] with the permission of Elsevier

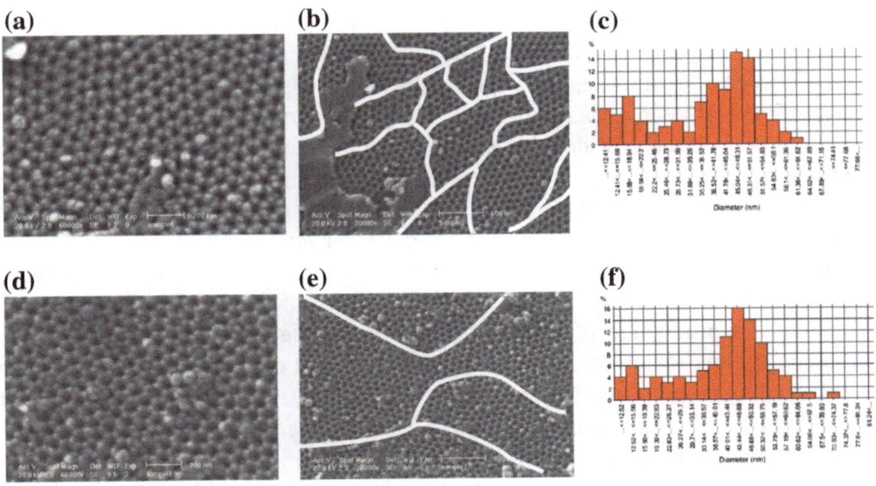

Fig. 5.8 **a** SEM image, **b** the identified domains and **c** the histogram of pore diameter, obtained from AAO film anodized for 0.5 h, 6 h and 10 min in the first, the second and the third steps, respectively. **d** SEM image, **e** the identified domains and **f** the histogram of pore diameter, obtained from AAO film anodized for 1 h, 6 h and 10 min in the first, the second and the third steps, respectively. Reproduced from [9] with permission from Elsevier

show a tighter pore size distribution which falls into a range of 40–45 nm, but also confirms a more uniform pore size distribution by increasing the first step anodization time which could be completely understood with the formation of better ordered barrier layer after dissolving the layers. Consequently, the ordered domain size also varies as a function of time. The ordered domain size of AAO was figured out from the regions on SEM images. Measurements reveal that the average domain size for three step anodized AAO templates with first step durations of 0.5

and 1 h are approximately 0.47 and 2.81 μm^2, respectively. On the other hand, using the equation: $\boldsymbol{D} = 0.52\ t^{0.7}$, in which D is the domain size in square micrometer and t is the first step anodization time in hours, the domain size of two step anodized samples with the same first step anodization times are estimated to be 0.308 and 0.52 μm^2, respectively. Thus, the larger areas are located in the ordered domain for the three step films. These estimations qualitatively explain that third step of anodic oxidation enhance the ordering of nanopores.

5.2.2.4 Effect of Voltage and Electrolyte on Pore Diameter of AAO Templates

Li et al. demonstrated a formula between interpore distance (D) and anodizing voltage (V) shown in Fig. 5.3f [18–20]:

$$D_{(nm)} = -1.7 + 2.81\ V_{(volts)} \tag{5.2}$$

The growth mechanism of anodic oxide films may be determined by a correlation between formation and dissolution processes depends on the types of electrolytes, temperature, anodizing voltage, etc. In the course of formation of the oxide coating, a large amount of heat is released. This phenomenon can be explained on the one hand by the exothermal reaction of aluminum with acid and aluminum oxidation and in the other hand by the Joule effect due to the high resistivity of the electrolyte column in the pores and of the barrier layer.

Anodizing electrolyte affects the growth mechanism in terms of its secondary dissolution capacity defined as the rate of dissolution of the layer produced during oxidation. For production of porous-type oxide films, electrolytes with a medium secondary dissolution are required in which the growth of oxide film thickness can be assumed as proportional to the amount of electricity consumed [21]. The major change in different electrolytes of this type is the *diameter of pores*, as it increases when the anodizing voltage increases.

Under the anodizing experiments, the growth rate of the oxide layer exceeds its dissolution rate, leading to the formation of a barrier type layer right on the substrate at the bottom and a porous-type layer at the top. The process of the development of pores requires the interaction between the electrolyte and surface and is currently understood using the so-called field-assisted dissolution (FAD) hypothesis which describes the initiation and propagation of pores at the barrier layer [22]. The effect of voltage on the diameter of pores is described by FAD hypothesis as increasing of the electrical field leads to a wide distribution electrical field and consequently it widens the pore diameter. Another major effect that we report here is that as the anodization voltage increases the non-uniformity of the film is increased. Shimuzu and co-workers [23] reported a similar behavior of the oxide films obtained in phosphoric acid and attributed the loss of planarity to wider pertuberances due to persistent crack-heal events.

The effect of the electrical field strength (or anodizing voltage) on pore diameter was studied by Nasirpouri and Peighambari [24, 25]. The electrical field or anodisation voltage can change depending on the electrolyte conductivity and its other transport properties, as discussed in Chap. 2. Depending upon the electrolyte type, a different potential of anodizing of Al has been chosen, as it includes: 18 V in sulphuric acid, 40 V in oxalic acid, and 150 V in phosphoric acid. Based on the applied anodizing voltage and also the type of electrolyte, pores with diameters may be produced up to 200 nm. Figure 5.9 illustrates AFM and SEM images of AAO films anodized in sulphuric, oxalic, and phosphoric acids under different voltages in a range of voltage between 20 and 170 V. These three types of electrolytes are categorized among those that lead to the formation of porous-type aluminum oxide films. Each electrolyte has a specific range of anodizing voltage, as shown in Fig. 5.9 [24, 25].

5.2.2.5 Topography and Kinetic Roughening of AAO Templates

Various studies have shown that the physical and chemical properties of nanostructure devices filled in porous AAO templates films are mainly exploited on top surface, i.e., magnetoresistance or magnetic properties [26, 27]. Therefore, the topography including surface morphology or roughness will influence the surface characteristics of the hybrid material AAO-based device. In this regard, several investigations are still conducted to understand the correlation between the preparation parameters and morphology of AAO films [28–33]. Some studies have recently reported the effects of specific fabrication parameters of AAO on the surface morphology of the produced films including substrate surface roughness, pore arrangement and structure. Electropolishing process is believed to significantly control the surface roughness of aluminum substrate by varying the process conditions such as solution temperature, applied potential and electropolishing time [34, 35]. Furthermore, it was found that there is a correlation between surface roughness achieved after electropolishing and pore size or density [36]. Although the flatness of the two-dimensional array was found to be dictated by electropolishing quality, but more importantly the pore size, spacing, and uniformity could be determined by the condition of the achieved surface following these preparations. During anodization, the barrier layer is developed first. The rough surface of Al sheet with large crests and deep troughs may disturb the electric field distribution and electrolyte concentration at local regions, and consequently barrier layers with nonuniform thickness may be formed. This indicates that even after anodizing an Al foil under otherwise identical conditions, barrier layers with different properties can be formed on substrates depending on whether an electropolishing step was taken or not. Effect of duration of dissolution stage has also some effects on the morphology of AAO films [37]. Very little investigations exist on the effect of other anodization parameters on the surface roughness of AAO films, an example can be the effect of anodization bath temperature [38].

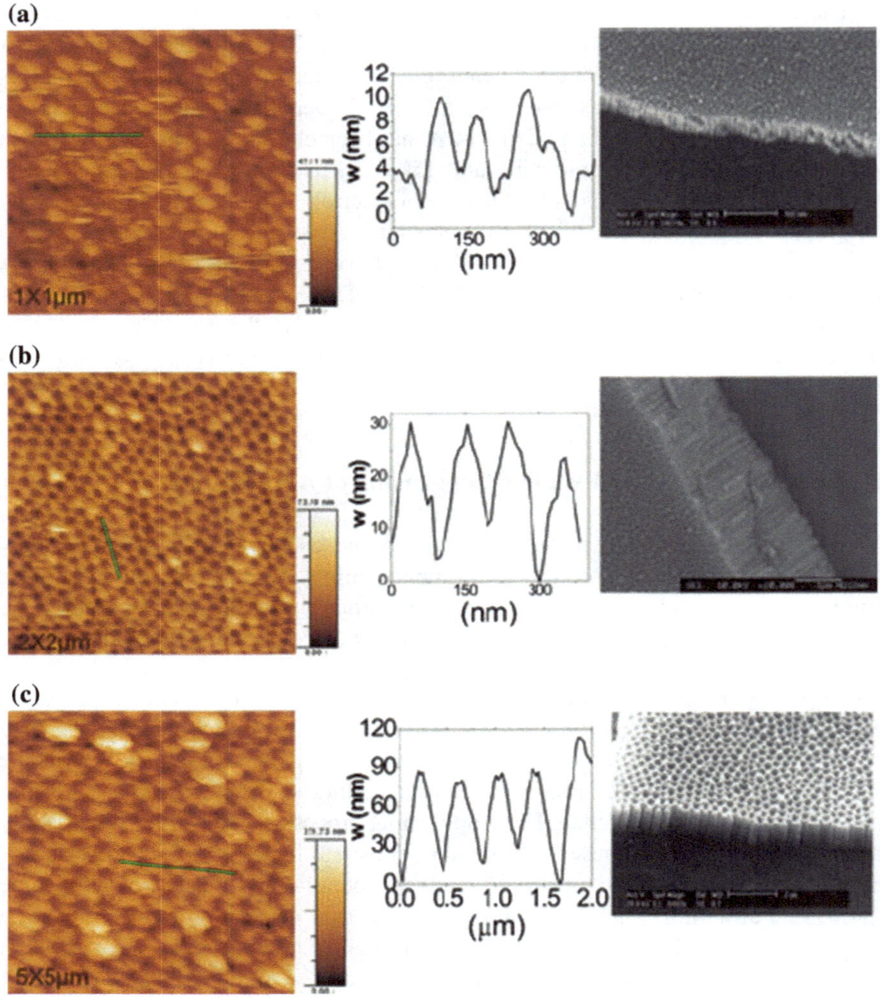

Fig. 5.9 AFM images, SEM images, and surface line scan or roughness profile of a highly ordered AAO produced after anodization for 60 min in the first step and 60 min in the second step in sulfuric acid under 18 V and b highly ordered AAO produced after anodization for 60 min in the first step and 60 min in the second step in oxalic acid under 40 V, and c highly ordered AAO produced after anodization for 60 min in the first step and 60 min in the second step in phosphoric acid under 150 V. Reproduced from [24, 25] with permission from Springer, 2016

Nasirpouri and Peighambari recently investigated the roughness variation of AAO templates in a systematic study [24]. RMS value of surface roughness was measured on the whole scan area obtained from AFM images. The variation of RMS measured for the surface roughness recorded at different stages of a typical

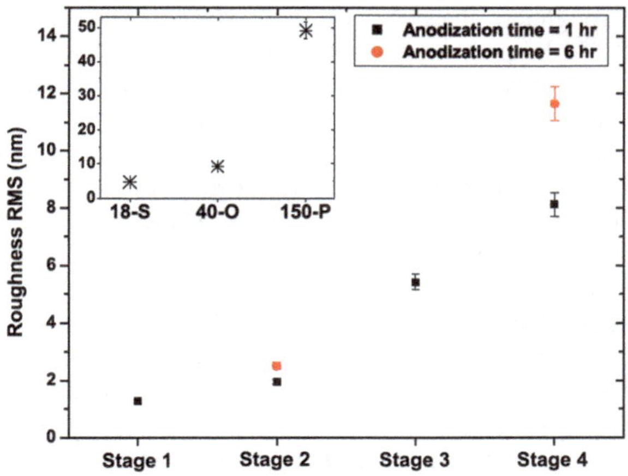

Fig. 5.10 The variation of RMS value of roughness measured on AAO templates using AFM images. Stage 1: electropolishing; Stage 2: first-step anodization; Stage 3: dissolution; and Stage 4: second-step anodic oxidation. For all samples except those mentioned on inset, anodic oxidation was performed in oxalic acid under 40 V. The inset shows RMS values of roughness measured on AFM images of highly ordered AAO produced in sulfuric acid under 18 V (18-S), oxalic acid under 40 V (40-O), and phosphoric acid under 150 V (150-P). Reproduced from [24] with permission from Springer, Copyright (2016)

two-step anodic oxidation process is illustrated in Fig. 5.10 indicating a dramatic increase of the surface roughness during four steps of the anodization process.

The lowest RMS value was obtained for the electropolished Al substrate. Electropolishing was conducted under the optimum condition generally used to produce highly ordered nanoporous aluminum oxide structures. After electropolishing, Al surface shows an RMS value of 1.30 nm which spontaneously occurs due to the formation of nanometer scale surface bumps and pits [7]. After anodizing the electropolished Al for the first time, the surface roughness slightly increases. The oxide layer grows on the surface controlled by a dynamic equilibrium between film growth and dissolution reactions described by FAD hypothesis. This has significant effect on the surface roughness. They assumed this behavior happening as a consequence of the formation of a non-uniform film on the cellular sections of the films as the film growth propagates. As a consequence of pore development, the electrical field and ionic current become concentrated in the barrier layer beneath the major pores. This leads to a successive ingress of O^{2-}/OH^{-} ions to form solid film at the metal-film interface and corresponding Al^{3+} ejection at the pore bottom-electrolyte interface. However, in the regions of current concentrations, a localized scalloping of the metal-film interface is developed. Non-uniform concentration of electrical current after the initiation of pores diminishes the surface uniformity of the oxide layer. This is what is exactly observed when we extend the anodization to obtain AAO films. Prolonging the anodization after electropolishing stage normally thickens the oxide layer and consequently gives rise to a greater

RMS value of roughness measured on the surface. The FAD theory given by Thompson [22] for the pore propagation is consistent discussed in above. At the initial stages of pore growth, a non-uniform disordered layer of AAO film develops on the surface with a typical roughness about 1.96 nm, while the ordering is obviously achieved after anodizing for 6 h exhibiting the formation of quasi-hexagonal arrays of pores with higher roughness, 2.93 nm.

The formation of the uniform cellular shape at the bottom of pores after an appropriate duration of anodization is proved when the AAO film grown is removed and the morphology of the barrier layer (or substrate) is studied. Figure 5.4c demonstrated that this hemi-spherical curvature retains on the surface which generates a remarkable surface roughness of 2.50 nm compared with the roughness of the electropolished Al before anodic oxidation. Montero-Moreno and co-workers [28, 29] based on the surface brightness studies have indicated that the dissolution of oxide layer in a mixture of phosphoric and chromic acid is associated with a significant increase in surface brightness due to the dissolution of layer from the pore walls. This evidence can qualitatively explain the evolution of roughness on the Al substrate after and before anodic oxidation.

Anodic oxidation for the second time develops the oxide films with a considerable increase either in the pore arrangement or in the surface roughness. The same behavior is observed for the roughness of the films after second time anodizing (or highly ordered) for different durations as of the films anodized once. This should be noted that the initial surface morphology not only directs the process towards highly ordering of the films, but also increases the surface non-uniformity at the walls of the pores, meaning that after dissolution of the first oxide layer, a patterned substrate is employed.

Therefore, regions of initial preferred pore development are established as a result of the geometry or curvature of the local film surface. A similar argument can be assumed for the surface roughness evolution of the oxide layer as given earlier in the previous section for the first-step anodic oxide film. In this regard, this was pointed out that either the funneling of pores near by the oxide surface or a variation of electrical field strength (average value of 8×10^6 V/cm and local increase to 2×10^7 V/cm) takes place. The change of electrical field has a remarkable effect on the pore growth of AAO but can also qualitatively explain the change of roughness due to the growth rate of the oxide film. The thickening of film would lead to variations of the strength of electrical field as the high dielectric effect of the oxide layer. We interpret this type of growth with the development of rough edges on the pore walls. In Fig. 5.11, Nasirpouri proposed a model behavior for the growth of AAO films based on our observations from the real surface topography measurements using AFM.

The schematic image is proposed here with the corresponding line scan taken from AFM image of surface after each stage of two-step anodic oxidation process. The line scans clearly show a considerable roughening takes place on the pore wall, as the two-step anodic oxidation process proceeds.

Kinetic roughening of AAO films is investigated using the dynamic scaling theory (see Chap. 3, Sect. 3.3.4). Nasirpouri and peighambari [24] have used two

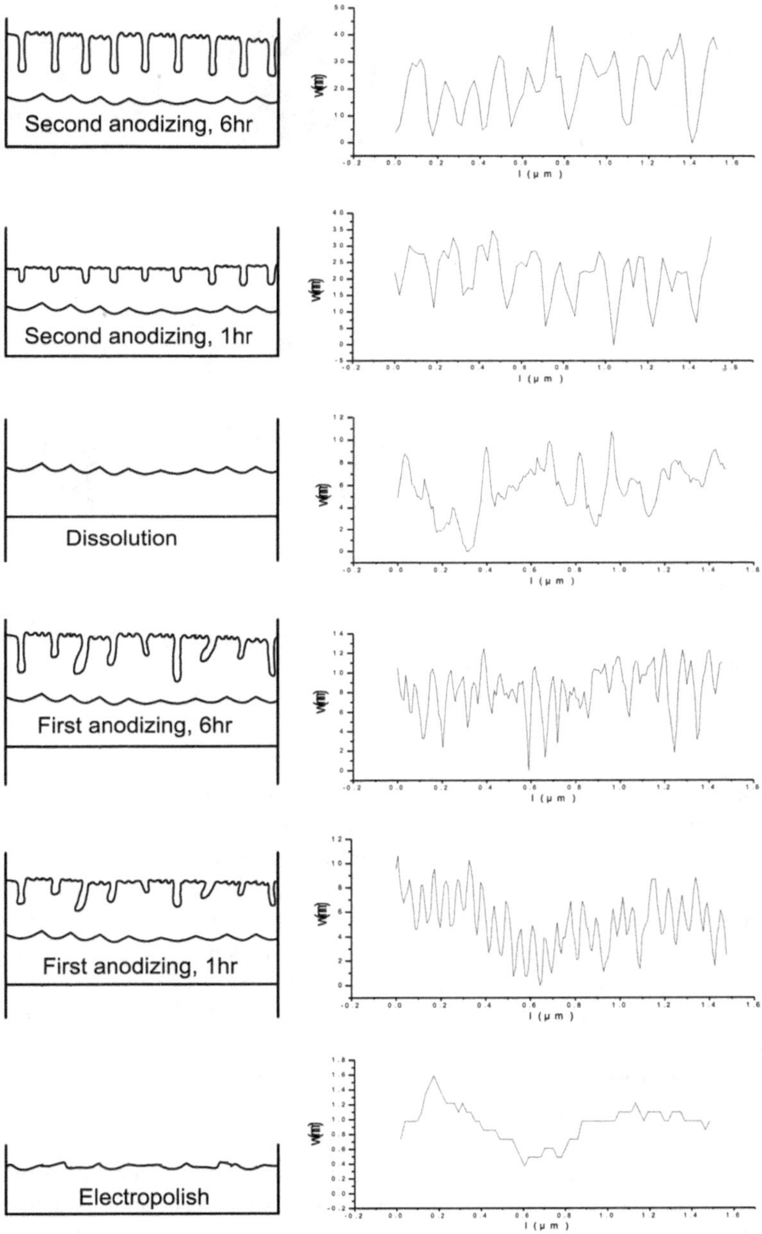

Fig. 5.11 A schematic model for roughness evolution of AAO templates during two-step anodic oxidation of aluminum including electropolishing, first-step anodization for different times (1 and 6 h), dissolution, and second-step anodization for different times (1 and 6 h). *Right*, the real line scan or roughness profile taken from the corresponding AFM images, shown before in Fig. 1; *left*, schematics driven based on the line scans. Reproduced from [24] with the permission of Springer

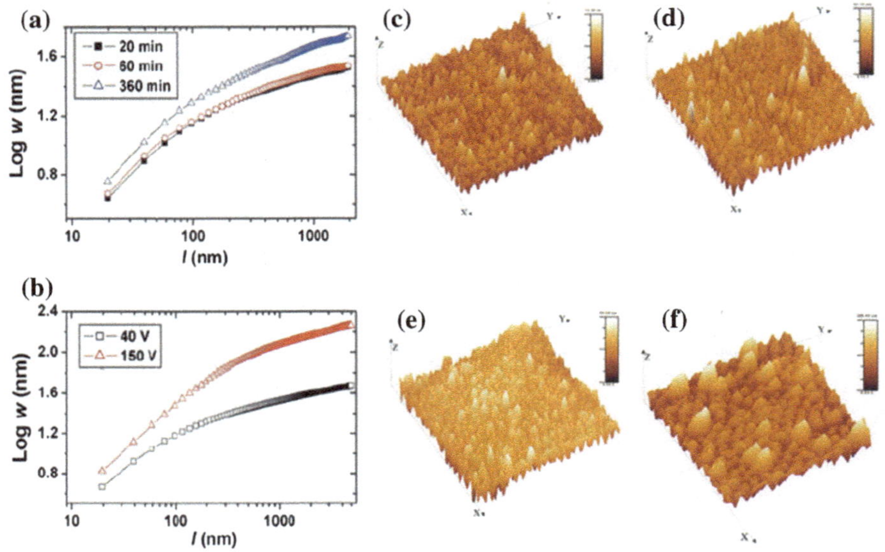

Fig. 5.12 Logarithmic scale of roughness as a function of scan length for **a** highly ordered AAO produced in oxalic acid under 40 V for different durations and **b** highly ordered AAO produced in oxalic acid under 40 V and in phosphoric acid under 150 V. Three dimensional AFM images of highly ordered AAO produced in oxalic acid under 40 V for **c** 20, **d** 60, and **e** 360 min. **f** A three-dimensional AFM image of highly ordered AAO produced in phosphoric acid under 150 V for 60 min. From [24] with permission from Springer

different sets of samples to scale the dynamics of roughness evolution. In the course of scaling, the effects of (1) the thickness of highly ordered AAO films and also (2) the diameter of nanopores (or hexagonal cell size) were investigated. For the former category, highly ordered AAO films anodized in oxalic acid with different duration of the second-step anodic oxidation were prepared to study the oxide layers with different thicknesses. The later effect has been investigated in the AAO films obtained from oxalic and phosphoric acids.

Figure 5.12a shows w(l, t) calculated from AFM data for a series of oxide films grown in oxalic acid under 40 V for 20, given film thickness or anodization time. In Fig. 5.12b, the variation of surface roughness as a function of length scale is illustrated for two different anodization electrolyte under different voltages exploiting different nanopore diameters. Equation 3.62 (see Chap. 3, Sect. 3.3.4) was used to fit the data shown in Fig. 5.12 for the calculation of the values of H (the gradient of the plot for small l), w_{sat} (the value of w for large l), and l_c (the cross-over between two regimes) for all the curves shown in Fig. 5.12. All the calculated data are summarized in Table 5.1.

As is evident, the variation of roughness is best described by the *anomalous* scaling. It is seen that the increase in anodization voltage significantly changes H, w_{sat}, and l_c. However, the effect of thickness or anodic oxidation time is only

Table 5.1 Scaling parameters obtained for AAO templates

Sample condition	w_{sat}	l_c	H
V = 150 v	216.6	1,643.1	0.6
V = 40 v	55.6	1,188.7	0.4
t = 20 min	35.0933	330.7184	0.5622
t = 20 min	36.3499	340.228	0.5648
t = 360 min	61.7493	577.1880	0.5686

From [24] with permission from Springer

significant on w_{sat} and l_c. These results are consistent with the direct observations shown by AFM. On the one hand, the change in Hurst parameter (H) is substantially due to the change of electrolyte and anodizing potential or current, as reported elsewhere [32] which in turn influences the mechanism of pore initiation and size based on FAD hypothesis. On the other hand, anodization time increases the roughness of AAO film since w_{sat} and l_c increases from 20 up to 360 min. H does not change as the composition of electrolyte and anodizing voltage are the same for different film thickness or anodizing time. One would expect that the roughness increases with second-step anodizing time in such a manner based on the anomalous scaling behavior.

There has been also an argument to perhaps interpret the cross-over length, lc, with the ordered nanopore domain size discussed earlier for AAO templates. The average area of an ordered domain has been described with an equation to be dependent on the duration of first-step anodic oxidation. The equation predicts a value of about 550 nm for the AAO studied for roughness scaling. Inside an ordered domain, the orientation of pore arrangement is assumed to be in one particular direction. Therefore, FAD theory which explains the change of roughness and pore size is applicable for an ideally grown hexagonal AAO structure. All the pores have the same structure inside all domains, but it may differ at the interfaces of the domains. When the length scale is greater than the domain size, the roughness is statured to the maximum value and becomes independent on the length.

5.3 Electrodeposition into Template

Template electrodeposition is known for years after successful achievements demonstrating that electrodeposition is capable of depositing materials in extremely confined and ordered nanopores of *templates*. First, Possin [39] in 1970 reported that metal wires with diameters as small as 400 Å were successfully fabricated using a method involving electroplating into etched particle tracks in mica. Consequently, W. D. Williams and N. Giordano expanded this technique to routinely producing wires as small as 80 Å [40]. Pioneering work by Martin [41] introduced the template synthesis as an efficient and versatile method of fabrication of one dimensional nanostructures like nanowires. This method involves the preparation of monodisperse nanoscopic structures of a desired material inside confined spaces of a host mesoporous membrane as templates [42–45]. Template electrodeposition has been consequently been used to produce a wide variety of materials in form of nanowires.

5.3.1 Template Electrodeposition: Electrochemical Consideration

5.3.1.1 General Mechanism

Nucleation and growth during a typical template electrodeposition process until a complete pore filling has a well-defined sequence, described first time by Whitney et al. [46]. Figure 5.13 schematically illustrates the variation of deposition as a function of deposition time (current transient) during a typical template electrodeposition process under a constant potential (potentiostatic electrodeposition). The curve depicts different regions each of that designates a peculiar nucleation or growth stage. A four stage process is well established for template electrodeposition. Stage (1) namely the pre-growth stage occurs when a sudden drop of current takes place once the potential is applied through the cell during which nucleation starts at the pore bottoms. The remarkable reduction of current at the beginning is attributed to the mass transport limitation. Consequently, a slightly increased current is observed at which the metal is growing in the pores, Stage (2). As the nanopores are filled, the current rises with a shadow slope. Finally, the nanowires smear out of the nanopores where a sharp increase of current appears on the current transient, Stage 3. This leads to the formation of hemispherical caps, originating from each nanowire, which calescence to form a cohort film covering the whole surface of the template, Stage (4). Thus, the effective cathode area increases and a rapidly increasing deposition current can be observed.

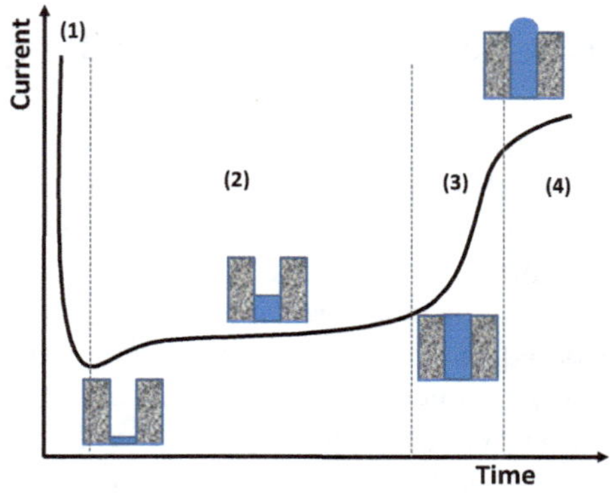

Fig. 5.13 A Schematic illustration of the growth stages during template electrodeposition

5.3.1.2 Electrochemical Implications

In terms of electrochemical principles present for the template electrodeposition, a nanoporous template could be compared with arrays of nanoelectrodes in the sense that small potential drops at low concentration electrolytes. Let's assume that the electrodeposition in controlled under diffusion (or mass transport reactions). A template working as an array of nanoelectrodes has a current response under two diffusion regimes including (i) time-dependent diffusion exhibiting linear or Cottrell diffusion and (ii) spherical diffusion for the longer times under the steady-state current values. This is accordance with the model suggested by Bond et al. [9, 10] in which diffusion controlled current density of microelectrodes were determined. Accordingly, at the initial times a recessed electrode obeys the Cottrell equation, whereas at the longer times it approaches the steady-state current.

The current transient for a single recessed electrode, i.e. a single nanopore in a template, can be described by the formula:

$$i = \frac{\pi n F c_b D r^2}{L + \pi r/4} \theta_3 \left\{ 0, \frac{Dt}{L + \pi r/4^2} \right\} \tag{5.3}$$

n is the number of transferred electrons, F is the Faraday's constant, c_b is the bulk concentration of the electroactive species, D is the diffusion coefficient, L is the pore length (recessed dimension), r pore radius. The term consisting of θ_3 can be expressed individually for short term and long term expansions [47] The steady-state diffusion current to a single micro- or nanoelectrodes is rapidly achieved. The current maintains at this steady-state value due to the high component of spherical diffusion at the same time as the diffusion zone continues to grow radially into the bulk solution [48].

For long expansion, the θ_3 is expressed by:

$$i = \frac{\pi n F c_b D r^2}{L + \pi r/4} \left\{ 1 + 2 \exp \left[-\frac{\pi^2 Dt}{(L + \pi r/4)^2} \right] + \cdots \right\} \tag{5.4}$$

and for the short term expansion is:

$$i = \pi n F c_b r^2 \sqrt{\frac{D}{\pi t}} \left\{ 1 + 2 \exp \left[-\frac{(L + \pi r/4)^2}{Dt} \right] + \cdots \right\} \tag{5.5}$$

Which is close to Cottrellian behavior [49].

For a template electrode consisting of an array of nanoelectrodes (or nanopores), the diffusion zones of the neighboring pores start to coalesce after a certain time. This provokes a change in the deposition current. A few cases for the deposition behavior of arrays of differently ordered inlaid microelectrodes were described by Scharifker [50, 51] explaining the change in the current due to coalescence of diffusion zones. This results in a decrease of the measured current.

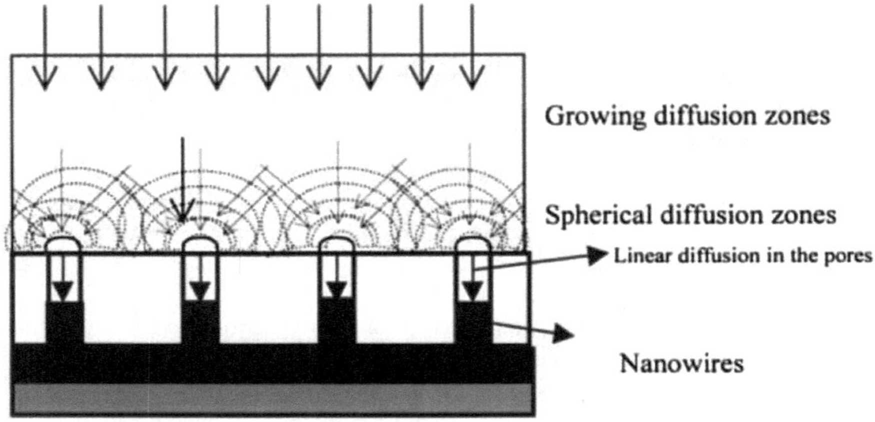

Fig. 5.14 Schematic view of the evolving growing diffusion zones as time elapses. The extended vertical line shows the direction of fluxes toward the diffusion zones. Adapted from [52] with the permission of Elsevier

Valizadeh et al. [52] have calculated the variation of current and the concentration profile on the mouth of pores and in the pores for an array of nanoelectrodes for an AAO template. Generally, the rate of electrode reaction is governed by the rate of the movement of electroactive species from the regions of high concentration (bulk concentration, c_b) to low concentration. For longer times, the diffusion of electroactive species towards the pores is assumed to take place by spherical diffusion along the radii of the nanopore electrode. Figure 5.14 shows a schematic view of the evolving diffusion zones. At the initial stage, the steady-state rate of mass transport is reached within a few milliseconds. In addition, at longer times, a spherical diffusion zone develops from each individual nanopore electrode and increases with the time.

Consequently, an extended three-dimensional diffusion volume from overlapping of neighbouring diffusion volume from each individual nanopore electrode is established near to the solution. Thus, more electroactive species diffuse in the direction normal to the region where the electrode surface including both the insulating and active surface area covers symmetrically with a spherical diffusion volume. Accordingly, electroactive species ions move rapidly towards the electrode surface by the radial diffusion from the bulk solution to the partially depleted diffusion zone. It is important to note that the diffusion-controlled limiting current at the pore array essentially facilitates the non-linear steady state, i.e. time independence at longer time. Since reduced electroactive species instantaneously established a continuous depleted region at the electrode surface and in the vicinity of the electrode, a concentration gradient is formed in the electrolyte between the nanopore mouth and bulk concentration. Thus, a continuous linear diffusion normal to the spherical diffusion zones at the electrode surface prevails. As a result, the concentration of the electroactive species in the mouth of the nanodes will differ from the concentration of electroactive species in the vicinity of the electrode. Consequently, for any constructed nanoelectrode whether it is planar, spherical, disc, or array of every geometry, the diffusion controlled limiting current is concentration dependent. This has been shown to affect the initial stages of nucleation of nanowires at the bottom of the pores as well as the morphology of nanowires grown in the first stages [53, 54].

5.3.2 Template Electrodeposition: Techniques

In general, template electrodeposition may be performed using three techniques namely direct current (DC), alternating current (AC) and pulse current (PE) deposition.

5.3.2.1 Direct Current Template Electrodeposition

DC electrodeposition is being frequently used for electrodeposition of nanowires into nanoporous template. Since the template is an insulating material, then in order to be used as the cathode in an electrodeposition cell it must become electrically conductive. For track-etched templates, there exists an advantage which is the open ended nanopores on either sides of the template. In this case, a metallic thin layer such as Au or Cu may be evaporated or sputter coated to cover the pores from one side, forming a conductive layer. For AAO templates, the situation is rather complicated. The anodic oxidation process leaves an insulating and rather thick barrier layer (1.2 nm/V of anodisation potential [55]). Considering anodizing potential range of periodically arranged AAO templates, i.e. 20 ± 195 V [7, 9, 24], the formation of a thick barrier layer (20–250 nm in thickness) is obvious. Thus, high electrodeposition potentials must be applied in order to transfer the electrons by tunneling through the barrier layer. This is associated with side reactions like hydrogen evolution in aqueous solutions and therefore is a very unstable process which does not provide uniform pore filling. Therefore, a similar procedure mentioned above for DC electrodeposition of nanowires in track-etched polymers must be conducted to deposit a metal film on top of an AAO film. However, the porous aluminium oxide layer has to be detached from the Al substrate in order to achieve open pores form the bottom side. Moreover, it is crucial to dissolve barrier layer from pore bottom to make pores open at both sides. This approach involves the transferring of the alumina film. This is only applicable for free-standing thick alumina films (>20 μm) and stable enough to be handled.

DC electrodeposition into conductive templates can be controlled by either *potentiostatic* or *galvanostatic* deposition methods, depending upon whether applied potential or current is adjusted precisely for the given material to be deposited. For more details about the instrumental information, see Chap. 3.

5.3.2.2 Alternating Current Template Electrodeposition

This method is solely applicable for the electrodeposition of materials into AAO templates which was first developed by Caboni [56]. This method has been traditionally used for coloring anodic alumina coatings for industrial applications like protective or decorative surfaces. Porous alumina remains on the aluminum substrate and the metal is deposited on the barrier layer at the pore tips by an alternating deposition potential. This method is based on using the barrier layer as a rectifying source for an alternative current (AC). This takes advantage from the natural growth of AAO films on aluminum substrate by an anodisation technique. The general mechanism for the growth of the oxide layer includes in particular Al^{3+} ions egress originating from Al metal lattice and oxygenated (OH^{-}/O^{2-}) ions ingress through the oxide layer under an adequate anodisation voltage. Thus, the ion transport rate affects the growth of the oxide layer and consequently the chemical composition. It is approved that the chemical composition of the oxide layer throughout its thickness is not uniform. In Al_2O_3/Al interface, the oxide layer is Al^{3+}-rich, whereas on the Al_2O_3/electrolyte interface, the concentration of oxygen ions is high. In this regard the oxide layer represents a p-n junction. The region close toAl_2O_3/electrolyte interface is the semiconductor type p (hole conductivity) and the region close to Al_2O_3/Al interface is semiconductor type n (electron conductivity) is formed. Therefore, Al/Al_2O_3 (AAO) template can be used directly as the cathode during an electrodeposition process. The barrier layer then can pass electrical current during cathodic half-cycles of a typical AC wave supply electrons required for reduction reaction of electroactive species. Alternating current is needed either because of the rectifying nature of the Al metal/oxide junction, or because of diffusion barriers associated with the deep nanopores [57].

Having explained this technique, it must be added that the barrier layer formed during the general anodic oxidation is so thick that the current cannot tunnel through it. In order to conduct electron transfer required for the electrochemical reduction reaction of nanowires, the thickness of the barrier layer must be reduced down to a minimum thickness. By thinning the barrier layer the quality and the uniformity of the electrodeposition process in the pores is significantly improved. Thinning of the barrier layer is performed by two methods including wet chemical etching and electrochemical mechanism. The chemical wet etching is carried out in used to either widen the pores or thin the barrier layer at the bottom of the pore. This can be carried out in oxalic or phosphoric acid at a little elevated temperature up to 35 °C [7, 58]. The second method involves the reduction of anodisation voltage at the final stage of production of AAO templates. This method was firstly used by O'sullivan et al. in 1970. The barrier layer can be thinned by consecutive lowering of the anodising voltage. If the sharp reduction of voltage is applied in a single step, the barrier layer just at the bottom of a few pores becomes thinner [59, 60]. In contrast, a gradual reduction of anodising voltage effectively decreases the thickness of barrier layer [61]. Provided that the voltage reduction process continues until zero then the AAO film is completely detached from Al substrate [62].

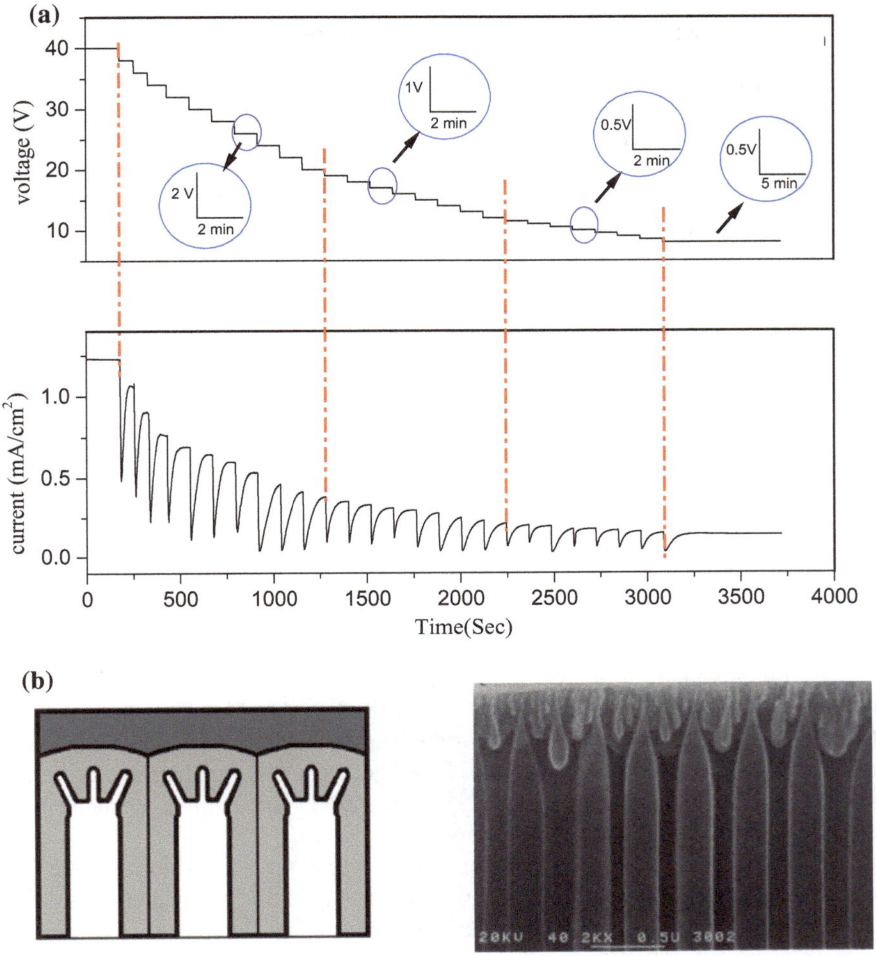

Fig. 5.15 **a** Reduction of anodisation voltage in a step-wise manner for thinning the barrier layer, Courtesy of *E. Pananhi* from his MSc thesis, **b** schematic picture and a SEM image of the modified barrier layer "routed" structure. Adapted from [63] with permission of Elsevier and [62] with the permission of Nature, copyright 2016

Figure 5.15a shows a typical step-wise reduction of anodising voltage as a function of time employed to thinning of AAO template grown in acid oxalic.

The mechanism of thinning of barrier layer can be explained by dissolving of the barrier layer leaving small pores. This finally leads to the formation of dendrite or divided barrier layer. This is shown in Fig. 5.15b, c. Using this method, the template electrodeposition is not limited by the thickness and size of the barrier oxide and its rectifying contact allows metal deposition by an alternating potential in a range of 10 V and more.

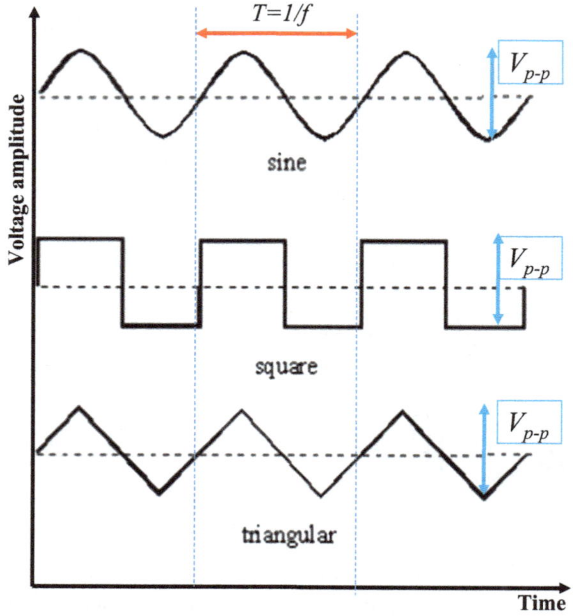

Fig. 5.16 Characteristics of AC signal generated by a power supply for template electrodeposition in AAO templates

General advantages of AC electrodeposition into AAO templates include: (1) direct method without removing the thin brittle AAO film from the substrate, (2) improvement of barrier layer uniformity during cathodic cycle by reducing its overall dissolution, (3) reducing the Joule heating, (4) avoiding concentration polarization in reverse cycles, and (5) capability of controlling mechanism by different AC parameters.

AC electrodeposition may be performed using three waveforms including sinusoidal, triangular and squared shown in Fig. 5.16. The most important parameters that affect AC template electrodeposition mechanism is the frequency *f*, voltage V_{p-p} (which is the peak-to-peak volatge), and the waveform. There is a narrow literature on the effects of these parameters on the electrodeposition mechanism [63–65] which requires further studies. To the author's knowledge, all of these parameters have significant effects on growth mechanisms, pore filling and in consequence on microstructure and other physical and chemical properties of nanowires. A brief description on our understanding in this respect is given here.

We have studied the growth mechanism of metal nanowires such as cobalt, cobalt phosphor, cobalt-nickel, nickel-manganese, etc. [66–70] on using current-transients with various AC voltages, frequencies and waveforms. Figure 5.17 schematically demonstrates the effect of frequency, voltage and waveform on typical current transients for AC electrodeposition of metal or alloy nanowires. The four-stage nucleation and growth mechanism obviously is likely observable for AC electrodeposition of nanowires in AAO.

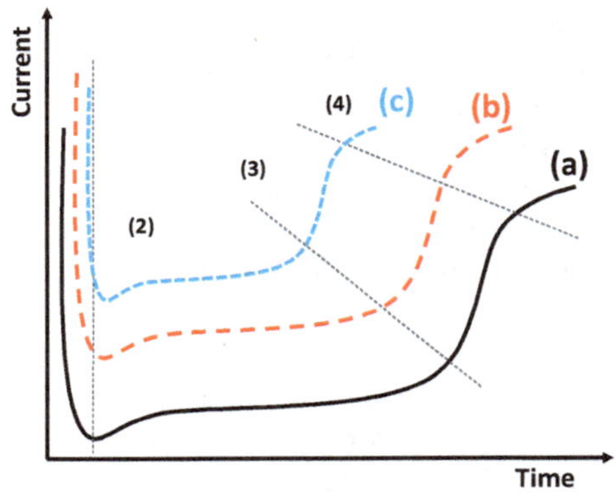

Fig. 5.17 Typical current transients for AC electrodepsotion of metal nanowires demonstrating 4 stages of nucleation and growth. **a** denotes low frequency (<0.2 kHz), low voltage (12 $V_{p\text{-}p}$), **b** medium frequency (0.2–0.5 kHz) and volatge (12–18 $V_{p\text{-}p}$), **c** high frequency (>0.5 kHz) and voltage (>20 $V_{p\text{-}p}$)

First on the effect of the electrodeposition frequency, it is seen that the deposition current increases with increasing frequency, which also shortens the filling time. This effect can be attributed to the dielectric properties of the thin aluminium oxide barrier layer at the bottom of the pores [71]. At low frequencies, the barrier layer creates a p–n junction, rendering the template conducting for electrodeposition during half of the sinusoidal period (half period pulses). At higher frequencies all polarization will be suppressed for two reasons. First, the response of the dielectric material changes upon increasing the frequency, i.e., there will be relatively no response to a high frequency field. This means that the relative dielectric constant (ε_r) will approach 1, when the frequency becomes very large [72]. Secondly, it has been reported that pulse electrodeposition in aqueous solutions postpones the hydrogen evolution [73]. This is the case here for electrodeposition of metallic or alloy nanowires into templates, as the higher the frequency the lower the hydrogen ad-atom adsorption and the lower the hydrogen evolution. Therefore, the electrodeposition process is accelerated at higher frequencies for both reasons. The morphological changes during electrodeposition of the nanowires by the frequency are also of critical importance. By increasing AC deposition frequency, ions have no sufficient time to relax and the mechanism will not follow diffusion controlled mechanism which is more common for electrodeposition of metal ions. This is also the case for hydrogen evolution reaction which is delayed making the growth of Co faster. At lower frequency, e.g. an order of 0.2 kHz, all pores are not completely filled and therefore, filling fraction is poor. The growth of the metal nanowires is facilitated with increasing frequency (e.g. up to about 0.5 kHz) owing to higher nucleation rate resulting in easier deposition of the metals into the pores and promoting homogeneous growth of nanowires. Current-time curves are consistent with this scenario. With a further increase of frequency (e.g. to about 1 kHz), surface growth is accelerated. As a general rule, we find that increasing the frequency makes the current at plateau increase while the filling factor and surface overgrowth change.

The influence of the electrodeposition voltage on the growth of nanowires exhibits a similar behavior as seen for the frequency. The only significant trend is a reduction of the growth rate as the deposition potential decreases which might be accompanied by the elimination of perfect four-stages of nucleation and growth on the current transients.

Waveform plays important role in AC electrodeposition of nanowires. In a study on cobalt nanowires, we observed that the four-stage nucleation and growth mechanism is clearly observed for the electrodeposition of nanowires in AAO using a sinusoidal wave form but it is only poorly resolved during growth with a square waveform. In another study reported by Yin et al. [65] the effects of different types of AC waves, including sine, square, triangle, and sawtooth waves, on the electrodeposition were also studied. From the experimental results, it was shown that a high filling ratio can be obtained using a sawtooth wave, triangle wave, or sine wave, while poor results were obtained using a square wave.

5.3.2.3 Pulse Current Template Electrodeposition

Pulse current electrodeposition is a major technique in electrdodepositing of materials such as nanowires in temapltes. Due to high AC voltages employed to electrodeposit into nanoporous templates, hydrogen evolution may take place which decays the deposition process. Therefore, pulse current template electrodeposition has more recently been introduced by Nielsch and coworkers [74] as a reliable method of electrodeposition into high aspect materials and which can compensate for the slow diffusion-driven transport in the pores. The rectifying properties of the barrier layer are here utilized to fill the pores of AAO template by pulsed electrodeposition. In contrast to alternating voltage pulses, in this techniques the current pulses are applied, which allow better control over the deposition parameters, such as deposition rate and ion concentration at the deposition interface. Figure 5.18 shows an example pulse current and voltages recorded during electrodeposition of nickel into an AAO template.

The deposition is based on modulated pulse signals in the microsecond range (Fig. 5.18). During the relatively long pulse of negative current the metal is deposited on the pore ground. In comparison to the direct current deposition, relatively high current densities are applied at the moment of metal deposition, which should increase the number of deposition centers in each pore. As is evident, after a short abrupt rise in the negative deposition voltage (8 ms, $I_{pulse} = \pm 70$ mA/cm^2), this measured absolute value of the signal increases a little further, which is due to the charging of the capacitance of the barrier layer and the depletion of metal ions near the deposition interface. The voltage signal varied between ±8 and ±12 V. After the deposition pulse, a short pulse of positive polarization (2 ms, $V_{pulse} = +3$ V) (Fig. 5.18b) follows to discharge the capacitance of the barrier layer and to immediately interrupt the electric field at the deposition interface. The current is limited to $I_{max} = \pm 70$ mA/cm^2 for both pulses. Consequently, this measured current signal (Fig. 5.18a) deviates slightly from the ideal form of an exponential discharging curve of a capacitor. As mentioned above, the positive pulse also repairs discontinuities in the barrier layer. The delay time, t_{off}, was varied

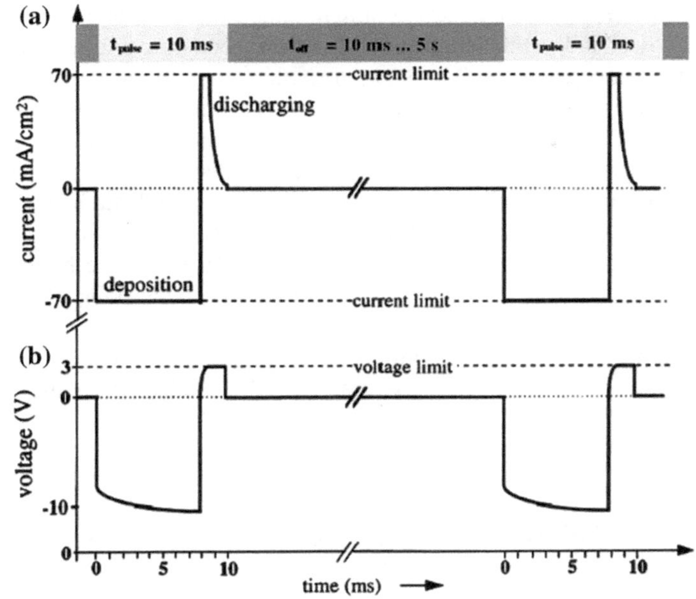

Fig. 5.18 Schematic diagram of the filling of porous alumina structure with nickel by pulsed electrodeposition. The deposition of nickel starts at the thinned barrier layer. Reproduced from [74] with permission of John Wiley and Sons, 2016

between 10 ms and 5 s. During this period the ion concentration recovers. Thus, the concentration of metal ions is high at the bottom of each pore when the subsequent deposition pulse appears. The delay time, t_{off}, improves the homogeneity of the deposition and limits the hydrogen evolution. The deposition was continued up to the drop in the deposition potential, which was due to the beginning of nickel deposition on top of the matrix structure [74].

5.4 Magnetic Nanowires

Ferromagnetic nanowire arrays have been paid much attention owing to their novel magnetic and transport properties because their dimensions approach the critical length scale in magnetism and electronics or spintronics [75]. The ultimate density of magnetic storage media depends on the minimum spacing between magnetic domains, their quality and the interaction between adjacent magnets. Miniaturization of the magnets can now be explored through the use of electrodeposited nanowires, whose typical dimensions may vary routinely from 500 down to 30 nm or even smaller. Generally, the factors determining the final magnetic response of nanowire arrays are, (i) the magnetic nature of individual nanowires, preferably having strong longitudinal magnetic anisotropy for data storage purposes, and (ii) the characteristics of the geometrical arrangement of the nanopore/nanowire array, which determines the strength of the magnetostatic interaction among neighboring nanowires [75]. Porous anodic aluminium oxide (AAO) is considered as particularly attractive template materials for fabricating nanowires, because the pore density is high, the pore distribution is uniform and the diameter of the pores is small [76]. However, due to very high surface pore density (typically 10^{-8}–10^{-12}/cm^2) of AAO templates, a considerable dipolar interaction occurs between nanowires that affects magnetic properties of the filled templates [77] The ideal result which is looked for, from the magnetic point of view, is the fabrication of controlled long range ordered arrays of nanowires with packing density as high as possible where individual nanowires should ideally keep its own magnetic state and consequently give rise to individualized response under the action of external magnetic field.

5.4.1 *Regular Metal and Alloy Magnetic Nanowires*

Regular magnetic nanowires, those with uniform and constant diameter along their total length, have potential applications in microwave electronics as tunable planar devices for very high frequencies [78, 79], storage devices with density beyond 2 Tb/in^2 [80] and biomedical sensors [81] for cancer treatments. Shape anisotropy of regular magnetic nanowires orients the magnetic moments along the nanowires' longitudinal axes forming a high density perpendicular magnetic recording media with small switching fields and high thermal stability. A hexagonal distribution of nanowires induces in-plane magnetic anisotropy with a large degree of magnitude, which is reasonably sufficient for performance required in the information storage media.

3d-Ferromagnetic metals including Co, Ni, Fe and their alloys have been electrodeposited into track-etched and AAO templates mainly using simple salt electrolytes. We will provide here a few examples to explain the general properties of electrodeposited magnetic nanowires.

5.4.1.1 Nickel Nanowires

Nickel nanowires have been electrodeposited via DC and AC techniques in PCTE and AAO templates and their magnetic and structural properties have been extensively studied.

Motoyama et al. [53] have investigated the template electrodeposition mechanism of nickel from a simple salt solution consisting of nickel sulphate and nickel chloride in PCTE with different diameters ranging from 15–200 nm. Figure 5.19 illustrates the current transients recorded during the DC electrodeposition into Pt-Pd coated PCTE templates. It is evident that all the current transients exhibit four stage

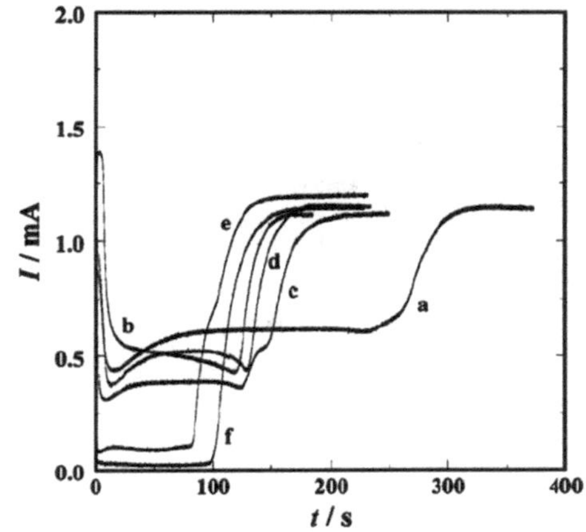

Fig. 5.19 Current transients recorded during electrodeposition of Ni nanowire arrays in PCTE templates with a pore diameter of: *a* 200, *b* 100, *c* 80, *d* 50 *e* 30, and *f* 15 nm under a cathodic potential of −1.0 V from a solution containing 1 M NiSO4, 0.2 M NiCl2, 0.6 M H3BO3 (pH 3.4). Reproduced from [53] with the permission of Elsevier

nucleation and growth mechanism even for the smallest diameter of 15 nm. The change of currents and times of each particular stage during the growth for different pore diameters can be easily interpreted by the cathodic surface area. For instance, the ratio of nominal cathodic surface area for a template with smaller nanopores less than 30 nm in diameter to that for 200 nm pore is less than 0.05, if it is assumed that all pores are effective for the electrodeposition. Thus, it is normal to expect that the measured current in Stage 2 is considerably small in the 15 and 30 nm than that in the 200 nm template. In addition, the beginning of Stage (3) is delayed for the growth in the 200 nm template in comparison with finer pores which is due to the longer pores of this template. The 200 nm PCTE templates used in this study have a thickness of 10 μm, while the others have a thickness of 6 μm. It has been mentioned that hydrogen evolution has no harmful effect on the growth of Ni nanowires [53].

Owing to their flexibility of fabrication and pre diameters, AAO templates have been filled with nickel and the magnetic properties of a range of Ni nanowires electrodeposited in AAO templates have been studied. One principle question about the magnetic properties of arrays of magnetic nanowires is their interaction within the array. Figure 5.6 shows a comparison between two hysteresis loops of Ni nanowire arrays, electrodeposited in AAO templates with different hexagonal order degree defined by the first anodization time. This implies that the ordering degree which can be tuned by first step anodisation, influences the magnetic properties. Clearly, the array with only 3 h of first anodization (lower order degree) presents an effective axial anisotropy less defined than that of the array that underwent 24 h of first anodization [75, 82] (Fig. 5.20).

Details of magnetic properties of nanowires with diameters less than domain wall have been explained by Kroll et al. [83]. The magnetisation curves for Ni and Fe nanowires within AAO membranes with an average pore diameter of 12 nm were studied. This figured behavior clearly indicates two stable orientations of the magnetic moments, namely pointing parallel and antiparallel to the long axis of the

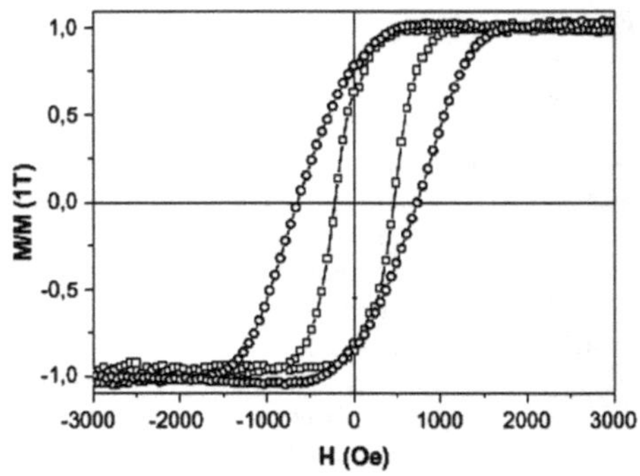

Fig. 5.20 Hysteresis loops of Ni nanowire arrays grown within nanoporous alumina film previously synthesized in oxalic acid solution (pore diameter = 35 nm and spacing = 105 nm). ○ 24 h and □ 3 h of first anodization times. Adapted from [75], Copy right (2016) with kind permission from of Elsevier

wire. These two orientations are separated by an energy barrier. The slight hysteresis that can be seen for the perpendicular measurement is probably due to a slight misalignment of the wires and due to the influence of the rather weak magnetocrystalline anisotropy. Furthermore, it is obvious that the hysteresis loops for the parallel measurement are slightly sheared. This can be attributed to dipolar interactions between the wires due to very high pore density of the template. Thus, it is concluded for Fe and Ni the easy direction for the magnetization lies preferentially parallel to the long axes of the wires.

Sellmyer and co-workers [84] investigated the magnetic properties of Ni nanowires electrodeposited into self-assembled porous alumina arrays. By anodizing aluminum in sulfuric acid and immersing the as-anodized template into phosphoric acid, called *pore-widening* process, for different lengths of time, a wide variety of pore diameters between 8 and 25 nm was obtained. Figure 5.21 illustrates that the coercivity measured along wire axis first increases with the wire diameter, reaches a maximum of 950 Oe near a diameter of 18 nm, and then decreases with further increase of wire diameter. The change of increasing trend of coercivity to decreasing one was suggested to be due to two modes of magnetisation, i.e. curling (vortex) and the coherent rotation (single domain) mode. The maximum of coercivity corresponds to a critical diameter of Ni at which the magnetisation reversal mechanism is switched. At diameters smaller than this critical size, the magnetisation takes place in single domain mode.

Samardak et al. have investigated the AC electrodeposition of nickel nanowires into AAO templates with different pore diameters (20 and 40 nm) [85, 86]. The microstructure and magnetic of Ni nanowires individually and embedded as array in the template were studied.

Figure 5.22 demonstrates the experimental data for nickel nanowire arrays embedded in AAO templates with diameters of 20 and 40 nm. In the array, spatial distribution and dimensional parameters (length and diameter) of Ni nanowires were studied by high resolution scanning electron microscopy (SEM) images and their spectral Fourier analysis. When magnetic field is applied perpendicular to the template plane, we found that $M_r/M_s = 0.9$ and $H_c = 600$ Oe in all samples irrespective to wire (or pore) diameter, as shown in Fig. 5.22a, b. It means that all

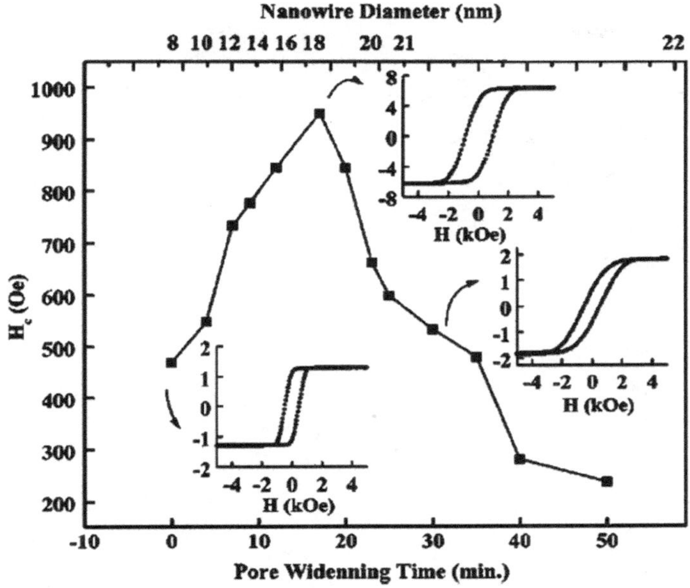

Fig. 5.21 Dependence of the coercivity versus the pore-widening time, i.e., nanowire diameter. The corresponding hysteresis loops for Ni nanowires with diameter of 8, 18, and 21 nm are shown as inset. Reprinted from [84], Copy right (2016) with kind permission from D.J. Sellmyer and American Physical Society (APS)

samples have strong perpendicular magnetic anisotropy with an easy axis aligned along longitudinal axis of the nanowires caused by shape anisotropy.

More complex behavior of magnetic properties was explored at the in-plane measurement geometry. A significant effect of spatial distribution of nanowires on coercive force and magnetization of magnetic arrays as maximum values of $M_r/M_s = 0.2$ and $H_c = 270$ Oe were found for all samples (Fig. 5.22c, d) which significantly smaller than in case of perpendicular anisotropy. As seen in Fig. 5.22e for 20 nm Ni nanowires electrodeposited by 14 $V_{AC\text{-}rms}$, 100 Hz, the Fourier spectrum indicates nearly hexagonal spatial distribution of nanowires which induced the sixfold in-plane configurational anisotropy in the array (Fig. 5.22c). The easy and hard axes have a periodicity of 60° in the substrate plane, in agreement with the hexagonal structure. A deviation of spatial distribution from hexagonal configuration can induce fourfold (Fig. 5.22d) for 40 nm Ni nanowires electrodeposited under 16 $V_{AC\text{-}rms}$, 50 Hz. In addition, it has been observed that a twofold in-plane configurational anisotropy persists for 20 nm Ni nanowires electrodeposited by 14 $V_{AC\text{-}rms}$, 100 Hz (c.f. Fig. 5.22a) and 40 nm Ni nanowires electrodeposited by 14 $V_{AC\text{-}rms}$, 50 Hz (c.f. Fig. 5.22b) [85, 86]

Ognev et al. [87] have studied the microstructure and magnetic domain structure of an individual Ni nanowire electrodeposited. The magnetic nanowires exhibit a strong shape anisotropy. However, the magnetoelastic anisotropy (MEA) induced

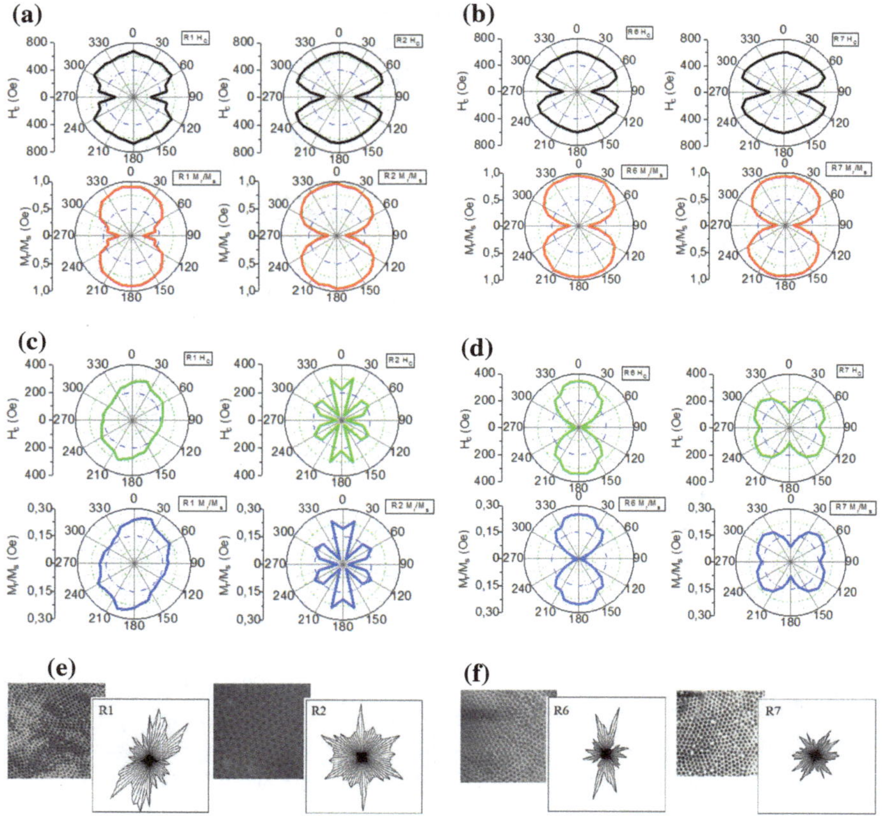

Fig. 5.22 Dependence of magnetic properties on spatial distribution of Ni nanowires with different diameters of 20 nm (**a**, **c**, **e**) and 40 nm (**b**, **d**, **f**) AC electrodeposited under different conditions. **a**, **b** Polar diagrams $H_c = f(\varphi)$ and $M_r/M_s = f(\varphi)$ when H is rotated out-of-plane. **c**, **d** Polar diagrams $H_c = f(\varphi)$ and $M_r/M_s = f(\varphi)$ when H is rotated in-plane. **e**, **f** SEM images and corresponding Fourier transform spectra for samples with different spatial distribution of Ni nanowires. Electrodeposition was carried out in bath was containing of $NiSO_4 \cdot H_2O$ 0.1 M and H_3BO_3 0.5 M with a pH value of 3.5 under different conditions of **a** 14 $V_{AC\text{-}rms}$, 100 Hz for 60 min **b** 14 $V_{AC\text{-}rms}$, 100 Hz for 90 min (**b**) 14 $V_{AC\text{-}rms}$, 50 Hz., **d** 16 $V_{AC\text{-}rms}$, 50 Hz. Reproduced from [85]

by internal and external strains can change the magnetization easy or hard axis in the NWs and hence the orientation of magnetic moments. In this study, Ni nanowires were released from AAO templates on a substrate. Figure 5.23 illustrates transmission electron micrographs (TEM) of Ni nanowires. Analysis of TEM images revealed that the unstrained Ni NWs have a polycrystalline structure with a grain size ranging from 10 to 15 nm (Fig. 5.23a). Selected area electron diffraction pattern (SAED) (Fig. 5.23b) obtained on unstrained parts of NWs demonstrates the characteristic circles indicating the polycrystallinity. In the strained parts of the NWs using the force of an AFM tip, it is observed that a sharp contrast is

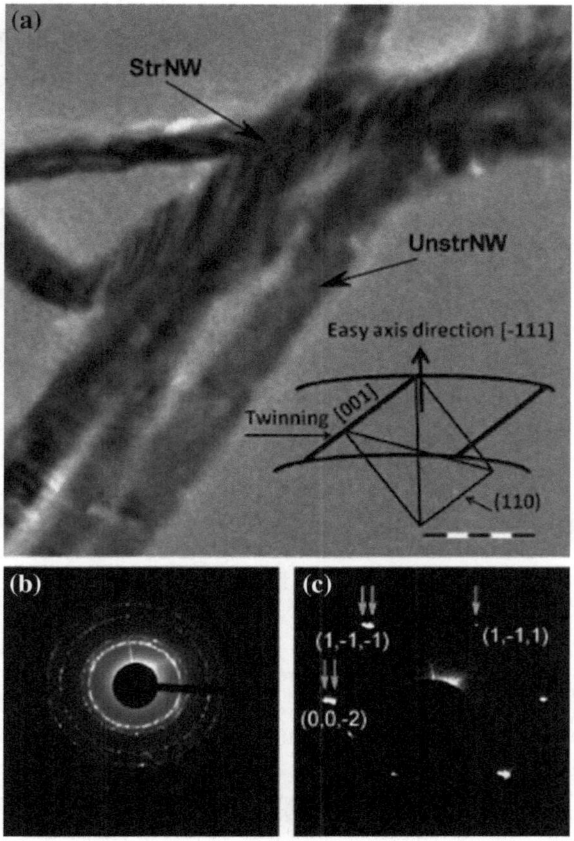

Fig. 5.23 **a** TEM image of Ni nanowires consisting of unstrained and strained parts. Arrows indicate the strained and unstrained nanowires. There is a scheme of (110) plane rotation with pointed orientation of easy axis along crystallographic axis $[\bar{1}11]$ in the strained part. SAED patterns of the unstrained (**b**) and strained (**c**) parts of a Ni NW. The scale bar of the TEM image is 40 nm long. Reproduced from [87] with permission of AIP publishing

distinguished in the image. This observation was interpreted with the crystal twinning due to the presence of some spots in the microdiffraction picture split. There are bright point maxima in the microdiffraction picture corresponding to the (110) *fcc* crystal plane (Fig. 5.23c). It implies that there exists a coarsening of grains as a result of its rotation and coalescence. Therefore, it is understood that the magnetocrystalline anisotropy (MCA) is induced to the stained parts of NWs.

In contrast, the energy of MCA in unstrained polycrystalline parts of NWs is negligible. SAED patterns were analyzed to define the easy axis direction of MCA. Deciphering of the experimental data (inset to Fig. 5.23a) implies that after the rearrangement occurred in the grains, the orientation of easy magnetocrystalline axis $[\bar{1}11]$ aligns in perpendicular direction to NW's long axis. In this case, the formation of quasi-periodic modulation of spin configuration is possible. Moreover, deformation-induced strains in NWs form MEA with an easy axis perpendicular to the tangent line in the nanowire curve.

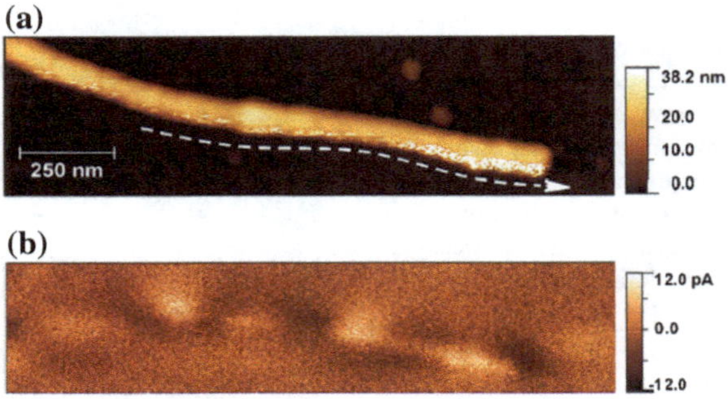

Fig. 5.24 **a** AFM and **b** MFM images of a strained nanowire. The *dashed line* with *arrow* indicates the movement trajectory of the AFM tip during deformation. Reproduced from [87] with the permission of AIP publishing, copyright 2016

The MFM investigation showed that images of unstrained parts of NWs seem to assume a single domain state, while the magnetic microstructure with altering contrast changes is realized in the deformed parts, as shown in Fig. 5.24.

It means that the strain induced MCA and MEA have changed the magnetic structure of the nanowire. As a result, some degree of magnetization rotation is observed which is not completely perpendicular to the NW axis, Fig. 5.24b. The magnesiation mechanism of single Ni nanowires was determined by the competition between magnetocrystalline anisotropy and magnetoelastic anisotropy.

5.4.1.2 Cobalt Nanowires

Cobalt nanowires are significantly influenced by shape anisotropy and temperature- and size dependent magnetocrystalline anisotropy. This orientation of the *c-axis* of an electrodeposited Co nanowire with *hcp* microstructure with respect to the long axis of the wire determines the magnetic properties of Co nanowires. For instance, if it is perpendicular to the wire axis, the easy magnetization axis of Co nanowires becomes perpendicular to the wire length, provided that one neglects the shape anisotropy. In contrast, Co nanowires with *fcc* microstructure exhibit an easy axis of magnetisation parallel to the wire long axis. In practice, the crystalline structure of Co nanowires electrodeposited in AAO templates show co-existence of *hcp* and *fcc* phases. However, many factors influence their content.

Paulus et al. [88] have reported a dependency of the formation of *fcc* phase on the diameter of Co nanowires. The content of *fcc* phase increases as the Co nanowire diameter is decreased. Thus, the shape anisotropy dominates in small Co nanowires even at low temperatures. At high or ambient temperatures, the presence of *fcc* phase does not influence the magnetic anisotropy of Co nanowires. At room temperature, however, the magnetic properties are predominated by the shape anisotropy. But there is seen a competition between shape anisotropy and magnetocrystalline anisotropy for cobalt nanowires at low temperatures can be observed. This can be explained by the strong temperature dependence of the magnetocrystalline anisotropy constant. The competition between the two types of the magnetic anisotropies is also controlled by the diameter of the wires. The effect of magnetocrystalline anisotropy becomes less significant even at low temperatures.

In other studies a relationship was established between the microstructure of Co nanowires and pH. Li et al. showed that Co nanowires electrodeposited in AAO with a diameter ranging 15–70 nm, have a *fcc* structure at pH = 2.7, a mixture of structures of *fcc* and *hcp* at pH = 3.5, and *hcp* structure at pH = 5.0. The effective perpendicular to plane (or along the nanowire axis) was stronger for Co nanowires with a microstructure consisting of fcc element. The coercivity of the fcc structure is also larger than that of the *hcp* structure for Co nanowires with a similar diameter [89]. The effects of the electrolytic bath acidity, or pH, on the magnetic properties in arrays of electrodeposited Co nanowires and their correlation with the crystalline properties have been also studied by other groups [90, 91]. Co-*hcp* phase with the c-axis oriented perpendicular to the wires was established for those nanowires electrodeposited at pH values of 3.8–4.0, and parallel to the wires for pH values ~6.0. This was shown to result in a competitive or extra component of magnetocrystalline anisotropy contribution to the total anisotropy field. Furthermore, the magnetocrystalline anisotropy is negligible of Co nanowires electrodeposited at a pH value of 2.0, indicating a lack of texture in the Co microstructure.

Scarani et al. [92] reported the effect of composition of the electrolyte on *hcp* or *fcc* phase formation in Co nanowires electrodeposited by DC technique in gold coated PCTE templates. In general, it was shown that low deposition rates were shown to be necessary to grow an *hcp* phase. The low deposition rates were obtained by lowering the ion concentration under low deposition voltage (or current). Figure 5.25 illustrates X-ray diffraction patterns of Co nanowires electrodeposited in PCTE template. Homogeneous Co nanowires grown with the high concentration electrolyte exhibited both the *fcc* and *hcp* phases. As is evident from XRD patterns, a relative percentage was obtained 70% *hcp* and 30% *fcc*. The formation of *fcc* Co phase was explained by tensile stress induced phase formation under high growth rates. The *fcc* phase has completely removed for Co nanowires electrodeposited in an inordinately low Co concentration electrolyte.

Another investigation by Kazadi et al. [93] reported the effect of addition of citrate $(NH_4)_2HC_6H_5O_7$ on the phase structure of Co nanowires. Figure 5.26a displays cyclic votammograms recorded in different citrate concentrations into PCTE template. The reduction peak corresponding to cobalt shifts to a more negative potential and a lower current by increasing the citrate concentration. This was attributed to the formation of complex ion between cobalt ions and citrate molecules. In addition, a decrease in the crystalline grain size was observed on the XRD patterns (Fig. 5.26b) by addition of citrate concentration. Furthermore, it reduces the maximum magnetic moment. From the electrochemical point of view, citrate has two major effects: slowing down the synthesis rate and diminishing the size of crystalline domains.

In addition, in the range of the synthesis conditions presented by Kazadi et al. [94], the magnetic behavior of nanowires electrodeposited from a simple cobalt sulphate salt bath have no dependence on the reduction potential and the electrolyte

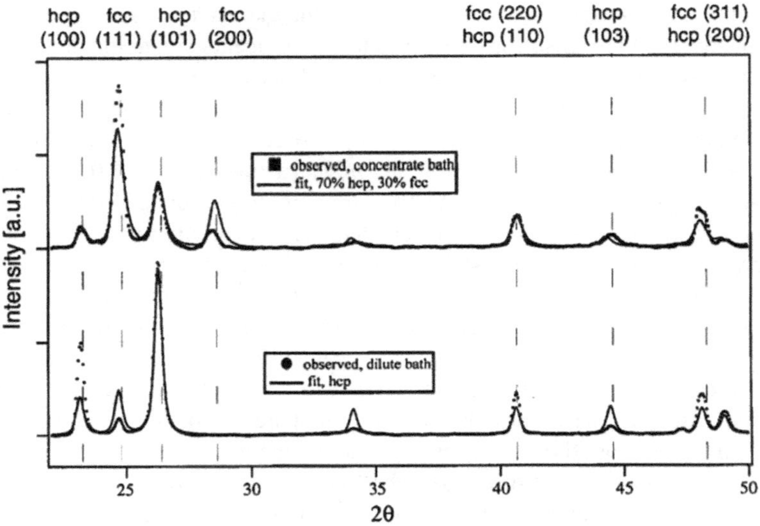

Fig. 5.25 X-ray diffraction patterns for Co nanowires produced with electrolytes of high (*top*) and low (*bottom*) concentrations. $CoSO_4$ concentration was 70–130 mM. Adapted from [92] with permission of Elsevier

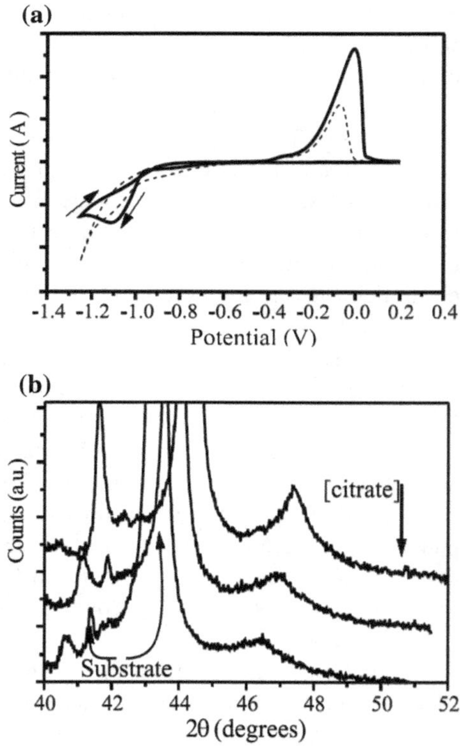

Fig. 5.26 **a** Cyclic voltammograms recorded from cobalt sulphate solution containing citrate compound with 0.25 g/50 mL (*solid line*) and 2.0 g/50 mL (*dash line*). **b** XRD patterns of Co nanowires electrodeposited into PCTE template with different citrate concentrations. Adapted from [93] with the permission of Elsevier, copyright 2016

concentration and the properties only depend on the geometry and aspect ratio of nanowires. It was shown that for Co nanowires with a diameter ranging 100–400 nm, the shortest wires exhibit an easy axis of magnetization parallel to the axes of the wires, and the longest wires perpendicular to the wire axis. This is easily explained by a simple magnetostatic model, which corresponds to competition between the dipolar interactions, demagnetizing field, and the Co magnetocrystalline anisotropy [95]. Dipole-dipole interaction (or magnetostatic interaction) may cause magnetic anisotropy be perpendicular to the wires. Under such circumstance, the effective anisotropy field is reduced by more closed packed nanowires. This has been shown for arrays of magnetic nanodots and nanowires. [96, 97].

5.4.1.3 Cobalt-Nickel Alloy Nanowires

As extensively discussed before, the electrodeposition of Co nanowires produces a polycrystalline structure with wires having many grains with random c-axis orientation. For perpendicular to plane magnetic storage application, many works have been carried out to rotate the c-axis of *hcp* Co grains parallel or perpendicular to the nanowire axis using electrodeposition parameters and solution conditioning as mentioned. Another method is to alloying cobalt with another element whose nanowire exhibit strong perpendicular to plane anisotropy. One possibility to tune the effective anisotropy factor (K_{eff}) is to produce binary alloy nanowires of cobalt and nickel [98–102]. Co-Ni Nanowires with tailored crystal structures can be produced by variation of electrodeposition conditions mainly through an adjusting of bath *pH* and/or temperature, amplitude of electroplating current, aspect ratio of template pores, amplitude and orientation of an external magnetic field during electrodeposition [103, 104]. A little attention has been paid to tune the crystal structure of nanowires by AC electrodeposition through the variation of frequency and waveform of electrodeposition current. A. Samardak and co-workers have investigated the effect of AC electrodeposition parameters on the microstructural and magnetic properties of CoNi nanowires [105].

CoNi nanowires were electrodeposited from an electrolyte containing 28 g/l $CoSO_4{\cdot}7H_2O$, 72 g/l $NiSO_4{\cdot}6H_2O$ and 30 g/l H_3BO_3 into AAO templates synthesized in oxalic acid. AAO templates have an average pore diameter of 40 ± 2 nm and length of 4 ± 0.2 μm and an interpore distance (from center to center) of 100 ± 4 nm. A typical TEM micrograph of isolated CoNi nanowires is shown in Fig. 5.27a which confirmed the diameter and quality of nanowires. Using electron energy-loss spectrometry (EELS) with atomic-level chemical analysis resolution, the uniformity of atomic-levels distribution of both Co and Ni was observed on EELS maps confirming the formation of Co-Ni binary alloy instead of Co and Ni segregated phases, as indicated in Fig. 5.27b. Selected area electron diffraction (SAED) revealed that crystal structure of NWs has a preferred plane orientation (texture) in dependence on the Co-Ni alloy composition. The SAED pattern for the nanowires electrodeposited at f = 800 Hz shows hexagonal symmetry illustrated in the inset in Fig. 5.27a, indicating that wires are polycrystalline

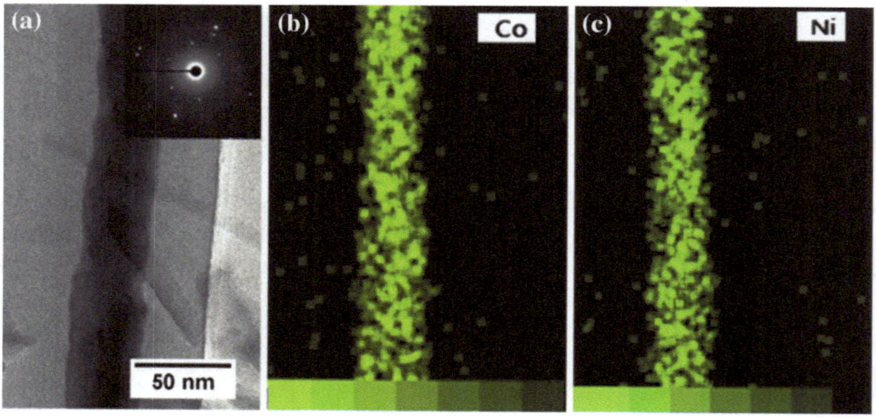

Fig. 5.27 **a** Dark field TEM image of $Co_{81}Ni_{19}$ nanowires electrodeposited by a sinusoidal waveform with a frequency of 800 Hz. The inset consists of a spot micro-diffraction pattern of a selected area of the same nanowire. **b** Atomic level EELS maps of Co (**b**) and Ni (**c**) atoms in the same nanowire. Reproduced from [105] with the permission of Elsevier, copyright 2016

with *hcp*-phase texture having c-axis perpendicular to the wire long axis. SAED patterns of other samples also reveal that electrodeposition at lower frequencies leads to polycrystalline NWs with *fcc-hcp* phase mixture.

Figure 5.28 illustrates XRD patterns of CoNi nanowire arrays electrodeposited under different waveform and frequency. The characteristics of nickel and cobalt diffraction peaks significantly depend upon the change of electrodeposition frequency and waveform. The main Bragg diffraction peaks corresponding to planes of cobalt and nickel are listed in Table 5.2.

The formation of polycrystalline Co-*hcp* showing a texture along [100], and [002] directions is observed for all the samples. This will have a strong effect on magnetocrystalline anisotropy.

Generally, a Bragg diffraction peak appeared at about 44.7° which corresponds to different crystalline scatterings coming from alloy nanowires. This consists of Co (111)-*fcc*, or Ni (111)-*fcc*, Al (002) and also Co (002)-*hcp*, overlapped on each other which individually could not be easily distinguished here. However, the existence of other diffraction peaks helps one understand the crystalline phase evolution in the NW arrays. Bragg peaks related to Ni (111), Ni (220) and Co(220) indices indicate that Co-Ni crystals have also been formed under different electrodeposition conditions in *fcc* phase. It is worth mentioning that Al also contributes to 002, 022 and 103 indices not indicated on the graph coming from aluminium substrate. With decreasing of deposition frequency, the intensity of Co(100)-*hcp* peak rises for both waveforms indicating that *hcp* phase is dominantly formed. For the nanowires electrodeposited at 800 Hz with square waveform, the peak (100) of *hcp*-Co disappears, but intensity of *fcc* Co-Ni peaks do not change. This was interpreted by the overlapping of *fcc* Co-Ni peaks with Al, whose contribution to

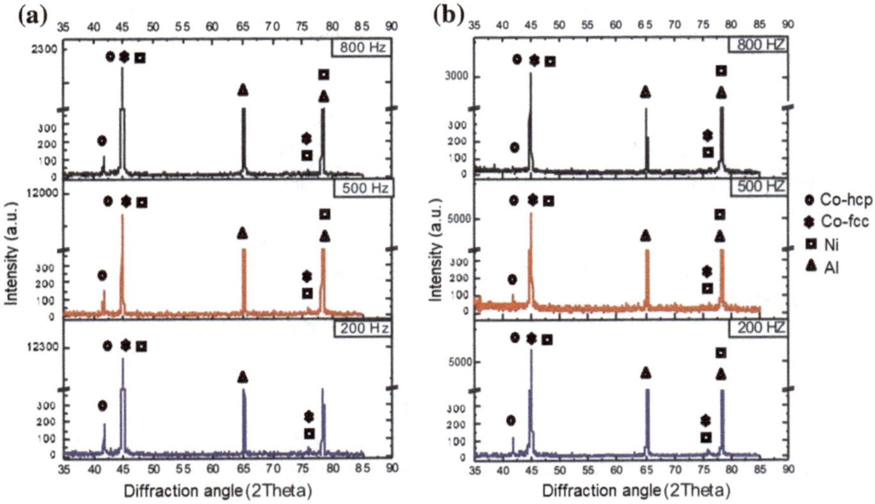

Fig. 5.28 X-ray diffraction patterns obtained for Co-Ni NWs electrodeposited at 18 V, frequency of 200, 500 and 800 Hz using **a** sinusoidal and **b** square waveforms. Main Bragg peaks for Co and Ni are labeled inside the panels, while estimated Al peaks are labeled outside the panels as a guide. Reproduced from [105] with the permission of Elsevier, copyright 2016

Table 5.2 Analysis of peaks observed for Co and Ni in XRD patterns for Co-Ni NW arrays electrodeposited in AAO

Phase	Co-hcp	Co-fcc	Co-fcc	Ni	Ni
2θ (degree)	41.3	44.7	76.1	44.7	76.1
Index	−100	−2	−220	−111	−220

Reproduced from [105] with the permission of Elsevier, copyright 2016

the peak intensity does not change from sample to sample. XRD results show a major change of crystalline structure from *fcc-hcp* to *hcp* phase in Co-Ni alloy nanowires with decreasing of the electrodeposition frequency. It is worth noting that the presence of Al peaks may mislead the precise interpretation of phase evolution with frequency change.

The composition of the nanowires was extracted from energy dispersive spectra (EDS) and summarized in Table 5.3. It is generally known that EDS could be used to analyze the chemical composition of metals and alloys, while the accuracy of the composition is highly dependent on many instrumental and experimental conditions on the atomic weight percentage level. A significant change was observed in the chemical composition of the alloy system and therefore, EDS could be satisfactorily used to evaluate the properties. Composition of Co-Ni NWs is very sensitive to the electrodeposition parameters. Nickel and cobalt content depends on electrodeposition waveform. Co content increases for both waveforms. For higher electrodeposition frequencies, i.e. 800 Hz, alloys become richer in cobalt.

Table 5.3 Composition of Co-Ni NW arrays electrodeposited in AAO under different conditions

Frequency, Hz	Sinusoidal signal			Square signal		
	200	500	800	200	500	800
Co, at.%	68.85	69.06	81.01	65.26	69.64	85.09
Ni, at.%	31.15	30.94	18.99	34.74	30.36	14.91

Reproduced from [105] with the permission of Elsevier, copyright 2016

Table 5.4 In-plane magnetic properties of Co-Ni NW arrays electrodeposited in AAO under different conditions

Frequency, Hz	Sinusoidal signal			Square signal		
	200	500	800	200	500	800
H_c, Oe	530	675	760	540	640	875
M_r/M_s	0.48	0.53	0.5	0.37	0.47	0.58

Reproduced from [105] with the permission of Elsevier, copyright 2016

Magnetic properties of Co-Ni electrodeposited using different AC parameters were investigated. Using the magnetization curves, the polar diagrams of H_c and M_r/M_s measured in the in-plane configuration were plotted. According to these diagrams, no particular anisotropic magnetic behavior inside the arrays was obtained. Magnetic data for the in-plane configuration are listed in Table 5.4. For both electrodeposition waveforms, coercivity (H_c) rises significantly when the deposition frequency increases. It can be linked to the mechanism of nucleation and growth during electrodeposition where the composition of nanowires becomes richer in cobalt and therefore it increases the coercivity. It also may be due to the higher nucleation rate leading to finer crystals which in turn is a reason for increasing of coercivity.

Once the magnetization direction changes to out of plane configuration, an anisotropic behavior of the magnetization reversal in CoNi nanowires has been realized. The magnetic hysteresis loops were measured when the magnetic field was applied out of sample plane and rotated in respect of the wire long axis as shown in Fig. 5.29. In case of out-of-plane rotation the easy axis declines from the parallel to perpendicular to the wire axis with increasing of electrodeposition frequencies.

Figures 5.30 and 5.31 show the dependence of the magnetic properties of CoNi nanowires on AC electrodeposition frequency. In these figures the polar diagrams of H_c and M_r/M_s at out-of-plane measurement configurations are presented. The behaviour of M_r/M_s in dependence on an angle θ between an applied magnetic field and the wire long axis enables one to visualize the direction of easy magnetization axis, indicated by arrows in Figs. 5.30 and 5.31. If to combine microstructure data with magnetic properties study, it was concluded that easy axis of magnetization matches c-axis of Co-*hcp* lattice in the samples deposited at 800 Hz only.

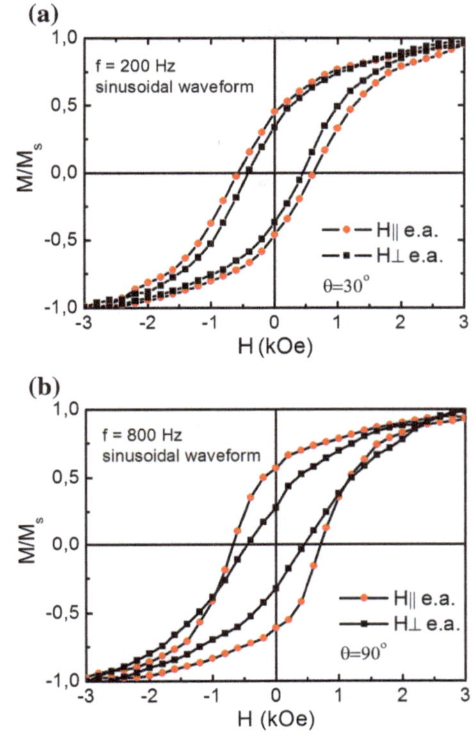

Fig. 5.29 Magnetic hysteresis loops for Co-Ni NWs electrodeposited at different frequencies. Reproduced from [105] with the permission of Elsevier, copyright 2016

Tables 5.5 describe magnetic properties of NWs embedded in AAO templates at the out of plane measurement configuration with respect to the easy axis (e.a.) of magnetization.

This study showed that any increase of electrodeposition frequency with waveforms of sinusoidal and square, the magnetisation easy axis rotates from parallel to perpendicular orientation to the wire axis with respect to *c-axis* of *hcp* phase of Co. As shown above with increase of deposition frequency the composition of Co-Ni alloy changes, Table 5.2. Consequently, it leads to change of phase from *fcc-hcp* mixture at $f = 200$ Hz to *hcp* at $f = 800$ Hz. This transformation occurs in accordance with the phase diagram for bulk Co-Ni alloy [105].

The microdiffraction image shown in the inset of Fig. 5.27a is similar to the diffraction pattern for a single crystal Co nanowire with *hcp* structure and perpendicular to the wire c-axis [106–108]. Authors claimed that easy axis of magnetization was aligned perpendicular to the wire long axis due to the competition between magnetocrystalline anisotropy K_{cry} and shape anisotropy K_{sh}, where $K_{cry} > K_{sh}$ at low temperatures only. Actually, at room temperature the constant of crystallographic anisotropy of bulk Co with *hcp* phase is $K_{cry\_hcp} = 4.1 \times 10^6$ erg/cm^3, which is smaller than $K_{sh} = \pi M_s^2 = 6.2 \times 10^6$ erg/cm^3 for an individual Co nanowire. It has to lead at normal conditions to the easy axis of magnetization

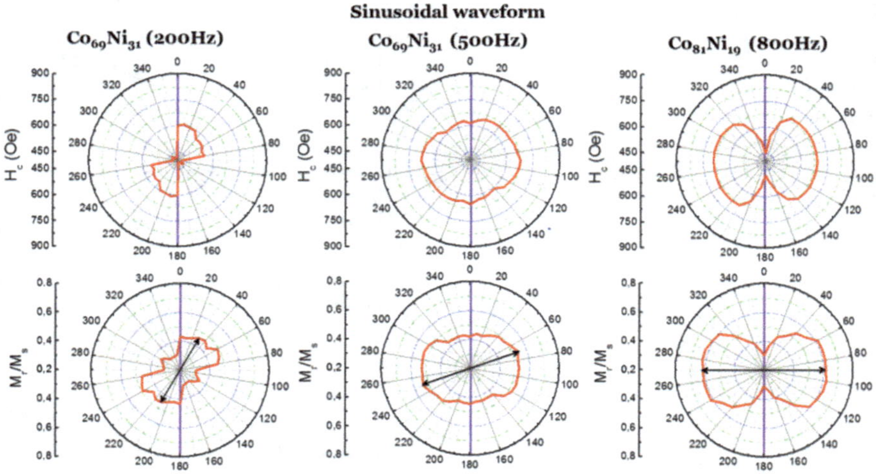

Fig. 5.30 Polar diagrams of H_c and M_r/M_s for CoNi nanowires electrodeposited using sinusoidal signal at different frequencies. Zero and 180° angles correspond to the magnetic field alignment parallel to the wire axis. *Arrows* roughly indicate the easy axis of magnetization. Reproduced from [105] with the permission of Elsevier, copyright 2016

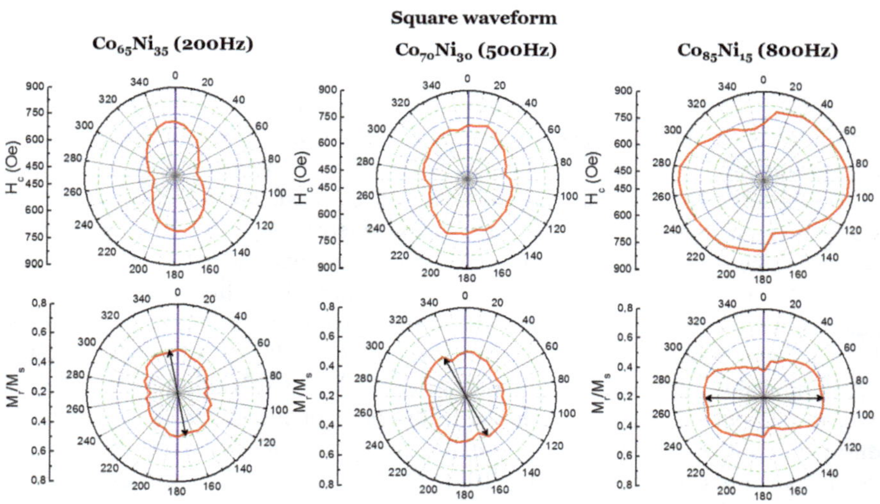

Fig. 5.31 Polar diagrams of H_c and M_r/M_s for CoNi nanowires electrodeposited using square signal at different frequencies. Zero and 180° angles correspond to the magnetic field alignment parallel to the wire axis. *Arrows* roughly indicate the easy axis of magnetization. Reproduced from [105] with the permission of Elsevier, copyright 2016

Table 5.5 Out of Plane magnetic properties of Co-Ni nanowire arrays electrodeposited in AAO under different conditions

Frequency, Hz	Sinusoidal signal			Square signal		
	200	500	800	200	500	800
H_c, Oe (H ‖ e.a.)	600	670	685	720	680	860
M_r/M_s(H ‖ e.a.)	0.47	0.55	0.63	0.5	0.52	0.61
H_c, Oe(H ⊥ e.a.)	425	595	475	535	600	680
M_r/M_s(H ⊥ e.a.)	0.35	0.43	0.3	0.38	0.46	0.39

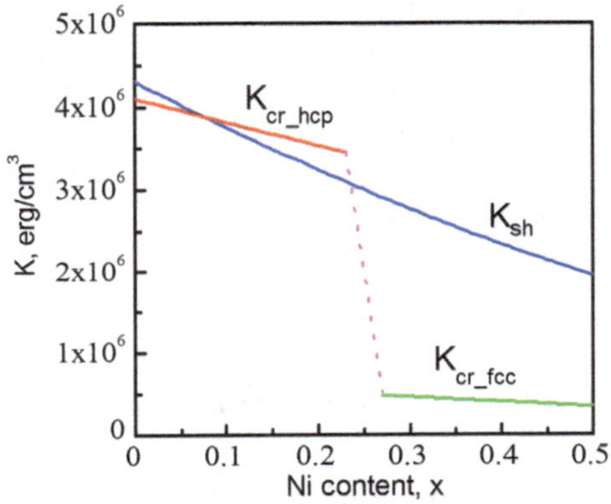

Fig. 5.32 Dependence of magnetic anisotropies on Ni content in Co-Ni alloy. Reproduced from [105] with the permission of Elsevier

aligned along the wire axis. However, in our case we have worked with the array of binary alloy nanowires, where magnetization depends on its composition and dipole-dipole interaction between the nanowires Consequently, shape anisotropy depends on magnetostatic coupling and changes with the content of Co-Ni alloy. It means that K_{sh} can be significantly smaller than K_{cry}.

In Fig. 5.32 the dependence of K_{sh} (blue bold line) for $Co_{1-x}Ni_x$ alloy is presented. This dependence was calculated as $K_{sh} = \pi(M_s(Ni)\cdot x + M_s(Co)\cdot(1 - x))^2(1 - 3p)$, where $M_s(Ni) = 500$ Gs, $M_s(Co) = 1400$ Gs, x is the Ni content in a Co-Ni nanowire and p is the porosity of membrane. The factor $(1 - 3p)$ takes into account the dipole-dipole interaction between wires and it was estimated from SEM images of membrane surface [108]. The behavior of magnetocrystalline anisotropy was defined in accordance with the phase diagram of Co-Ni binary alloy. According to this diagram, if the Ni content is below 25 at.%, then *hcp* structure is dominant. At the Ni content higher than 27 at.% *fcc* phase is prevalent. The mixture of *hcp* and *fcc* is realized at the Ni content lying between 25 and 27 at.%. The red line in Fig. 5.32 indicates K_{cry_hcp}, which correspond to *hcp* phase of Co. At 27 at. % of Ni content, Co-Ni alloy has *fcc* structure associated with significantly smaller magnetocrystalline anisotropy marked as K_{cry_fcc} in Fig. 5.32. As our previous study showed that for pure Co and Co-Ni alloy nanowires with the very low Ni content ($x < 0.07$) the easy axis of magnetization aligned along the wire axis.

A case, when the easy axis of magnetization aligned perpendicular to the wire long axis, corresponds to a region where $K_{sh} < K_{cry\_hcp}$. As seen from Fig. 7, this case is realized for the low Ni content (0.07 < x < 0.24) in accordance with observed behavior of easy axis. In alloys with x > 0.24 $K_{sh} > K_{cry\_fcc}$ and the easy axis tends to diverge from perpendicular to the wire direction and finally to be along the wire axis at higher Ni content [85]. This also agrees well with the data of Figs. 5.30 and 5.31.

Electrodeposition frequency variation leads to change of atomic percentage of Co and Ni. Rising in the frequency assists to increase of Co content. The crystal structure changes from *fcc-hcp* mixture to *hcp* phase with increase of Co percentage. Simultaneously easy axis of magnetization rotates from the being parallel to the perpendicular direction to nanowire long axis. This is discussed as result of competition the magnetocrystalline anisotropy of *hcp* crystallites formed with c-axes perpendicular to the wire axis and the shape magnetic anisotropy of the Co-Ni nanowire. The value of coercive force is sensitive to the electrodeposition signal waveform. Our results open up a way for tuning electrodeposition of binary alloy nanowires using frequency and waveform manipulation of AC current [105].

5.4.2 Core-Shell Nanowire Arrays

5.4.2.1 Core-Shell Structure in Co-P Nanowires

Nasirpouri et al. have recently exploited the possibility of formation of core-shell nanowires via AC electrodeposition into AAO templates [109]. In this work, the formation of an unusual core-shell structure in Co-P nanowires with high coercivity values comparable to that of cobalt nanowires, along with perpendicular to plane anisotropy. The interpretation was based on a careful analysis of experimental magnetization data, including hysteresis loops and magnetic torque measurements, as will be discussed here.

Co-P alloy nanowires were AC-electrodeposited into the AAO templates synthesized in oxalic acid at room temperature in an electrolyte consisting of 0.1 M $CoSO_4{\cdot}7H_2O$, 0.5 M boric acid and 0, 5, 15, and 25 g/l NaH_2PO_2 at a pH close to 4. A sinusoidal waveform was employed with voltage amplitude ranging from 12 to 15 V and a frequency ranging from 50 to 400 Hz. Prior to electrodeposition the AAO template was sonicated for 10 min in the electrolyte to facilitate wetting of the nanopores. Table 5.6 illustrates the chemical composition and magnetic properties of the electrodeposited Co-P nanowires.

XRD pattern of $Co_{100-x}P_x$ (X < 10%) nanowires embedded in AAO and electrodeposited at 15 V and 200 Hz using the sinusoidal waveform is shown in Fig. 5.33. For Co nanowires (x = 0), *hcp* crystalline structure is observed with clear (001), (200) and (110) Bragg diffraction peaks. Co (200) peaks are overlapped with a Bragg peak belonging to the aluminum substrate. Aluminum peaks are annihilated in templates with higher filling fraction. The electrodeposited nanowires containing

Table 5.6 Phosphor content and the corresponding coercivity and saturation fields of Co-P nanowires electrodeposited with a 200 Hz 15 V sine wave from different composition solutions

$NaH_2PO_2{\cdot}H_2O$, g/l	Phosphor content x, wt%	Parallel coercivity $H_c(\|\|)$, Oe	Perpendicular coercivity $H_c(\perp)$, Oe	Parallel saturation field $H_s(\|\|)$, Oe	Perpendicular saturation field $H_s(\perp)$, Oe
0	0	330 ± 20	865 ± 30	9000 ± 100	8600 ± 100
5	6.8	1260 ± 35	2380 ± 50	8500 ± 100	7000 ± 100
15	8.2	920 ± 30	1545 ± 40	6800 ± 80	6600 ± 80
25	9.3	645 ± 30	1005 ± 40	5700 ± 80	5600 ± 80

⊥ and || denote perpendicular and parallel directions to AAO substrate. Reproduced from [109] with the permission of AIP publishing

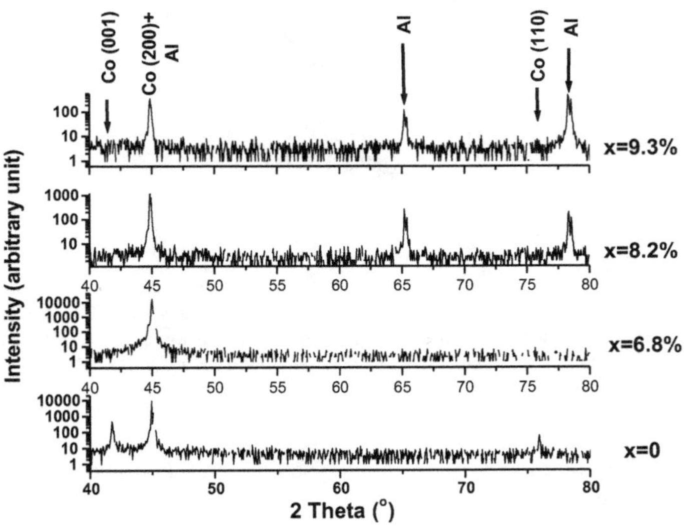

Fig. 5.33 X-ray diffraction patterns obtained from an array of $Co_{100-x}P_x$ nanowires with different phosphorcontent (x) electrodeposited in AAO templates using a 15 V, 200 Hz sinusoidal waveform from solutions containing 0, 5, 15 and 25 g/l sodium hypophosphate. Reproduced from [109] with the permission of AIP publishing

phosphor tend to lose crystallinity as (110) and (100) Bragg diffraction peaks disappear.

The disappearance of the characteristic Bragg peaks alone is not enough to confirm the presence of an amorphous phase. As it will be discussed later, if just an amorphous phase was present only, a broad halo would be observed in selected area diffraction (SAED) patterns. However, we also observe bright diffraction spots corresponding to the presence of a polycrystalline phase. This observation may be ascribed to the electrodeposition of Co-P nanowire arrays under non-equilibrium

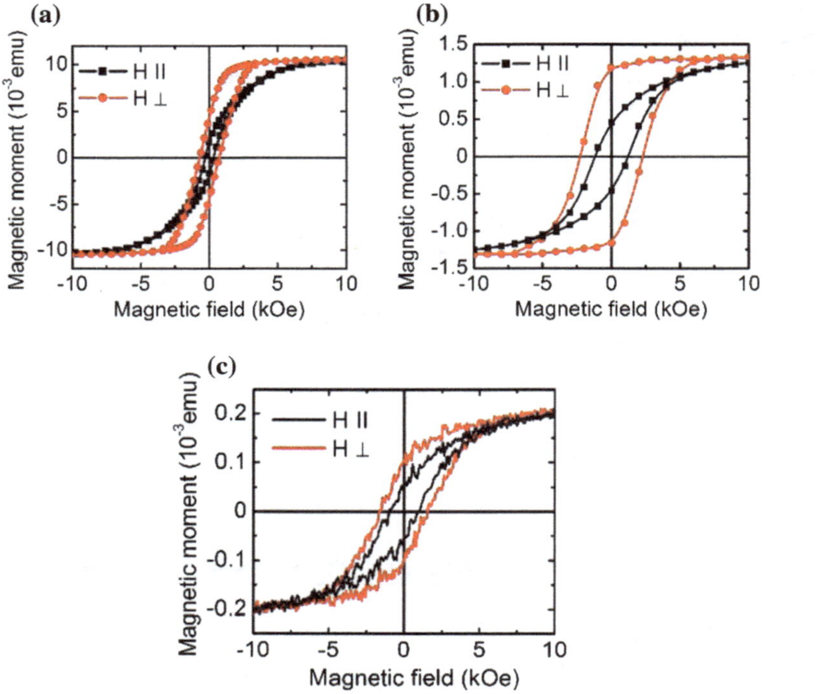

Fig. 5.34 Room-temperature hysteresis loops measured for arrays of **a** Co, **b** $Co_{93.2}P_{6.8}$ and **c** $Co_{91.8}P_{8.2}$ nanowires electrodeposited in AAO templates using a 15 V, 200 Hz sinusoidal waveform.⊥ and || denote perpendicular and parallel directions to AAO substrate. Reproduced from [109] with the permission of AIP publishing

conditions where the ratio between polycrystalline and amorphous phases is controlled by the P composition.

Figure 5.34 shows the magnetic hysteresis loops measured for $Co_{100-x}P_x$ nanowires with $x = 0$, 6.8 and 8.2 wt% for an external magnetic field applied parallel (||) and perpendicular (⊥) to the plane of the AAO template. For pure Co nanowires (Fig. 5.34a) the magnetic moment at saturation is 10.5 ± 0.1 μemu and the coercivity in different geometries is $H_c(||) = 330 \pm 20$ Oe and $H_c(\perp) = 865 \pm 30$ Oe. The addition of phosphor leads to variation of magnetic properties of Co-based alloy nanowires as shown in Tables 5.6 and 5.7. In Fig. 5.34b, the saturation magnetic moment decreases compared with that of pure Co nanowires while the coercivity significantly increases for both magnetic configurations. Further increase of phosphor content, however, significantly reduces the saturation magnetic moment (to 200 ± 5 μemu, which is 50 time smaller than that of Co wires) as well as the coercivity. The perpendicular to plane magnetic anisotropy in conjunction with the high coercivity makes Co-P systems very interesting, not only for high density data storage applications, but also for designing microwave filters, whose absorption bands can be selected as a function of the materials within the arrays.

As shown in Fig. 5.34, for Co and Co-P nanowire arrays the M-H loops exhibit large hysteresis along the parallel and perpendicular directions. All samples are more easily magnetized along the long axis of the nanowires rather than perpendicular to it due to the strong shape anisotropy. As expected from the theoretical predictions for bulk materials, the saturation field can be defined as $2\pi M_s$, (M_s is the magnetisation at saturation) if the shape anisotropy only exists. For Co and CoP nanowires H_s = 8792 Oe (700 kA m^{-1}) and 5100–6900 Oe (405–550 kA m^{-1}) [110]. This shows that theoretical predictions are consistent with experimental values recorded in Table 5.4 and Fig. 5.34. The reduction in saturation field with the increase of P content can be attributed to the reduced value of Ms. Magnetocrystalline anisotropy of CoP nanowires is weak or even absent due to the presence of an amorphous phase in the crystal structure. Small reduction in saturation field can be attributed to the likely presence of other in-plane anisotropy fields such as magnetocrystalline anisotropy, magnetostatic interactions between wires, and/ora reduction of Ms in nanostructures. The magnetostatic interactions between nanowires in an array are necessarily negligible as there is no significant evidence of sheared hysteresis loops [111].

As shown in Table 5.4, the difference in values of $H_c(\|)$ and $H_c(\perp)$ for $Co_{93.2}P_{6.8}$ nanowires is much larger than for pure Co nanowires. Assuming that the amorphisation of the crystalline structure in Co-P nanowires, and the loss of magnetocrystalline anisotropy, is solely responsible for this change in magnetic anisotropy, Co-P nanowires would be expected to exhibit stronger perpendicular to plane magnetic anisotropy than Co nanowires. However, we find that the addition of phosphor does not lead to this outcome.

Table 5.7 summarizes some data extracted from the magnetic hysteresis loops as well as torque measurements, along with the morphological characteristics of Co and $Co_{93.2}P_{6.8}$ samples. Data for the volume of nanowires, V_w, and saturation moment, m_s, allow us to estimate the saturation magnetization according to $M_s = m_s/V_w$ and compare it with literature values for Co-P alloys. If one knows the magnetic torque L, magnetic material volume V_w and saturation magnetization M_s, then it is possible to estimate the perpendicular magnetic anisotropy field as $H_{a\perp} = 2L/(V_w M_s)$ and to compare with $H_{a\perp}$ found from the hysteresis loops using $H_{a\perp} = \int \left[\left(M_\perp(H) - M_\|(H)\right)/M_s\right] dH$ ($K_\perp = M_s H_{a\perp}/2$). Various estimates of the magnetization are listed in Table 5.7 based on different assumptions about the volume occupied by the ferromagnetic nanowires. The influence of the addition of P in Co-P alloys manifests itself in a maximum for H_c and M_r/M_s at x = 6.8 wt% of P. The value of $H_{a\perp} = 3800 \pm 200$ Oe for $Co_{93.2}P_{6.8}$ nanowires is almost the same as for pure Co wires $H_{a\perp} = 3700 \pm 200$ Oe. We conclude that $Co_{93.2}P_{6.8}$ nanowires have optimal hysteresis properties for practical applications in data storage and microwave devices, for instance, filters for wireless communication and automotive systems, where a demand exists for smaller sizes and broader bandwidths. The so-called magnetic "nanowired" substrates (MNWS) [109], consisting

Table 5.7 Experimental magnetic data and morphological characterization for Co and Co-P nanowires

	Co	$Co_{93.2}P_{6.8}$
Saturation moment m_s (emu)	1.10E−02	1.30E−03
Torque L (erg)	31.1	2.1
Anisotropy field from loops $H_{a\perp}$ (Oe)	3700 ± 200	3800 ± 200
Pore diameter d (nm)	30 ± 2	30 ± 2
Core diameter D_1 (nm)	30 ± 2	5 ± 1
First shell layer diameter D_2 (nm)	...	20 ± 2
Second shell layer diameter D_3 (nm)	...	30 ± 2
Porosity v	0.095	0.095
Substrate square S (mm^2)	16 ± 1	12 ± 1
Pore volume (cm^3)	7.6E−06	5.70E−06
Core volume V_1 (cm^3)	7.6E−06	0.16E−06
First shell layer volume V_2 (cm^3)	...	2.53E−06
Second shell layer volume V_3 (cm^3)	...	5.70E-06
M_s of V_1 (Gs)	1450 ± 150	8200 ± 700
M_s of V_2 (Gs)		500 ± 100
M_s of V_3 (Gs)		230 ± 40
M, (literature) (Gs) [14]		800
M, (literature) (Gs) [15]	1420	1100
Anisotropy field from torque of V_1 (Oe)	5800 ± 800	52000 ± 2000
Anisotropy field from torque of V_2 (Oe)		3200 ± 800
Anisotropy held from torque of V_3 (Oe)		1400 ± 400

Reproduced from [109] with the permission of AIP publishing

of nanowires embedded in a porous template, are ideal candidates for this application due to their desirable materials properties and the possibility to build tunable nanomagnetic devices applicable for the desired frequency range. The advantages of CoP materials compared to classical ferrites are a higher operation frequency, a higher saturation magnetization and resonant frequency, observed without any applied external DC field [112, 113].

The magnetic properties of nanowires as a function of the addition of P have been analyzed by the microstructural features of electrodeposited nanowires. Figure 5.35 shows TEM images taken from typical Co and $Co_{93.2}P_{6.8}$ nanowires and the corresponding SAED patterns. For pure Co nanowires we observe a uniform polycrystalline structure, an interpretation supported by the SAED pattern (Fig. 5.36a). This is consistent with previous report on Co-based alloy nanowires with fine grain polycrystalline nature [114]. However, this is not the case for $Co_{93.2}P_{6.8}$ nanowires, when one observes a form of coaxial structure. As seen in Fig. 5.35b, the wire consists of a core and two shell layers composed of various materials with different image contrast. The corresponding SAED pattern

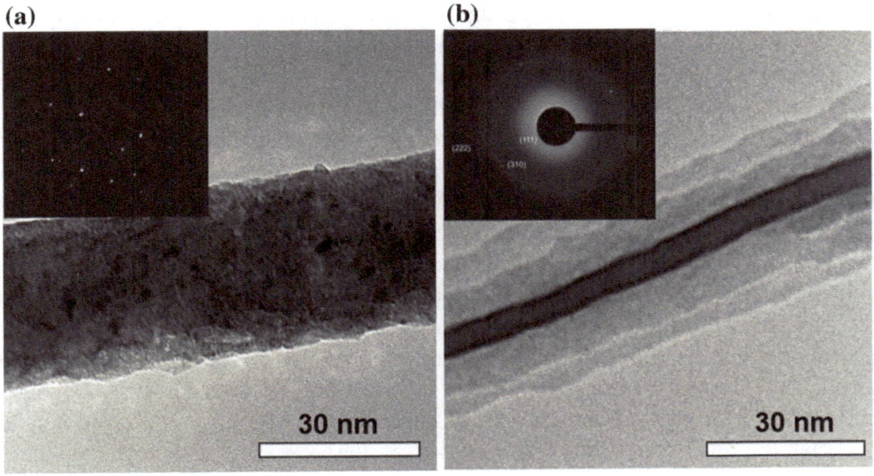

Fig. 5.35 Typical TEM images and selected area diffraction patterns for **a** Co and **b** $Co_{93.2}P_{6.8}$ nanowires electrodeposited in AAO templates using a 15 V, 200 Hz sinusoidal waveform. Reproduced from [109] with the permission of AIP publishing

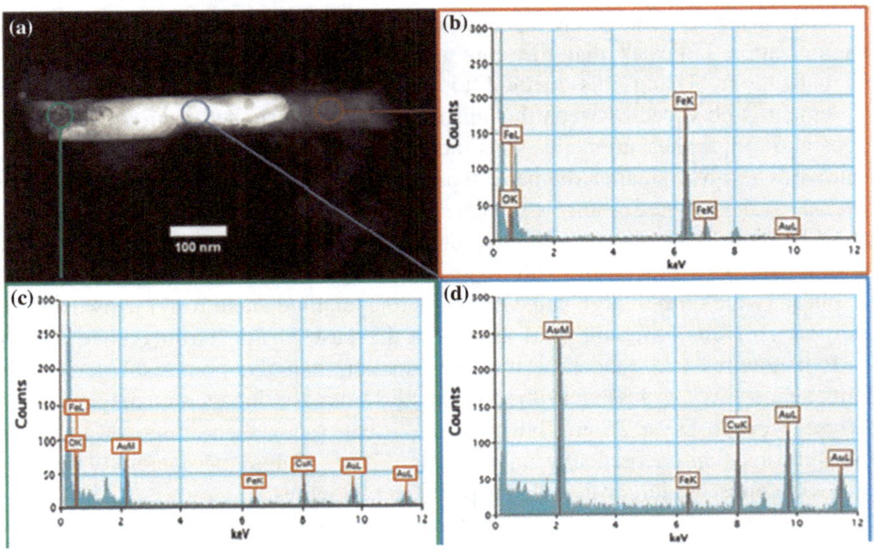

Fig. 5.36 **a** Scanning transmission electron micrograph of a single Fe@Au nanowire, EDX spectrum of **a** bare core shown in *red colour*, **b** partially coveredcore and **c** Fe@Au core-shell region shown in *blue*. Reproduced from [118] with the permission of Elsevier and M. Vazquez

demonstrates a few bright spots on the halo background, pointing out an existence of the mixture of crystalline and amorphous phases.

To estimate M_s for Co and Co-P nanowires, we have calculated the volume occupied by nanowires on a substrate with porosity $\nu = 0.095$ as $V_w = LS\nu(D_i/d)^2$, where $L = 5$ µm is the pore length (assumed equal to the nanowire length), S is the substrate area, D_i is the assumed diameter of a ferromagnetic nanowire and d is the pore diameter (30 nm for the samples). As seen in Table 5.7 for Co nanowires, the experimental value of M_s is in good agreement with literature data, indicating that the pores are fully and uniformly filled with pure Co. The inferred values of anisotropy, measured from hysteresis loops and magnetic torque measurements, are not consistent. This may be because the magnetization reversal for nanowires with this diameter is not limited to coherent magnetization rotation and mainly occurs through the domain wall nucleation and propagation [76, 100, 115].

In case of $Co_{93.2}P_{6.8}$ nanowires we assume that the volume V_w of ferromagnetic material is equal to: (1) the volume, V_1, of the core with the actual diameter $D_1 = 5$ nm or (2) the volume, V_2, of the core together with the first shell layer having $D_2 = 20$ nm or (3) the volume V_3 of the whole coaxial nanowire with $D_3 = 30$ nm. Three scenarios for the estimations were compared and summarized in Table 5.2. The first assumption, when $V_w = V_1$, was discarded, because it yielded physically meaningless value $M_s \sim 8200$ Gs. The value of M_s calculated using an assumption $V_w = V_2$ is closest to the literature data, though still smaller. This might be expected for a coaxial structure where the core has larger magnetization that the first shell layer. In general, the discrepancy between the experimental and literature magnetization indicates that the core and the shell have significantly different magnetizations. Using this method to estimate the volume fraction, the good agreement is observed between the anisotropy field determined from the hysteresis loops and torque measurements. This indicates that magnetization reversal in these nanowires (with a smaller diameter than for Co nanowires with inhomogeneous reversal of the magnetization) is mediated by the uniform (coherent) rotation of the magnetization. It also helps to explain the sharp rise of H_c for Co-P nanowires. On the one hand, CoP nanowires with the smallest P content have the largest fraction of *fcc* phase (which has a large value of magnetocrystalline anisotropy). However, the addition of a small amount of P leads to a decrease of the ferromagnetic core in Co-P nanowires and rendering it much smaller than the pore diameter. If the diameter of NWs is less than the critical diameter for incoherent magnetization reversal (for Co $D_{cr} = 25$ nm, but for Co-P it can vary), then domain wall nucleation becomes an energetically unfavorable process. Thus, the coercivity of Co-P nanowires must be larger than H_c in pure Co nanowires.

Hence the analysis indicates that the core, and the adjacent shell layer, are ferromagnetic, and the outer shell layer is weakly magnetic or possibly nonmagnetic (e.g, Co_2P) as deduced from TEM SAED as discussed above. A few diffraction spots are seen in these core/shell structures which might possibly arise from a crystalline Co_2P phase which has an orthorhombic lattice and is a Pauli paramagnet [116]. It was suggested that the core consists of a Co-P alloy with a higher saturation magnetization than the first shell layer, which is possibly a mixture of crystalline and amorphous Co-P phases. For Co-P with x = 8.2 and 9.3 wt% the coaxial structure was still clearly observed, although the core becomes thinner, leading to a reduced nanowire magnetization and coercive field.

As the magnetocrystalline anisotropy of CoP nanowires is weak or even absent due to the presence of an amorphous phase in the crystal structure, there is a simple and reliable method to estimate the effective anisotropy of a nanowire array. This method is based on the assumption that, due to the dipolar interaction between nanowires, the shape anisotropy of nanowires is reduced by factor ($1–3\,\vartheta$), where ϑ is the porosity of the template [102, 105]. For the self-ordering regimes of pores formation ϑ is 0.1 [117]. If one assumes that for Co-P the effective diameter of magnetic nanowires decreases, then produces proportionally to the diameter squared. For $Co_{93.2}P_{6.8}$ nanowires, the effective diameter of the ferromagnetic core is about 5 nm. It means that ϑ decreases 36 times compared to the pure Co nanowires with diameter of 30 nm. As a result, the value of $3\,\vartheta$ becomes very small. Thus, magnetostatic interaction between the $Co_{93.2}P_{6.8}$ nanowires in the array is negligible. This conclusion is supported by the hysteresis loops, which are not significantly sheared as evidence for dipolar–dipolar interaction.

5.4.2.2 Core-Shell TM@Au Nanowires

In a recent report by Ovejero et al. [118] successful electrochemical synthesizing of TM@Au core–shell magnetic nanowires into AAO templates was mentioned for biological functionalization application as Au protects the harmful effects of TM elements (TM stands for transition metal). In this work, Fe, Ni, CoFe were electrodeposited as the core material. The method of electrodeposition involves an optimized two-step potentiostatic deposition. In the first step, inner walls of nanopores in AAO were carefully electroplated with Au forming nanotubular structure. In the second step TM were filled inside the AU nanotubes. Potentiostatic electrodeposition technique at different overpotentials was used to control the morphology of Au. It is crucial to achieve a continuous Au film electrodeposited on the inner walls of AAO nanopores. Au electrodeposition was performed in an electrolyte consisting of 0.93 g/l $HAuCl_4.3H_2O$ and 30 g/l H_3BO_3 in a range of potential (−0.35 to −1.25 V). It was shown that the optimum condition of the formation of a complete Au film involves the electrodeposition at −1.0 V. However, it is not clear why the Au film is only deposited on the inner walls in the form of tube. This needs further research to understand the electrodeposition mechanism of Au nanotubes on AAO pore walls. Figure 5.36 illustrates an individual Fe@Au core-shell nanowire with EDS spectra recorded at different sections of the wire.

The presence of Au film as a shell encapsulating the magnetic has been shown to change the magnetic properties of TM core-shell nanowires compared with TM nanowires. The magnetisation curves for TM@Au and TM nanowire arrays embedded in AAO templates are shown in Fig. 5.37. The sheared hysteresis loops formed due to the presence of a strong magnetostatic interaction between the wires

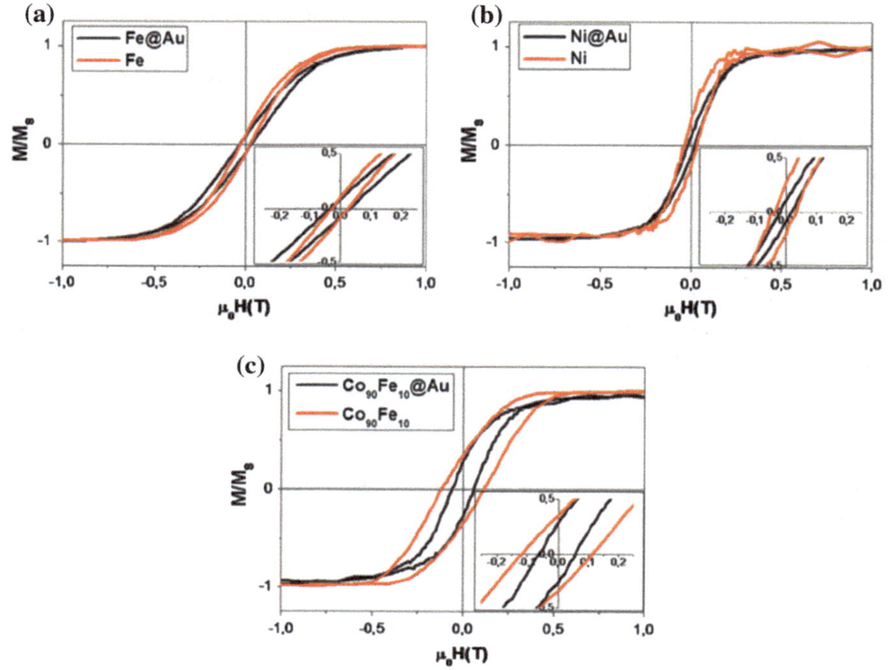

Fig. 5.37 Magnetic hysteresis loops of **a** Fe and Fe@Au, **b** CoFe and CoFe@Au and **c** Ni and Ni@Au nanowires. Adapted from [118] with the permission of Elsevier and M Vazquez

in the array. For Fe@Au and Ni@Au core-shell nanowires, except a small reduction of susceptibility values, the magnetic properties are nearly similar to the Fe and Ni nanowires. This has slightly enhanced the magnetic anisotropy of the core-shell nanowires. However, for $Co_{90}Fe_{10}$@Au nanowires, the coercivity has decreased from 0.12 to 0.06 T and the initial susceptibility has increased. This was explained to be due to an enhanced axial magnetic anisotropy likely by the shape of core-shell nanowires.

5.4.3 *Multilayered Nanowires*

The discovery of giant magnetoresistance (GMR) by Albert Fert and co-workers [119] in artificially fabricated Fe/Cr multilayers leading to win the Noble prize of 2007 in physics, has pioneered a great deal of research effort to enhance the physics behind this phenomena and develop novel fabrication techniques. In a very earlier time in 1921 by Bund, electrodeposition had been a technique for depositing metal multilayers by successive deposition from two different electrolytes [120]. The more recent progresses of the electrodeposition of multilayers [120–124] revealed the strengths of this deposition technique so that W. Schwarzacher and coworkers at Bristol discovered current-in-plane CIP-GMR in CoNiCu/Cu multilayers in 1993 [125]. Several review articles and book chapters have already been published on the electrodeposited multilayers. For more details see [126–128].

In the CIP geometry, it is crucial that the anti-ferromagnetic coupling takes place between two ferromagnetic layers through nonmagnetic layer whose optimum width must not exceed the electron mean free path (MFP) [129, 130] for efficient GMR devices. Valet and Fert [131] theoretically proposed another geometry namely perpendicular to plane (CPP) in which the spin diffusion length (SDL) becomes the critical length which is generally an order greater than MFP. However, extremely low resistance in a direction perpendicular to the plane of multilayers makes it very difficult to exploit efficient devices for applications. Electrodeposition of multilayered nanowires has been a breakthrough for CPP-GMR device research and technology demonstrated by Piraux et al. [132] and Blondel et al. [133]. These nanowires provide very small cross sections that facilitate magnetoresistance measurements along wire axis as well as enables electron transport almost perpendicular to the interfaces of layers. Therefore, an ideal CPP geometry is realised.

5.4.3.1 Electrodeposition Techniques of Multilayered Nanowires

In general two major techniques have been known for the electrodeposition of magnetic multilayered nanowires including *dual bath* and *single bath* method. The former has a relatively longer history as its application goes back to the Bund's report in1921 on Cu/Ni multilayers. This method involves the electrodeposition process of each layer in separate electrolytes and the sample exchanged between the two baths. The later, the single bath method, is performed in one distinct electrolyte in which ions of all the metals to be electrochemically reduced exist. The composition of the film is so modulated by switching deposition potential or current between two values each of which corresponds to that of each layer. To justify this method, one need to consider the formal equilibrium potential and mechanism of metal or alloy to be reduced in each layer. For A/B multilayer system where A is less noble than B, B is reduced at a potential more positive than the reduction potential of A forming a pure layer of B. When the potential is switched to the reduction potential of A, co-electrodeposition of A-B certainly takes place. For a graphical explanation of this process, See Chap. 3, Sect. 3.2.6.1: alloy electrodeposition. However, the magnitude of B deposited in A-B alloy is controlled as low as possible by choosing the deposition mechanism to be diffusion limited process. One solution is to keep the concentration of B as low as possible in the electrolyte, the resulting alloy will be predominantly A-rich. First reports on the template electrodeposition of magnetic multilayered nanowires systems were demonstrated to have used the single bath method technique for the electrodeposition into PCTE membranes. However, other reports exist for AAO or other templates later.

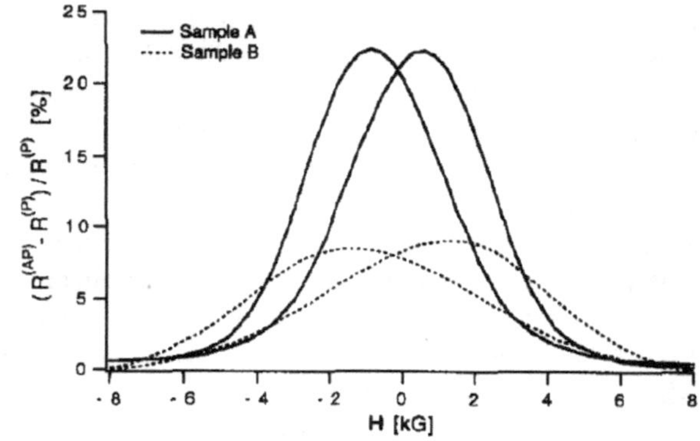

Fig. 5.38 CPP-MR measured at 20 K for multilayered Co/Cu nanowires with a diameter of 80 nm wires into PCTE. *Solid line:* Sample (A), 8 nm of layers thickness, produced by a single bath technique. *Dotted line*: Sample (B), 10 nm of layer thickness, produced by the dual bath method. Reprinted from [134], Copy right (2007) with kind permission of Elsevier

Blondel et al. [134] compared GMR values in Co/Cu multilayred nanowires electrodeposited by single and dual bath methods. In spite of neglecting the co-deposition of Cu with Co, it was pointed out that the CPP-GMR values of the nanowires electrodeposited by dual bath are much lower than those by single bath method. This was attributed to the incorporation of impurities and non-smooth interfaces of multilayers during exchanging the electrolytes. Figure 5.38 illustrates the result of this study on the variation of magnetoresistance (MR) as a function of applied magnetic field.

5.4.3.2 Electrodeposition Mechanism of Ni/Cu Multilayered Nanowires from Single Bath

The optimum range of electrodeposition potentials of the ferromagnetic (Ni) layers and non-magnetic (Cu) was determined using the cyclic voltammetry method. Figure 5.39 shows typical cyclic voltammograms recorded at different scan rates from 2.3 M nickel sulphamate ($Ni(SO_3NH2)_2$), 0.05 M copper sulphate ($CuSO_4$) and 0.5 M boric acid solution at a pH of about 2 at room temperature into polyester track etched template (PETE). Multilayered nanowires meeting within the template PETE show. The data extracted from the cyclic voltammograms are summarized in Table 5.8.

According to Fig. 5.39a, the cathodic reduction peak of copper commences from a potential of about −0.05 V_{SCE} and terminates at a potential of −0.45 V. However, the cathodic peak position of copper shifts towards more negative values by increasing the potential scan rate as a consequence of concentration gradient of copper electroactive ions at the bottom of the pores. This behavior may represent diffusion-limited electrodeposition mechanism of copper. It should be noted that in the electrodeposition of multilayers the more noble metal must be deposited under

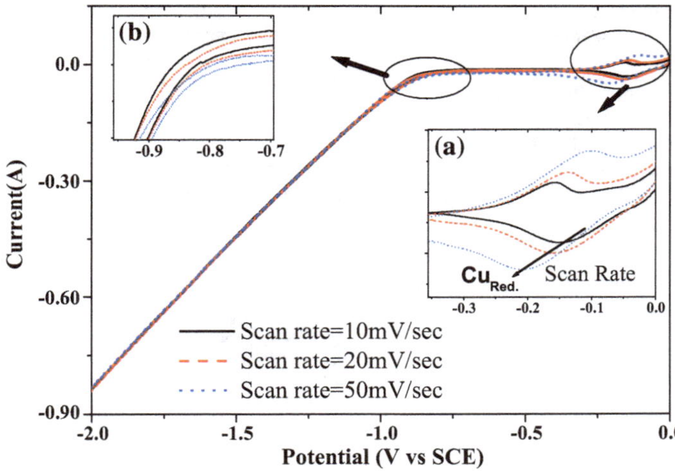

Fig. 5.39 Typical cyclic voltammograms recorded from 2.3 M nickel sulphamate (Ni ($SO_3NH2)_2$), 0.05 M copper sulphate ($CuSO_4$) and 0.5 M boric acid solution at a pH of about 2 at room temperature into polyester track etched template (PETE) with a pore diameter of 100 nm. **a, b** Insets show the reduction regions of copper and nickel respectively. From [137]. Total cathode surface was 1 cm^2

Table 5.8 Data extracted from the cyclic voltammograms shown in Fig. 5.39, (HRR = hydrogen reduction reaction). From [137]

Metal	$I_{p\ (A)}$	$E_p(V_{SCE})$	$E_{initial}\ (V_{SCE})$	$E_{final}\ (V_{SCE})$	Scan rate (mV/s)
Cu	−4.12E-04	−0.05866	−0.05	−0.35	10
Cu	−5.00E-04	−0.07755	−0.05	−0.4	20
Cu	−6.16E-04	−0.1016	−0.05	−0.45	50
Ni	–	–	−0.8	HRR	10
Ni	–	–	−0.8	HRR	20
Ni	–	–	−0.8	HRR	50

diffusion controlled process in order to limit its content in the less noble layers. The voltammogram shown in Fig. 5.39b indicates the potential range of the reduction of nickel which starts from −0.8 V_{SCE}. There is no clearly a sharp reduction peak for nickel because of hydrogen reduction (HRR) occurring and even being intensified by nickel reduction. In addition, a dissolution (or oxidation) peak on reverse scan is observed which corresponds to the dissolution of deposited nickel.

Based on to the characteristics obtained from the cycle voltammograms, the deposition potentials to achieve multilayers of copper and nickel (Ni rich) could be chosen. Copper potential must be chosen such that in addition to be deposited under diffusion control, the dissolution of nickel occurs at minimum. On this basis, a potential of −0.4 V may be used for depositing copper layer. The deposition

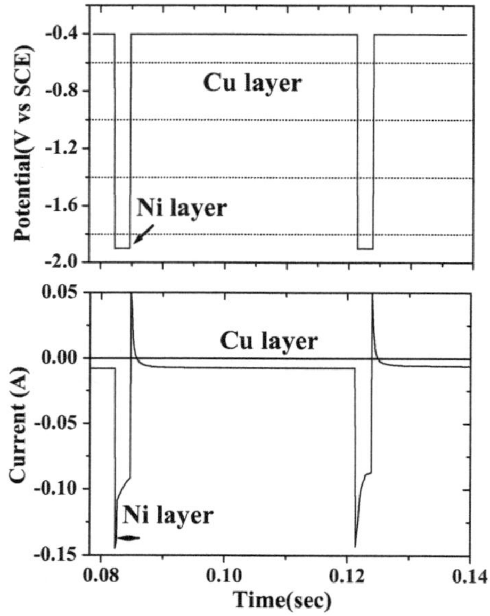

Fig. 5.40 *Top panel* modulating potential plot applied for the electrodeposition of Ni/Cu nanowires, and *bottom panel* current-transient recorded during the electrodeposition [137]

potential of nickel must be chosen with adequately high overpotential to control the deposition of copper at its diffusion limited current. High nickel sulphamate concentration used eliminates any concentration gradients even at high overpotentials. As a result, the percentage of copper in the nickel layer is reduced. Selected potentials are consistent with those reported elsewhere [135, 136].

Figure 5.40 illustrates the modulating potential plot used for the electrodeposition of multilayered Ni-Cu/Cu nanowires and corresponding typical current transient curve recorded during the first potential modulations. When the potential is switched from the deposition potential of nickel to that of copper a sharp drop happens likely due to either the depletion of copper ions or dissolution of nickel film (positive current is seen) at the bottom of the pore. However, it increases up to a plateau corresponding to the diffusion limiting current of copper. For nickel layers, the current reaches a maximum upon switching followed by a fall which is a typical nucleation and growth process.

Electrodeposition of Ni-Cu/Cu nanowires into PETE templates followed the general four stage mechanism of nucleation and growth of template electrodeposition processes. Figure 5.41 shows the variation of current recorded at the end of each modulating potential for copper layer.

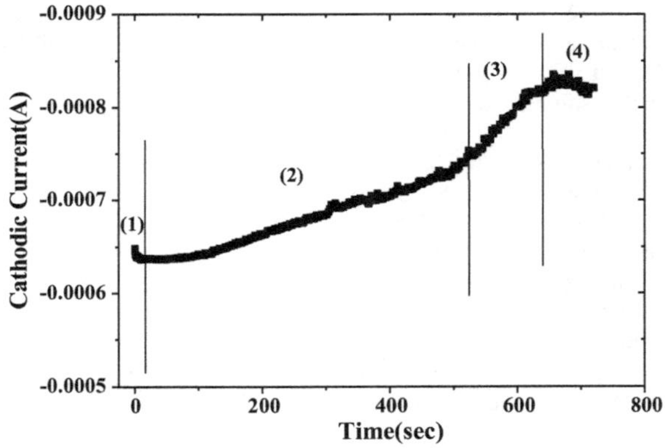

Fig. 5.41 A typical current transient for deposition of Cu layers during electrodeposition of multilayered NiCu/Cu nanowires in PETE [137]

5.4.3.3 Electrodeposition Mechanism of CoNiCu/Cu Multilayered Nanowires from Single Bath

In this section the electrochemical deposition mechanism of ternary Co-Ni-Cu system into PETE templates is described. Figure 5.42 illustrates cyclic voltammograms recorded from two solutions including solution (1) 2.3 M nickel sulphamate ($Ni(SO_3NH_2)_2$), 0.41 M cobalt sulphate ($CoSO_4$), 0.05 M copper sulphate ($CuSO_4$) and 0.5 M boric acid solution at a pH of about 2 and solution (2) M nickel sulphamate ($Ni(SO_3NH_2)_2$), 0.41 M cobalt sulphate ($CoSO_4$), 0.5 M copper sulphate ($CuSO_4$) and 0.5 M boric acid solution at a pH of about 2 at room temperature into polyester track etched template (PETE). It is observed that the potentials of reduction and oxidation of copper and nickel are almost the same for different copper ion concentrations. However, the peak current of copper has significantly increased by increasing the copper sulphate concentration. Figure 5.42c shows the variation of the copper reduction peak current as a function of root square of potential scan rate ($\nu^{1/2}$). According to Delahay's equation it shows that the electrodeposition mechanism of copper from Co-Ni-Cu ternary solution is diffusion controlled.

At lower concentration of copper ions, the limiting current of the deposition of copper is quite low such that it is off-limit of the current resolution of most conventional potentiostats/galvanostats. Thus it is clearly observed that to use the higher concentration of copper ions in solution to avoid structural defect in the multilayered nanowires [137]. It was found out that higher GMR values are obtained for multilayers nanowires electrodeposited from solution (2). The behaviour is quite similar for the electrodeposition into PETE and PCTE templates. Figure 5.43 shows current transients recorded during electrodeposition of CoNiCu/Cu nanowires with different bi-layer thicknesses.

The Co-Ni-Cu current was recorded at the end of the deposition of each Co-Ni-Cu layer, just before switching to the potential for Cu deposition. Each stage can be recognized from the scanning electron micrographs of grown nanowires in

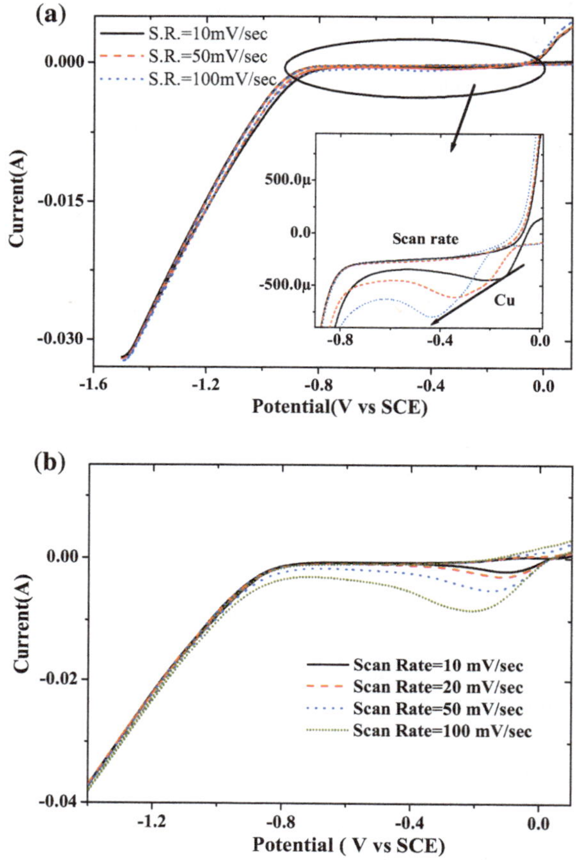

Fig. 5.42 Cyclic voltammograms recorded from **a** solution 1: (1) 2.3 M nickel sulphamate (Ni $(SO_3NH2)_2$), 0.41 M cobalt sulphate ($CoSO_4$), 0.05 M copper sulphate ($CuSO_4$) and 0.5 M boric acid solution at a pH of about 2 and solution (2) M nickel sulphamate (Ni $(SO_3NH2)_2$), 0.41 M cobalt sulphate ($CoSO_4$), 0.5 M copper sulphate ($CuSO_4$) and 0.5 M boric acid solution at a pH of about 2 [137]

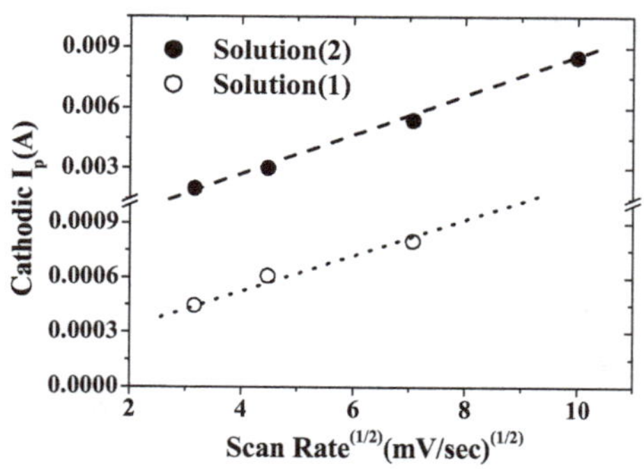

Fig. 5.43 Variation of cathodic reduction peak of copper as a function of root square of potential scan voltage as according to the Delahay's equation for diffusion-controlled electrodeposition process [137]

the template. For example, SEM images of top surface of the PETE template (with a nominal thickness of 6 μm) are shown in Fig. 5.44. For a nominal wire length of less than 4 μm, only nucleation and little amount of growth take place at the bottom of pores. Extension of deposition time will be accompanied by growth of wires within nanopores. Growth of nanowires continues until the nominal wire length exceeds 4 μm (Fig. 5.44a). Consequently, an overgrowth can occur on the template, seen in Fig. 5.44b.

Figure 5.45 shows the Co-Ni-Cu current as a function of time during deposition of (a) Co-Ni-Cu (6 nm)/Cu (6 nm) multilayered nanowires in polycarbonate membranes and (b) Co-Ni-Cu (3.3 nm)/Cu (3.3 nm) superlattice nanowires in polyester membranes. The current transients recorded in both types of templates show the four distinct growth stages with some difference from the general 4-stage mechanism. These are (1): initial nucleation and growth of the nanowires at the ends of the pores, (2): growth of wires within the pores, (3) pore filling and (4) overgrowth. Stage (2) may be subdivided into a time during which the current

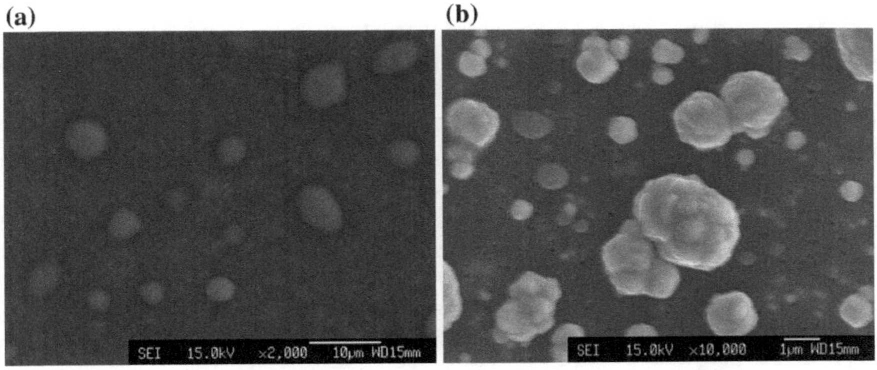

Fig. 5.44 SEM images taken from *top* surface of a PETE template after electrodeposition of **a** 4 μm and **b** 7 μm of CoNiCu/Cu nanowires [137]

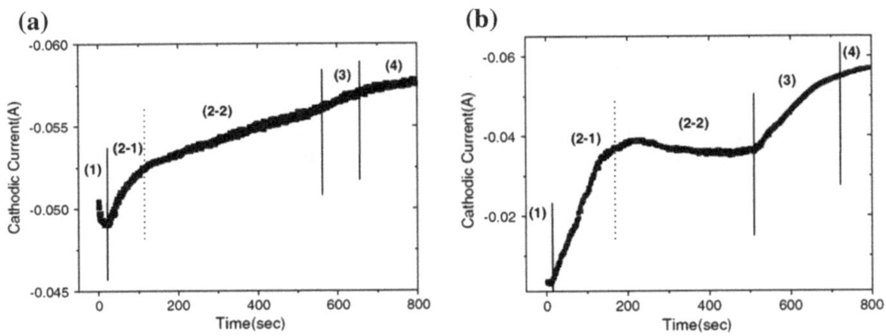

Fig. 5.45 Co-Ni-Cu current as a function of time during deposition of **a** Co-Ni-Cu (3.3 nm)/Cu (3.3 nm)-superlattice nanowires in polyester membranes and **b** Co-Ni-Cu (6 nm)/Cu (6 nm)-superlattice nanowires in polycarbonate and. Reproduced from [139] with the permission of Elsevier

increases (2-1) and a time during which the current changes much more slowly (2-2). A probable explanation of the increasing current during Stage (2-1) is that poor wetting leads to a delay in nucleation in some pores, so that the number of pores in which growth takes place and therefore also the current rise gradually [138]. The increase in current during Stage (2-1) and its duration are significantly greater for the polycarbonate membranes than for the polyester ones, consistent with poorer wetting in the former case. According to Fokkink et al. [138], poor pore-wetting also leads to a smearing out of the transition to bulk growth and it is noticeable that Stage (3) is much longer for the polycarbonate than for the polyester membranes. Current transients also show that the ratio of the steady state ratio of Cu to the Co-Ni-Cu current is larger for the polyester than the polycarbonate membranes. This ratio determines the Cu content of the Co-Ni-Cu layers, which is therefore likely to be greater for the nanowires grown in polyester membranes [139].

5.4.3.4 GMR and Magnetism in Electrodeposited Multilayered Nanowires

Let's begin this section with the first nanowire composition in which CPP-GMR was explored. In general, the magnetoresistance (MR) is defined as:

$$\%\mathrm{MR} = \frac{R(H) - R(H_{\max})}{R(H_{\max})} \times 100, \tag{5.6}$$

in which R is the resistance, H is the applied field and $H_{\max}$ the maximum applied field.

Ansermet and co-workers reported $\sim 14\%$ MR at room temperature and $\sim 18\%$ MR at 20 K for $\mathrm{Co_{5nm}/Cu_{5nm}}$ nanowires electrodeposited in PCTE membranes with a nominal average pore diameter of 80 nm and a length of 6 μm [132]. Piraux et al. obtained quite similar values of MR results for $\mathrm{Co_{7nm}/Cu_{3nm}}$ nanowires electrodeposited in PCTE membranes comprising pores with 40 nm average diameters and 10 μm lengths, i.e. 15 and 19% MR for at room temperature and 4.2 K, respectively. They have clearly shown the dissolution of cobalt layer during the deposition of copper layer illustrating relatively rough and irregular interlayer interfaces [131]. Later, Liu et al. [119] exploited a CPP-MR of 11 and 22% at room temperature and 5 K, respectively for $\mathrm{Co_{5nm}/Cu_{0.8nm}}$ nanowires electrodeposited in PCTE templates. A relatively higher GMR values may be explained by their modification of their method of electrodeposition. Major modifications were (1) a more precise estimation of layer thickness by calculating the current efficiencies of Co and Cu and consideration of hydrogen evolution during Co deposition and (2) application of a few seconds open-circuit rest potential during the transition between Co to Cu potentials to avoid severe cobalt dissolution during copper deposition cycle which could damage the interfaces and leave the Co thickness inaccurate [58, 140] reported a comparative study between Co/Cu nanowires electrodeposited in PCTE and AAO. The nanowires electrodeposited in AAO exhibited greater values of GMR than those electrodeposited in PCTE.

Co-Ni-Cu/Cu was an alternative system to Co/Ni because it was found that Ni appears to reduce cobalt dissolution during the deposition of copper layers. Blondel et al. reported Co-Ni-Cu/Cu nanowires electrodeposited into PCTE from a suphamate bath [141] exhibiting of CPP-GMR values of 15–20% depending on bilayer thickness. More recent study by Heydon et al. [142] showed that the composition of Co-Ni-Cu/Cu depends on some re-dissolution of the less noble metal like Co or Ni and hydrogen reduction reaction as well as galvanic replacement of copper with less noble metal during a off-time pulse. The electrodeposition of CoNiCu/Cu nanowires were also successfully carried out in AAO template [143] exhibiting CPP-GMR with very large values of about 55% at room temperature and 115% at 77 K, the highest recorded GMR value so far. These high values were relatively ascribed to the enhanced diffusion of electroactive species along ordered AAO nanopores leading to more uniform interlayer interface with less roughness (Fig. 5.46).

More recently, Nasirpouri et al. reported CPP-GMR in CoNiCu/Cu nanowires electrodeposited in PETE templates and compared with those electrodeposited in PCTE. The electrodeposition mechanism of these nanowires were compared as explain in previous section. Figure 5.47 compares TEM images of Co-Ni-Cu (3.3 nm)/Cu (3.3 nm) superlattice nanowires electrodeposited in PCTE and PETE templates, following dissolution of the templates. The imaged nanowires seem to have lengths of 4–5 μm, which approaches the maximum expected for a membrane thickness of 6 μm, but shorter fragments are also seen, perhaps resulting from wire breakages during the dissolution of the template. The releasing procedure of nanowires from PETE template is much more difficult than for that of PCTE, template which are easily dissolved in chloroform at room temperature, and the sample of Fig. 5.47a clearly still includes a significant amount of residual polymer

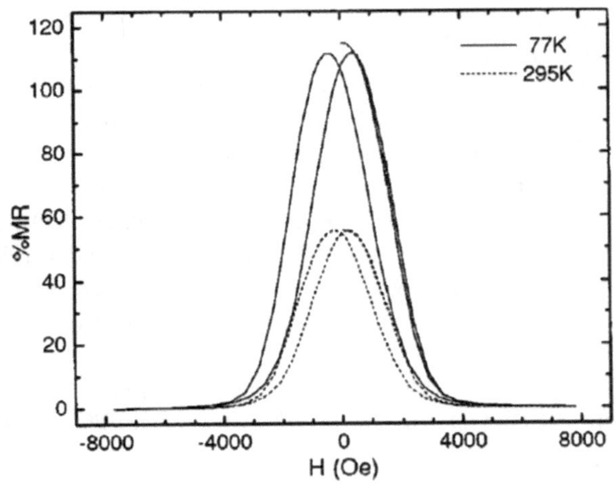

Fig. 5.46 Current perpendicular to plane giant magnetoresistance measured for multilayered nanowire with nominal layer thicknesses 54 Å Co-Ni-Cu/21 Å Cu at 77 K (*solid line*) and room temperature (*dashed line*). Reproduced from [143] with the permission of AIP Publishing, copyright 2016

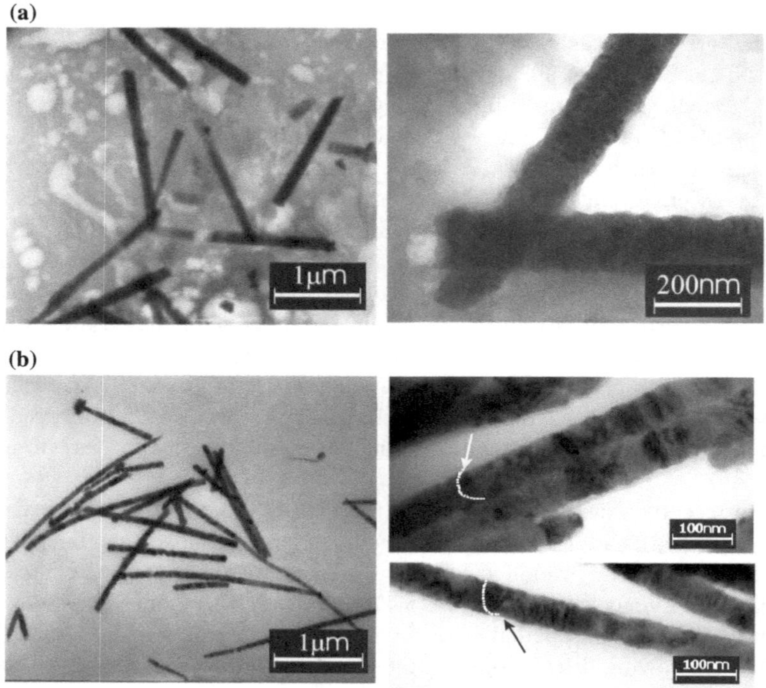

Fig. 5.47 Bright-field TEM images of Co-Ni-Cu (3.3 nm)/Cu (3.3 nm) superlattice nanowires electrodeposited in PCTE and PETE templates, following dissolution of the templates

whereas the nanowires released from PCTE seem brighter with our any residual materials. Higher magnification images of Co-Ni-Cu (3.3 nm)/Cu (3.3 nm) superlattice nanowires electrodeposited in PETE and PCTE are also shown in Fig. 5.47. The formation of successive parallel interfaces are obvious for both types of nanowires while the layer interfaces are not flat. Surface tension considerations suggest that the evaporated Au contact at the start of a pore is unlikely to be flat. As this contact forms the initial growth surface for the nanowire, the layer interfaces are likely to be curved [139]. This is more intensified for PCTE templates owing to their small pore mouth size (30 nm).

Figure 5.48a shows the measured percentage magnetoresistance (%MR) for series of Co-Ni-Cu/Cu superlattice nanowires electrodeposited in PETE with fixed nominal Co-Ni-Cu layer thickness $t_{Co\text{-}Ni\text{-}Cu}$ and varying Cu layer thickness t_{Cu}. In each case, the %MR curve is broad, with changes in the MR continuing to be recorded even for $H > 2$ k Oe. There is also hysteresis (as well as some drift) in the MR data. For comparison, Fig. 5.48b shows the magnetization curve measured at 300 K with the applied field in the plane of the membranes, as for the MR measurements, for the Co-Ni-Cu (3.3 nm)/Cu (2 nm) sample. Comparison of

Fig. 5.48a, b shows that the maximum %MR occurs close to $H = H_c$ (the coercive field).

Note that from Fig. 5.48a the maximum measured %MR *increases* with decreasing t_{Cu} as is expected according to the Valet-Fert (VF) model for CPP-GMR, because decreasing t_{Cu} decreases the component of the resistance that is independent of magnetization [144]. This simple model will only apply, however, if decreasing t_{Cu} does not influence e.g. the effective $t_{Co\text{-}Ni\text{-}Cu}$. For example, if interlayer roughness or pinholes lead to strong ferromagnetic coupling between the Co-Ni-Cu layers when t_{Cu} is reduced, the VF model will break down. Hence the observation of increasing %MR with decreasing t_{Cu} is an important criterion by which the structural quality of superlattice nanowires may be judged.

Figure 5.49 shows that the VF model does not apply to the Co-Ni-Cu/Cu superlattice nanowires that we electrodeposited in polycarbonate membranes. For these nanowires, like those electrodeposited in polyester membranes, the real layer

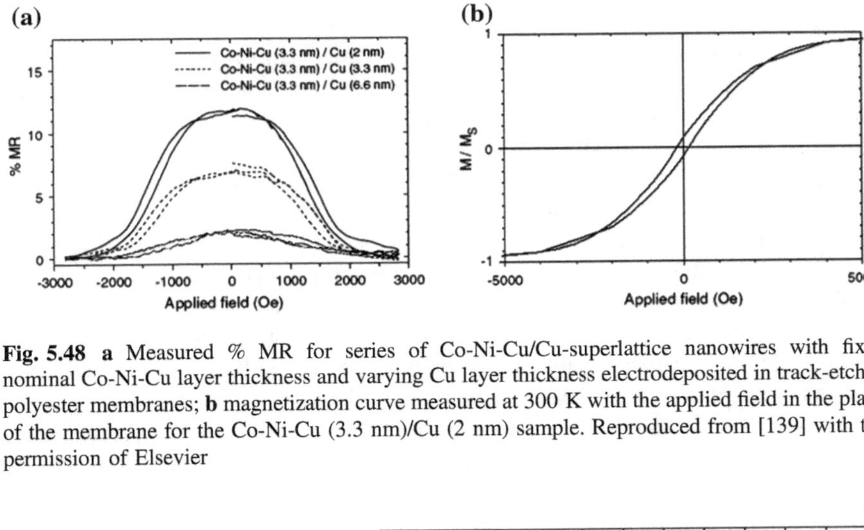

Fig. 5.48 **a** Measured % MR for series of Co-Ni-Cu/Cu-superlattice nanowires with fixed nominal Co-Ni-Cu layer thickness and varying Cu layer thickness electrodeposited in track-etched polyester membranes; **b** magnetization curve measured at 300 K with the applied field in the plane of the membrane for the Co-Ni-Cu (3.3 nm)/Cu (2 nm) sample. Reproduced from [139] with the permission of Elsevier

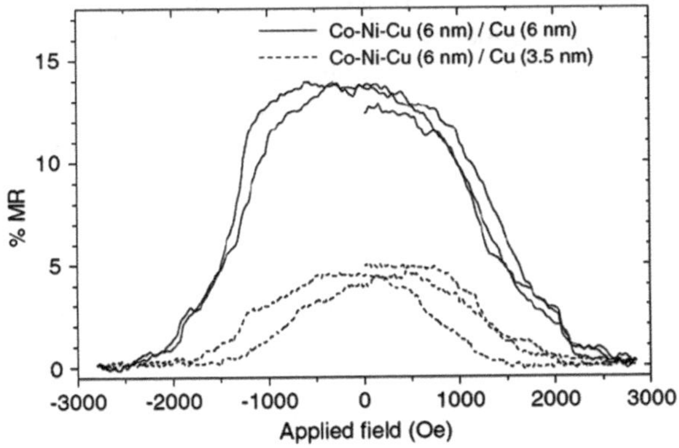

Fig. 5.49 Measured %MR for a series of Co-Ni-Cu/Cu-superlattice nanowires with fixed nominal Co-Ni-Cu layer thickness and varying Cu layer thickness electrodeposited in track-etched polycarbonate membranes. Reproduced from [139] with the permission of Elsevier

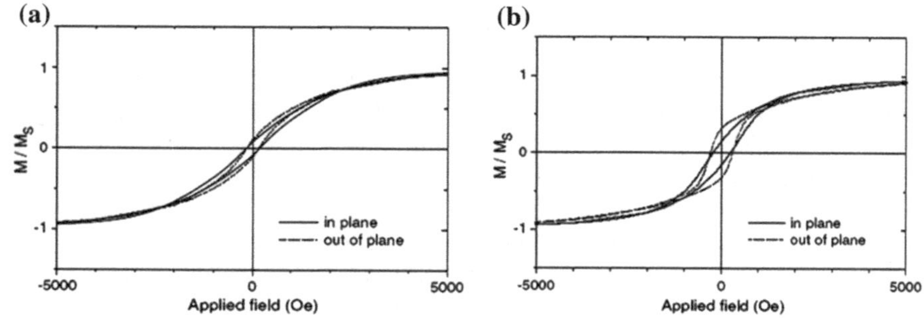

Fig. 5.50 **a** Hysteresis loops measured at 300 K with the applied field in and out of the plane of the membrane for Co-Ni-Cu (3.3 nm)/Cu (3.3 nm)-superlattice nanowires electrodeposited in track-etched polyester; **b** hysteresis loops measured at 300 K with the applied field in and out of the plane of the membrane for Co-Ni-Cu (6 nm)/Cu (6 nm)-superlattice nanowires electrodeposited in track-etched polycarbonate. Reproduced from [139] with the permission of Elsevier

thicknesses appear to be close to the nominal layer thicknesses, since TEM measurements of Co-Ni-Cu (6 nm)/Cu (6 nm) superlattice nanowires gave an estimated repeat distance of 9.0 ($\pm$0.4) nm, which is in reasonable agreement with the nominal value of 12 ($\pm$2.5) nm. Although the general shape of the MR curves is similar to Fig. 2a, the maximum measured %MR *decreases* between $t_{\mathrm{Cu}} = 6$ nm and $t_{\mathrm{Cu}} = 3.5$ nm. This suggests that, at least by this criterion, the structural quality of Co-Ni-Cu/Cu multilayer nanowires electrodeposited in polyester membranes is superior to that of similar nanowires electrodeposited in polycarbonate membranes.

Although the shape anisotropy of the individual layers would make the direction parallel to the superlattice nanowires' long axis (out of the plane of the membrane) a hard axis, ferromagnetic coupling between the layers would make this direction an easyaxis. Figure 5.50 shows hysteresis loops measured with the applied field parallel and perpendicular to the plane of the membrane. For Co-Ni-Cu (6 nm)/Cu (6 nm) superlattice nanowires electrodeposited in polycarbonate membranes there is clearly an easy axis parallel to the nanowires' long axis, suggesting ferromagnetic coupling, whereas in the case of Co-Ni-Cu (3.3 nm)/Cu (3.3 nm) superlattice nanowires electrodeposited in polyester membranes there is no obvious hard or easy direction. Current transients also show that the ratio of the steady state ratio of Cu to the Co-Ni-Cu current is larger for the polyester than the polycarbonate membranes. This ratio determines the Cu content of the Co-Ni-Cu layers, which is therefore likely to be greater for the nanowires grown in polyester membranes. This could be one explanation for the observation that the maximum measured %MR for the polycarbonate membranes ($\sim$14% for $t_{Cu} = 6$ nm) is greater than that measured for polyester membranes ($\sim$12% for $t_{Cu} = 2$ nm) [27, 139].

Other superlattice nanowires including Ni-Fe/Cu [114, 133, 145–147] and Ni/Cu [148–151] have been studies. For instance, multilayered permalloy/copper ($Ni_{80}Fe_{20}$/Cu) nanowire arrays having a bilayer thickness ranging 5–10 nm were firstly demonstrated with a CPP-MR of 10% at ambient temperature [133]. Further studies in PCTE and lithographically patterned templates [133, 145–147] have been reported indicating of precisely layered structure of Ni-Fe/Cu nanowires with CPP-MR ratios as high as 80% at 4.2 K.

Nickel/copper multilayered nanowires have been studied as well. In previous section, the electrodeposition mechanism of NiCu system was explained. It is surprising that NiCu/Cu multilayered nanowires were shown to exhibit different microstructures and consequently magnetoresistance properties. For NiCu/Cu multilayers GMR magnitudes up to 4% have been reported [152], but they are not comparable with other multilayers systems. Robinson and Schwarzacher [150]. found that while having actual multilayered structure, exchange coupling between the ferromagnetic layers are influenced by the formation of individual nickel islands and intermix Ni/Cu interface. This has been even used to produce NiCu alloy nanowires through rapid modulating potentials [136].

5.4.4 Diameter Modulated Nanowires

More recent development of magnetic nanowires have been achieved by electrodeposition of magnetic materials into a particularly confined templates. As has been shown in former sections on the nanowires, any changes in diameter would change the properties and as a consequence of a controlled manner applied to diameter changes of nanowires or tubes, namely diameter modulated nanowires, should give rise to novel physics and technology. In the magnetic nanowires such modulation in diameter would strangely affect the motion and lock of magnetic domain walls and their magnetisation reversal.

AAO templates with uniform and smooth diameters were described earlier in this chapter. A controlled modification of diameter in form of the alternating modulations, has been exploited [153–156] using a combination of “mild” and “hard” anodisation of aluminum anodization. Figure 5.51 shows a modulated AAO template [161].

Electrodeposition of magnetic nanowires of ferromagnetic metal and alloys has been successfully carried out into these diameter modulated AAO templates with a wide variety of diameter profiles along nanowires. For permalloy nanowires with modulated diameters it has been shown to reverse their magnetisation via the nucleation and propagation of a vortex domain wall. Vortex domains are initially nucleated in the nanowire ends and in the transition region between the two

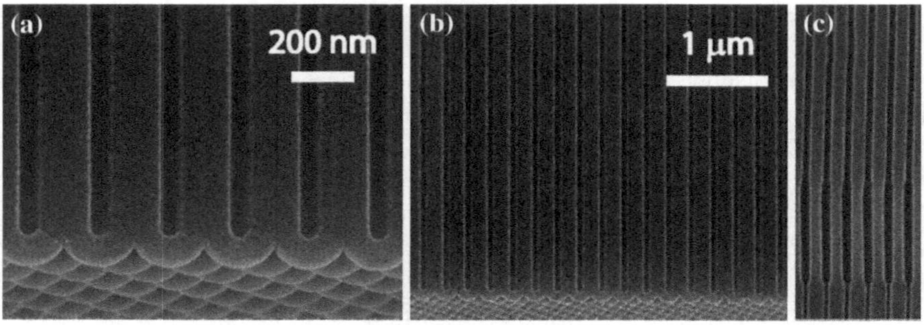

Fig. 5.51 Scanning electron micrographs of a modulated porous anodic alumina template. The membrane is displayed in section; pores are broken in halves along their z axis: **a** end of the pores with hemispherical barrier layer; **b** the last 4 _m of a perfectly ordered "hard" segment; the pore diameter is absolutely constant and homogeneous; **c** a section of membrane displaying two changes in diameter, from "hard" anodization (*narrow pores*) to "mild" (*thicker pores*) and back to "hard"; the interfaces between the different segments are sharply defined. The scale bar is common for panels **b** and **c**. Reproduced from [161] with the permission of American Chemical Society, copyright 2016

diameters in order to minimize magnetostatic energy generated by surfaces perpendicular to the initial magnetization of the nanowires [162]. Magnetic behavior of Co nanowires with round shape and squared shape diameter modulations strongly exhibited mgnetisation reversal dependence on nanowire shape developing magnetostatic interactions at the ends of the wider segments in modulated wires. Research on this subject is ongoing and further investigations are required.

References

1. S.K. Chakarvarti, J. Vetter, Nucl. Instr. Meth. Phys. Res. B **62**, 109 (1991)
2. D. Dobrev, J. Vetter, N. Angert, Nucl. Instr. Meth. Phys. Res. B **149**, 207 (1999)
3. E. Ferain, R. Legras, Nucl. Instr. Meth. Phys. Res. B **174**, 116 (2001)
4. E. Ferain, R. Legras, Radiat. Meas. **34**, 585 (2001)
5. H. Masuda, K. Fukuda, Science **268**, 1466 (1995)
6. H. Chik, J.M. Xu, Mater. Sci. Eng. R **43**, 103 (2004)
7. M. Ghorbani, F. Nasirpouri, A. Irajizad, A. Saedi, Mater. Des. **27**(10), 983–988 (2006)
8. Reprint copyright 2016, with the permission of Elsevier
9. F. Nasirpouri, M. Abdollahzadeh, M.J. Almasi et al., Curr. Appl. Phys. **9**(1), S91–S94 (2009)
10. Reprint copyright, 2016, with the permission of Elsevier
11. O. Jessensky, F. Muller, U. Gosele, Appl. Phys. Lett. **72**(10), 1173–1175 (1998)
12. K. Shimizu, K. Kobayashi, P. Skeldon, G.E. Thompson, G.C. Wood, Corros. Sci. **39**, 701–718 (1997)
13. F. Li, L. Zhang, Metzger RM. Chem. Mater. **10**, 2470–2480 (1998)
14. Y.C. Sui, B.Z. Cui, L. Martmez, R. Perez, D.J. Sellmyer, Thin Solid Films **406**, 64–69 (2002)
15. T. Xu, G. Zangari, Metzger RM. Nano Lett. **2**(1), 137–141 (2002)

16. J.I. Paredes, A. Martinez-Alonso, Tascon JMD. Micropor Mesopor Mat **65**, 93–126 (2003)
17. Y.C. Sui, J.M. Saniger, Mater. Lett. **48**:127–136. 15–18 (2001)
18. A.P. Li, F. Muller, A. Birner, K. Nielsch, U. Gosele, J. Vac. Sci. Technol. A **17**, 1428 (1999)
19. A.P. Li, F. Muller, A. Birner, K. Nielsch, U. Gosele, J. Appl. Phys. **84**, 6023 (1998)
20. A.P. Li, F. Muller, U. Gosele, Electrochem. Solid-State Lett. **3**, 131 (2000)
21. T. Biestek, J. Weber, *Electrolytic and chemical conversion coatings* (Portcullis Press Limited, Warsaw, 1976)
22. G.E. Thompson, Thin Solid Films **297**, 192–201 (1997)
23. K. Shimizu, K. Kobayashi, G.E. Thompson, G.C. Wood, Philos. Mag. A **66**, 643 (1992)
24. F. Nasirpouri, S.M. Peighambari, Ionics **19**(3), 535–542 (2013)
25. Reprint copyright (2016) with the permission of Springer
26. Z. Ye, H. Liu, Z. Luo, H.-G. Lee, W. Wu, D.G. Naugle, I. Lyuksyutov, Nanotechnology **2**, 045704 (2009)
27. F. Nasirpouri, IEEE Trans. Magn. **47**(8), 2015–2021 (2011)
28. J.M. Montero-Moreno, M. Sarret, C. Müller, Microporous Mesoporous Mater. **136**(68–74), 22 (2010)
29. J.M. Montero-Moreno, M. Sarret, C. Müller, Surf. Coat. Technol. **201**(6352–6357), 23 (2007)
30. K.M. Alam, A.P. Singh, S.C. Bodepudi, S. Pramanik, Surf. Sci. **605**, 441–449 (2011)
31. S. Yue, Y. Zhang, J. Du, Mater. Chem. Phys. **128**(187–190), 25 (2011)
32. Microelectronic Engineering. 97, Sept 2012, pp. 147–149 (2012)
33. H. Yan, P. Lemmens, D. Wulferding, J. Shi, K.D. Becker, C. Lin, A. Lak, M. Schilling, Mater. Chem. Phys. **135**(206), 21121–21126 (2012)
34. R.E. Ricker, A.E. Miller, D.F. Yue, G. Banerjee, S. Bandyopadhyay, J. Electron. Mater. **25** (10), 1585–1592 28 (1996)
35. M.T. Wu, I.C. Leu, M.H. Hon, J. Vac, Sci. Technol. B **20**, 776–782 (2002)
36. C.-U. Yu, C.-C. Hu, A. Bai, Y.-F. Yang, Surf. Coat. Technol. **201**, 7259–7265 (2007)
37. M.H. Rahimi, S. Saramad, S.H. Tabaian, S.P. Marashi, A. Zolfaghari, M. Mohammadalinezhad, Appl. Surf. Sci. **256**, 12–16 (2009)
38. G.D. Sulka, K.G. Parkola, Electrochim. Acta 52:1880–1888 (2007)
39. G.E. Possin, Rev. Sci. Instrum. **41**, 772 (1970)
40. W.D. Williams, N. Giordano, Rev. Sci. Instrum. **55**, 410 (1984)
41. R.M. Penner, C.R. Martin, Anal. Chem. **59**, 2625 (1987)
42. C.J. Brumlik, C.R. Martin, J. Am. Chem. Soc. **113**, 3174 (1991)
43. C.R. Martin, Science **266**, 1961 (1994)
44. C.R. Martin, Chem. Mater. **8**, 1739 (1996)
45. S.A. Sapp, B.B. Lakshmi, C.R. Martin, Adv. Mater. **11**, 402 (1999)
46. T.M. Whitney, J.S. Jiang, P.C. Searson, C.L. Chien, Science **261**, 1316 (1993)
47. A.M. Bond, D. Luscombe, K.B. Oldham, C.G. Zoski, J. Electroanal. Chem. **249**, 1–16 (1988)
48. M. Granstrom, J.C. Carlberg, O. Inganas, Polymer **36**, 3191 (1995)
49. F.S. Fedorov, I. Monch, C. Mickel et al., J. Electrochem. Soc. **160**(1), D13–D19 (2013)
50. B.R. Scharifker, J. Electroanal. Chem. **240**, 61 (1988)
51. B. Scharifker, G. Hills, J. Electronal. Chem. **130**, 81 (1981)
52. S. Valizadeh, J.M. George, P. Leisner, L. Hultman, Electrochim. Acta **47**, 865–874 (2001)
53. M. Motoyama, Y. Fukunaka, T. Sakka, Y.H. Ogata, S. Kikuchi, J. Electroanal. Chem. **584**, 84–91 (2005)
54. M. Motoyama, Y. Fukunaka, T. Sakka, Y.H. Ogata, Electrochim. Acta **53**, 205–212 (2007)
55. F. Keller, M.S. Hunter, D.L. Robinson, J. Electrochem. Soc. **100**, 411 (1952)
56. V. Caboni, Italian Patent 339 232, 1936. Langbein-Pfandhauser GmbH, German Patent 741 753 (1940)
57. M. Metzger, V.V. Konovalov, M. Sun, T. Xu, G. Zangari, B. Xu, M. Benakli, W.D. Doyle, IEEE Trans. Magn. **36**, 30 (2000)

58. T. Ohgai, X. Hoffer, A. Fabian, L. Gravier, J.-P. Ansermet, J. Mater. Chem. **13**, 2530 (2003)
59. H. Takahashi, M. Nagayama, H. Akahori, A. Kitahara, J. Electron Microscopy **22**, 149–157 (1973)
60. R.C. Furneaux, Thompson GE abd Wood GC. Corr. Sci. **18**, 853–881 (1978)
61. O'sullivan, G. Wood, Proceedings of the Royal Society of London. Ser. A, Math. Phys. Sci. 511–543 (1970)
62. R. Furneaux, W. Rigby, A. Davidson, Nature **337**, 147–149 (1989)
63. A. Saedi, M. Ghorbani, Mater. Chem. Phys. **91**, 417–423 (2005)
64. S.M. Peighambari, MSc Thesis: Electrodeposition of Co-base alloy nanowire arrays effect of AC deposition parameters, Sahand University of Technology, 2011
65. F. Nasirpouri, S.M. Peighambari, S.J. Bending, E.V. Sukovatitsina et al., J. Spintronics Mag. Nanomaterials **1**(1), 23–27 (2012)
66. M.A. Kashi, A. Ramazani, A. Khayyatian, J. Phys. D Appl. Phys. **39**, 4130 (2006)
67. A.J. Yin et al., Appl. Phys. Lett. **79**, 1039 (2001)
68. E. Panahi Danaei, MSc Thesis: Effect of electrolyte concentration pH, content and AC electrodeposition parameters on growth of CoNi alloy nanowires, Sahand University of Technology, 2013
69. M. Nadi, MSc Thesis: Engineering composition and microstructure of CoNi nanowires by AC electrodeposition, Sahand University of Technology, 2012
70. Sarshar-noshar, A., MSc thesis: AC electrodeposition of NiMn nanowires in anodic aluminium oxide nanopores", Sahand University of Technology, 2014
71. M. Abdollahzadeh, MSc Thesis "Fabrication of amorphous magnetic nanowire arrays in alumina nanoporous template", Sahand University of Technology, 2007
72. S. Beheshti, MSc Thesis: "AC Electrodeposition of ZnO in AAO template", Sahand University of Technology, 2013
73. H. Birey, J. Appl. Phys. **49**, 2898 (1978)
74. B.Z. Katsenelenbaum, High-frequency Electrodynamics, John Wiley & Sons
75. F. Nasirpouri, A.R. Nogaret, *Nanomagnetism and Spintronics* (World Sci Pub Co, Singapore, 2011)
76. M. Hernandez-Velez, Thin Solid Films **495**, 51 (2006)
77. M.S. Chandrasekar, M. Pushpavanam, Electrochim. Acta **53**, 3313–3322 (2008)
78. K. Nielsch, F. Müller, A.-P. Li, U. Gösele, Adv. Mater. **12**, 582–586 (2000)
79. D.J. Sellmyer, M. Zheng, R. Skomski, J. Phys. Condens. Matter **13**, R433 (2001)
80. V. Raposo, J.M. Garcia, J.M. Gonzalez, M. Vazquez, J. Magn. Magn. Mater. **222**, 227 (2000)
81. J. Spiegel, I. Huynen, Solid State Phenom. **152–153**, 389 (2009)
82. S.Y. Chou, Proc. IEEE **85**, 652 (1997)
83. T. Shimizu, K. Aoki, Y. Tanaka, T. Terui, S. Shingubara, Jap. J. Appl. Phys. **50**, 06GE01 (2011)
84. H. Schmidt, Appl. Organomet. Chem. **15**(3311), 2 (2001)
85. M. Vazquez, K. Nielsch, P. Vargas, J. Velazquez, D. Navas, K. Pirota, M. Hernandez-Velez, E. Vogel, J. Cartes, R.B. Wehrspohn, U. Gosele, Phys. B **343**, 395 (2004)
86. M. Kroll, W.J. Blau, D. Grandjean, R.E. Benfield, F. Luis, P.M. Paulus, L.J. de Jongh, J. Magn. Magn. Mater. **249**, 241 (2002)
87. M. Zheng, L. Menon, H. Zheng, Y. Liu, S. Bandyopadhyay, R.D. Kirby, D.J. Sellmyer, Phys. Rev. B **62**, 12282 (2000)
88. A.S. Samardak, E.V. Sukovatitsina, A.V. Ognev, L.A. Chebotkevich, R. Mahmoodi, S.M. Peighambari, M.G. Hosseini, F. Nasirpouri, J. Phys. Conf. Ser. **345**(1), 012011 (2012)
89. A.S. Samardak, E.V. Sukovatitsina, A.V. Ognev, L.A. Chebotkevich, R. Mahmoodi, S.M. Peighambari, M.G. Hosseini, F. Nasirpouri, Physics Procedia **22**, 549–556 (2011)
90. A.V. Ognev, A.S. Samardak, L.A. Chebotkevich, E.V. Sukovatitsina, M.E. Stebliy, E.B. Modin, E.V. Pustovalov, V.S. Plotnikov, F. Nasirpouri, J. Appl. Phys. **113**(17), 17A334 (2013). Reprint copyright, 2016, with the permission of AIP publishing

91. P.M. Paulus, F. Luis, M. Kroll, G. Schmid, L.J. de Jongh, J. Magn. Magn. Mater. **224**, 180 (2001)
92. F. Li, T. Wang, L. Ren, J. Sun, J. Phys. Condens. Matter **16**, 8053 (2004)
93. M. Darques, A. Encinas, L. Vila, L. Piraux, J. Phys. Condens. Matter **16**, S2279 (2004)
94. M. Darques, A. Encinas, L. Vila, L. Piraux, J. Phys. D Appl. Phys. **37**, 1411 (2004)
95. V. Scarani, B. Doudin, J.P. Ansermet, J. Magn. Magn. Mater. **205**, 241 (1999)
96. A. Kazadi Mukenga Bantu, J. Rivas, G. Zaragoza, M.A. Lopez-Quintela, M.C. Blanco, J. Non-crystalline Solids **287**, 5 (2001)
97. A. Kazadi Mukenga Bantu, J. Rivas, G. Zaragoza, M.A. Lopez-Quintela, M.C. Blanco, J. Appl. Phys. **89**, 3393 (2001)
98. J. Rivas, A. Kazadi Mukenga Bantu, G. Zaragoza, M.C. Blanco, M.A. Lopez-Quintela, J. Magn. Magn. Mater. **249**, 220 (2002)
99. A. Encinas-Oropesa, M. Demand, L. Piraux, U. Ebels, I. Huynen, J. Appl. Phys. **89**, 6704 (2001)
100. F. Nasirpouri, A. Nogaret, S.J. Bending, IEEE Trans. Magn. **47**(12), 4695–4700 (2011)
101. W.O. Rosan, L.G. Vivas, K.R. Pirota, A. Asenjo, M. Vazquez, J. Magn. Magn. Mat. **324**, 3679–3682 (2012)
102. A. Pereira, C. Gallardo, A.P. Espejo, J. Briones, L.G. Vivas, M. Vázquez, J.C. Denardin, J. Escrig, J. Nanopart. Res. **14**, 2041 (2013)
103. L.G. Vivas, M. Vazquez, J. Escrig, S. Allende, D. Altbir, D.C. Leitao, J.P. Araujo, Phys. Rev. B **85**, 035439 (2012)
104. R.S. Iskhakov, S.V. Komogortsev, L.A. Chekanova, A.D. Balaev, V.A. Yuzova, O.V. Semenova, Tech. Phys. Lett. **29**(4), 263–266 (2003)
105. V. Vega, T. Böhnert, S. Martens, M. Waleczek, J.M. Montero-Moreno, D. Görlitz, V.M. Prida, K. Nielsch, Nanotechnology **23**, 465709 (2012)
106. D. Kaur, D.K. Pandya, S. Chaudhary, J. Electrochem. Soc. **159**, D713-D716 (2012). KR Pirota, J. Appl. Phys. **109**, 083919 (2011)
107. S. Thongmee, H.L. Pang, J.B. Yi, J. Ding, J.Y. Lin, L.H. Van, Acta Mater. **57**, 2482–2487 (2009)
108. A.S. Samardak, F. Nasirpouri, M. Nadi, E.V. Sukovatitsina, A.V. Ognev, L.A. Chebotkevich, S.V. Komogortsev, J. Magn. Magn. Mater. **383**, 94–99 (2015). Reprint copyright, 2016, with the permission of Elsevier
109. V. Langlais, S. Arrii, L. Pontonnier, G. Tourillon, Scr. Mater. **44**, 1315–1319 (2001)
110. H.N. Hu, H.Y. Chen, S.Y. Yu, J.L. Chen, G.H. Wu, F.B. Meng, J.P. Qu, Y.X. Li, H. Zhu, J. Q. Xiao, J. Magn. Magn. Mat. **295**, 257–262 (2005)
111. V. Vega, V.M. Prida, J.A. García, M. Vazquez, Phys. Stat. Solidi a **208**, 553 (2011)
112. F. Nasirpouri, S.M. Peighambari, A.S. Samardak, A.V. Ognev, E.V. Sukovatitsina, E.B. Modin, L.A. Chebotkevich, S.V. Komogortsev, S.J. Bending, J. Appl. Phys. **117**(17), 17E715 (2016). Reprint copyright, 2016, with the permission of AIP publishing
113. G. Rivero et al., J. Appl. Phys. **69**, 5454 (1991). K. Huller et al., J. Magn. Magn. Mater. **53**, 103 (1985)
114. G. Kartopu, O. Yalçın, M. Es-Souni, A.C. Başaran, J. Appl. Phys. **103**, 093915 (2008)
115. M. Darques, J.D. Torre Medina, L. Piraux, L. Cagnon, I. Huynen, Nanotechnology **21**, 145208 (2010)
116. J. De La Torre Medina, J. Spiegel, M. Darques, L. Piraux, I. Huynen, Appl. Phys. Lett. **96**, 072508 (2010)
117. L. Maurice, D. Imho, P. Etienne, O. Durand, S. Dubois, L. Piraux, J.M. George, P. Galtier, J. Magn. Magn. Mater. **184**, 1 (1998)
118. K.R. Pirota et al., J. Appl. Phys. **109**, 083919 (2011)
119. H. Hou, Q. Peng, S. Zhang, Q. Guo, Y. Xie, Eur. J. Inorg. Chem. 2625–2630 (2005)
120. K. Nielsch et al., Nano Lett. **2**, 677 (2002)
121. J. Ovejeron, C. Bran, M.P. Morales, M. Vazquez, E. Vilanova, J. Kosel, J. Magn. Magn. Mater. **389**, 144–147 (2005)

122. M.N. Baibich, J.M. Broto, A. Fert, F. Nguyen Van Dau, F. Petroff, Phys. Rev. Lett. **61**, 2472 (1988)
123. W. Blum, Trans. Amer. Electrochem. Soc. **40**, 307 (1921)
124. U. Cohen, F.B. Koch, R. Sard, J. Electrochem. Soc. **130**, 1987 (1983)
125. D. Tench, J. White, Metall. Trans. A **15A**, 2039 (1984)
126. J. Yahalom, D.F. Tessier, R.S. Timsit, A.M. Rosenfeld, D.F. Mitchell, P.T. Robinson, J. Mater. Res. **4**, 755 (1989)
127. D.S. Lashmore, M.P. Dariel, J. Electrochem. Soc. **135**, 1288 (1988)
128. K. Attenborough, R. Hart, S.J. Lane, D.S. Lashmore, C. Younes, W. Schwarzacher, Appl. Phys. Lett. **53**, 2144 (1993)
129. W. Schwarzacher, D.S. Lashmore, IEEE Trans. Magn. **32**, 3133 (1996)
130. G. Nabiyouni, in *Nanomagnetism and Spintronics*, ed. by F. Nasirpouri, A. Nogaret, World Scienctific Pub., (2011)
131. I. Bakonyi, L. Péter, Prog. Mater Sci. **55**, 107–245 (2010)
132. Q. Yang, P. Holody, S.F. Lee, L.L. Henry, R. Loloee, P.A. Schroerder, W.P. Pratt Jr., J. Bass, Phys. Rev. Lett. **72**, 3274 (1994)
133. S.D. Steenwyk, S.Y. Hsu, R. Loloee, J. Bass, W.P. Pratt, J. Magn. Magn. Mater. **170**, L1 (1997)
134. T. Valet, A. Fert, Phys. Rev. B. **48**, 7099 (1993)
135. L. Piraux, J.M. George, J.F. Despres, C. Leroy, E. Ferain, R. Legras, K. Ounadjela, A. Fert, Appl. Phys. Lett. **65**, 2484 (1994)
136. A. Blondel, J.P. Meier, B. Doudin, J-Ph Ansermet, Appl. Phys. Lett. **65**, 3019 (1994)
137. A. Blondel, B. Doudin, J-Ph Ansermet, J. Magn. Magn. Mater. **165**, 34 (1997)
138. I. Kazeminezhad, W. Schwarzacher, J. Magn. Magn. Mater. **226–230**, 1650–1652 (2001)
139. I. Kazeminezhad, H.I. Blythe, W. Schwarzacher, Appl. Phys. Lett. **78**, 1014 (2001)
140. F. Nasirpouri, PhD Thesis: Electrodeposition of multilayered nanowires and magnetic magnetotransport properties of nanomagnetic structures, Sharif University of Technology, 2006
141. C. Schönenberger, B.M.I. Zande, L.G.J. Fokkink, M. Henny, C. Schmid, M. Krüger, A. Bachtold, R. Huber, H. Birk, U. Staufer, J. Phys. Chem. B **101**, 5497 (1997)
142. F. Nasirpouri, P. Southern, M. Ghorbani, A. Irajizad, W. Schwarzacher, J. Magn. Magn. Mater. **308**(1), 35–39 (2007). Reprint copyright 2016 with the permission of Elsevier
143. K. Liu, K. Nagodawithana, P. Searson, C.L. Chien, Phys. Rev. B **51**, 7381 (1995)
144. A. Blondel, J. Meier, B. Doudin, J-Ph Ansermet, K. Attenborough, P. Evans, R. Hart, G. Nabiyouni, W. Schwarzacher, J. Magn. Magn. Mater. **148**, 317 (1995)
145. G.P. Heydon, S.R. Hoon, A.N. Farley, S.L. Tomlinson, M.S. Valera, K. Attenborough, W. Schwarzacher, J. Phys. D Appl. Phys. **30**, 1083 (1997)
146. P.R. Evans, G. Yi, W. Schwarzacher, Appl. Phys. Lett. **76**, 481 (2000)
147. L. Piraux, S. Dubois, A. Fert, J. Magn. Magn. Mater. **159**, L287 (1996)
148. S. Dubois, C. Marchal, J.M. Beuken, L. Piraux, J.L. Duvail, A. Fert, J.M. George, J.L. Maurice, Appl. Phys. Lett. **70**, 396 (1997)
149. J.L. Duvail, S. Dubois, L. Piraux, A. Vaures, A. Fert, D. Adam, M. Champagne, F. Rousseaux, D. Decanini, J. Appl. Phys. **84**, 6359 (1998)
150. S. Dubois, J.L. Duvail, A. Fert, J.M. George, J.L. Maurice, L. Piraux, J. Appl. Phys. **81**, 4569 (1997)
151. M. Chen, L. Sun, J.E. Bonevich, D.H. Reich, C.L. Chien, P.C. Searson, Appl. Phys. Lett. **82**, 3310 (2003)
152. M. Chen, P.C. Searson, C.L. Chien, J. Appl. Phys. **93**, 8253 (2003)
153. A. Robinson, W. Schwarzacher, J. Appl. Phys. **93**, 7250 (2003)
154. A. Robinson, P. Southern, W. Schwarzacher, J. Phys. Conf. Ser. **17**, 16 (2003)
155. I. Bakonyi, E. Toth-Kfidfir, T. Becsei, J. Toth, T. Tarnoczi, A. Czirfiki, I. Gerocs, G. Nabiyouni, W. Schwarzacher, J. Magn. Magn. Mater. **156**, 347–349 (1996)
156. W. Lee, K. Schwirn, M. Steinhart, E. Pippel, R. Scholz, U. Gosele, Nat. Mater. **5**, 741–747 (2006)

157. W. Lee, R. Scholz, G sele, U. Nano Lett. **8**, 2155–2160 (2008)
158. W. Lee, K. Schwirn, M. Steinhart, E. Pippel, R. Scholz, U. Gosele, Nat. Nanotechnol. **3**, 234–239 (2008)
159. M.S. Salem, P. Sergelius, R.M. Corona, J. Escrig, D. Gorlitz, K. Nielsch, Nanoscale **5**, 3941–3947 (2013)
160. K. Pitzschel, J.M. Montero Moreno, J. Escrig, O. Albrecht, K. Nielsch, J. Bachmann, ACS Nano **3**, 3463–3468 (2009)
161. I. Minguez-Bacho, S. Rodriguez-López, M. Vázquez, M. Hernández-Vélez, K. Nielsch, Nanotechnology **25**(14), 145301 (2014)

Chapter 6
Electrodeposited Nanocrystalline Films and Coatings

Abstract Electrocrystallisation is a major stage of electrodeposition of any materials in which ad-atoms adsorbed on the surface form crystalline nuclei under favoured electrochemical conditions. Provided that the driving force applied, the electrocrystallisation leads to the formation of nano-grained deposits, or nanocrystalline deposits. Two major approaches are known to obtain nanograin electrodeposits direct current electrodeposition using grain refining additives and pulse electrodeposition. We emphasize on the pulse electrodeposition in this chapter, thus, principles and important particulars will be discussed. Ni nanocrystalline films electrodeposited by direct, and pulse techniques will be compared.

6.1 Introduction

Bulk nanostructured materials or coatings refer to a particular category of bulk materials whose constituents or components are one or more of 0D, 1D or 2D nanostructures. One major group of bulk nanostructured materials is related to nanocrystalline coatings. Nanocrystalline coatings having a micro-structural length or grain size of up to about 100 nm possess unique chemical, physical or mechanical properties. The role of grain boundaries in nanocrystalline coatings are inevitable for instance more than 50 vol.% of atoms of a nanocrystalline material with a grain size below 10–20 nm incorporate into grain boundaries or interfacial phases. In this chapter, the principles and recent progresses of electrodeposition of nanocrystalline coatings and nanocomposite coatings are explained.

6.2 Electrodeposition of Nanocrystalline Films

Many advantages of the electrodeposition process for depositing of nanocrystalline coatings such as low cost, flexibility of production, industrial mitigation based on existing electroplating technologies, convenient of control over crystal grain size,

F. Nasirpouri, *Electrodeposition of Nanostructured Materials*,
Springer Series in Surface Sciences 62, DOI 10.1007/978-3-319-44920-3_6

microstructure, roughness, etc., capable of producing compact pore free dense coatings and economic production costs. As we have discussed in Chap. 3, the nucleation and growth of electrodeposits which determines crystallinity and morphology of structures, could be readily controlled via electrode potential, current density, electrolyte concentration, agitation, pH, temperature, etc. There have been two general methods to produce metal or alloy electrodeposits with nanocrystalline structures including (i) DC electrodeposition from electrolytes contacting grain refining additives such as saccharin ($C_7H_4NO_3S$) [1–3], and (ii) pulse electrodeposition [4, 5].

The most admitted method of the pulse electrodeposition is carried out under galvanostatic mode. In the DC electrodeposition, a constant current is applied throughout the whole deposition process, whereas in the pulse method, the current is switched between different magnitudes as a function of time. The duration of current pulses varies for different pulse current techniques. Many advantages have been shown for the pulse electrodeposition such as the capability of producing smoother, less porous, more uniform films by thickness, finer grained electrodeposits, no need to use additives in electrolyte bath in most cases, and the composition is the bath requires less metal salt. Since pulse electrodeposition technique has been shown to produce more uniform and finer nanograins which are not attainable by DC electrodeposition, we focus this section to pulse electrodeposition and its application to nanocrystalline coatings in particular nickel owing to its particular interest for many functional and mechanical applications.

6.3 Pulse Electrodeposition

6.3.1 Electrochemical Implications of Pulse Current Technique

Let's first consider a simple electrode reaction which is controlled by two modes: galvanostatic and potentiostatic. Figure 6.1a, b illustrate the ideal and real response of potential of a system during a potentiostatic and galvanostatic process, respectively. For a potentiostatic process (Fig. 6.1a), an increase of current at the beginning followed by a drop is observed. The initial increase owes to the capacitance of the double layer charging. For a galvanostatic process (Fig. 6.1b) a dramatic increase of voltage at the beginning is followed by a vertical increase due to the concentration polarization. The capacity of the double layer limits the slope of the potential increase. Therefore, it imposes a delay time in an order of a few microseconds before the charge transfer reaction occurs.

Figure 6.1c shows an equivalent circuit for a simple electrode reaction. In a potentiostatic circuit the most serious problem is the ohmic resistance, R_c, between electrode and Luggin capillary. This resistance prevents the application of potentiostatic conditions, i.e. $\eta = constant$, because η_L can only be externally controlled,

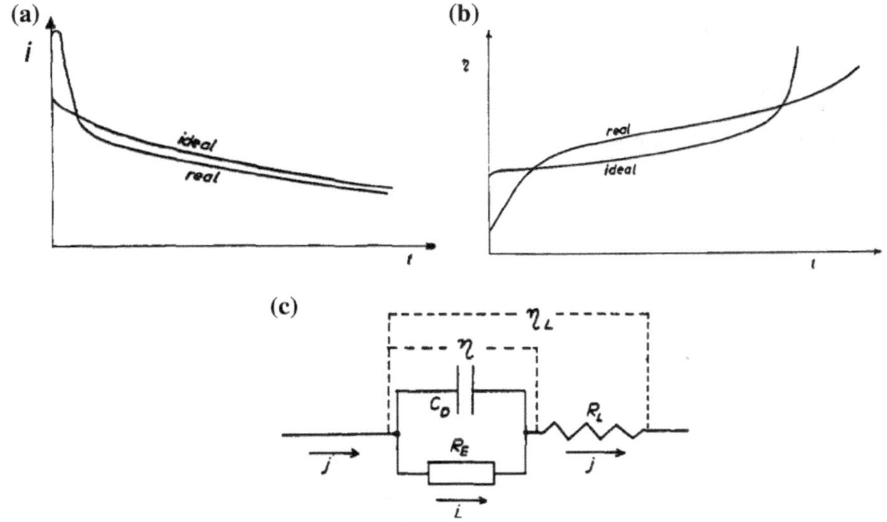

Fig. 6.1 **a** Current response transients for real and ideal systems in a potentiostatic circuit, **b** potential response transients for real and ideal systems in a galvanostatic, **c** the electrical equivalent circuit for a simple electrode reaction. Adapted from [6] with the permission of the American Chemical Society (ACS), copyright 2016

see Fig. 6.1c. For galvanostatic case, the ohmic voltage drop is neglected when R_L remains constant during the transient. The charging time of the double layer, however, is relatively long here, because the charging current is limited by the constant current applied to the system. Thus, the double layer capacity is a limit of application of the galvanostatic method which, in turn, makes greater changes of concentration at the interphase.

Now let's consider a double pulse technique as shown in Fig. 6.2a. A short (microseconds) at very high constant current pulse is superimposed upon a longer galvanostatic pulse. At short time a very high current charges the double layer associated with a linear overvoltage increase until it the current is switched. The overvoltage should remain constant owing to the charge transfer reaction until the concentration polarization becomes noticeable. Figure 6.2b illustrates the schematical responses of overvoltage for a double pulse. This can be well balanced by compensating the ohmic resistances. Using pulse galvanostatic technique, a potentiostatic condition as a balancing condition is applied. Therefore, it is a combination of both methods avoiding the difficulties of feedback in the potentiostatic method and the limitation of the charging current for a single galvanostatic impulse [6].

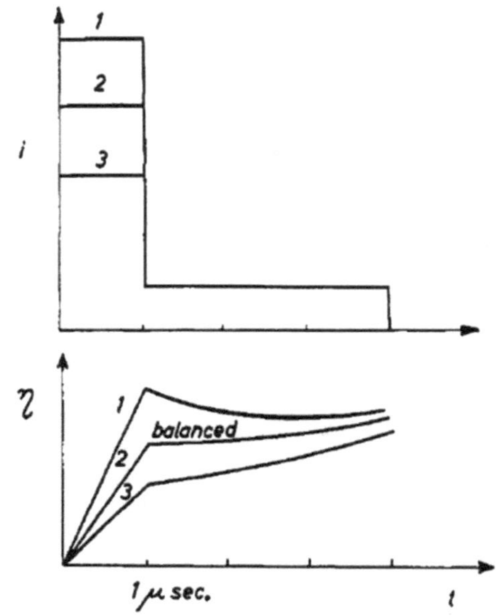

Fig. 6.2 Schematic picture of a double pulse galvanostatic technique along with the overvoltage responses. Reproduced from [6] with the permission of the American Chemical Society, Copyright 2016

6.3.2 *Principles of Pulse Electrodeposition*

Thanks to advances in microelectronics and control systems, pulse electrodeposition has become a predominant technique for deposition of nanocrystalline metal and alloy films and coatings. In principle, the electrical pulses are the created according to certain waveforms by modulating signal amplitude as a function of time. The electrical pulses are generally divided into major groups: (i) unipolar or single polarity pulses in which all pulses are in one direction and (ii) bipolar or double polarity pulses in which the pulses are both positive and negative. In the pulse electrodeposition processes, current pulses are employed. Three variables of primary importance defining the pulse modulation system include: peak current density, i_p, current 'on' time, t_{on} and current 'off' time, t_{off}. Current values can be cathodic and anodic with pulse on and off times ranging from microseconds to miliseconds. This is usually determined by the instrumental capabilities which limit the pulse parameters. A wide variety of pulse waveforms can be generated as shown in Fig. 6.3. The simplest pulse waveform is called direct pulse in which the current modulates between a cathodic current and zero current. If an anodic is applied this appears a pulse reverse technique. Each variable plays a role for the electrodeposition process which will be addressed in this chapter. Table 6.1 summarizes the most important effects of each variable [7].

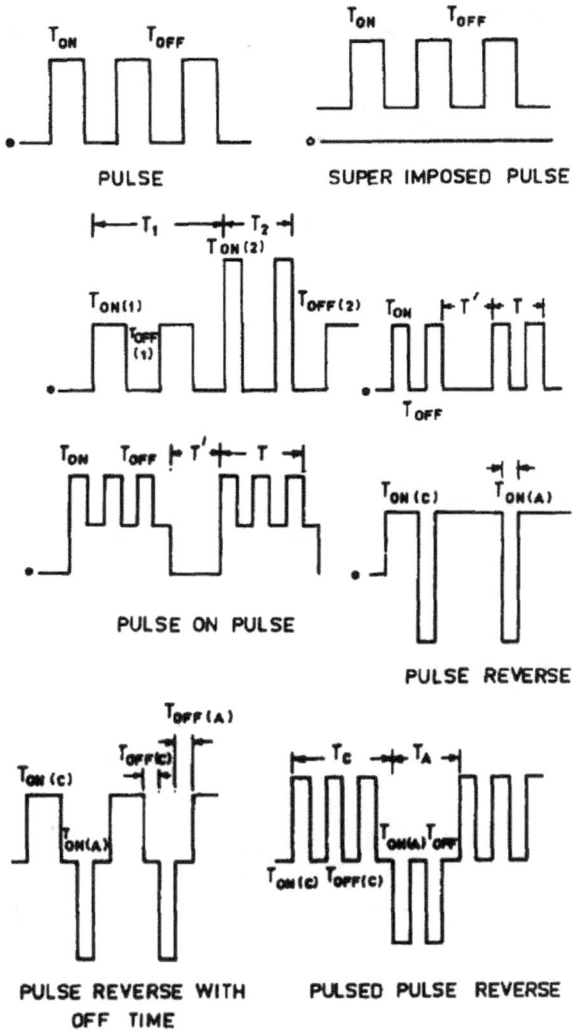

Fig. 6.3 Possible waveforms for different pulse electrodeposition techniques. Reproduced from [8] with the permission of Elsevier

The average current density i_a in pulse and pulse reverse electrodeposition is expressed by:

$$i_a = i_p \cdot \gamma \tag{6.1}$$

in which γ is duty cycle as is defined as:

$$\gamma = \frac{t_{on}}{t_{on} + t_{off}} \times 100 \tag{6.2}$$

Table 6.1 Possible effects of pulse plating on deposit structure. Reproduced from [7] with the permission of Elsevier

Time interval	Conditions that differ from DC plating	Phenomena affected
On-time	Double layer charging	Nucleation rate
	Overvoltage	Growth mechanism (e.g. dendrites)
	Concentration profile near electrode	Electrode reaction mechanism
	Adsorption (ions, additives, hydrogen)	Codeposition rate (H, alloy elements) Additive reactions
Off-time	Double layer discharge	Surface diffusion
	Potential relaxation	Surface recrystallisation
	Concentration profile relaxation	Corrosion, displacement reactions
	Desorption (additives, ions, hydrogen)	Passivation hydrogen diffusion
Pulse reverse-time	Anodic potential	(Selective) metal dissolution
	Sign change of double layer charge	Hydrogen reoxidation
	Concentration profile near electrode	Additive oxidation
	Desorption/adsorption (additives, ions)	Passivation

γ is in a range between 0 and 100%. Another parameter is the pulse frequency f which is expressed by:

$$f = \frac{1}{t_{on} + t_{off}} \tag{6.3}$$

6.3.3 *Electrochemical Aspects of Pulse Electrodeposition*

It was explained in Sect. 6.3.1 that in a typical electrode reaction under galvanostatic the time spent to charge the double layer is a limiting factor and rise the overpotential to an extent at which the charge transfer reaction takes place. In pulse electrodeposition, the on and off times are limited by the rate of charging and discharging, respectively, of the electrical double layer at the cathode-electrolyte interface. Let's consider a simple direct pulse (square waveform) electrodeposition process as is shown in Fig. 6.4a. At the beginning of each pulse the total current i_{total} applied in each pulse i_p must support two processes: a capacitive current i_C (which charges the double layer), and a faradaic current, i_F (which corresponds to the rate of metal deposition):

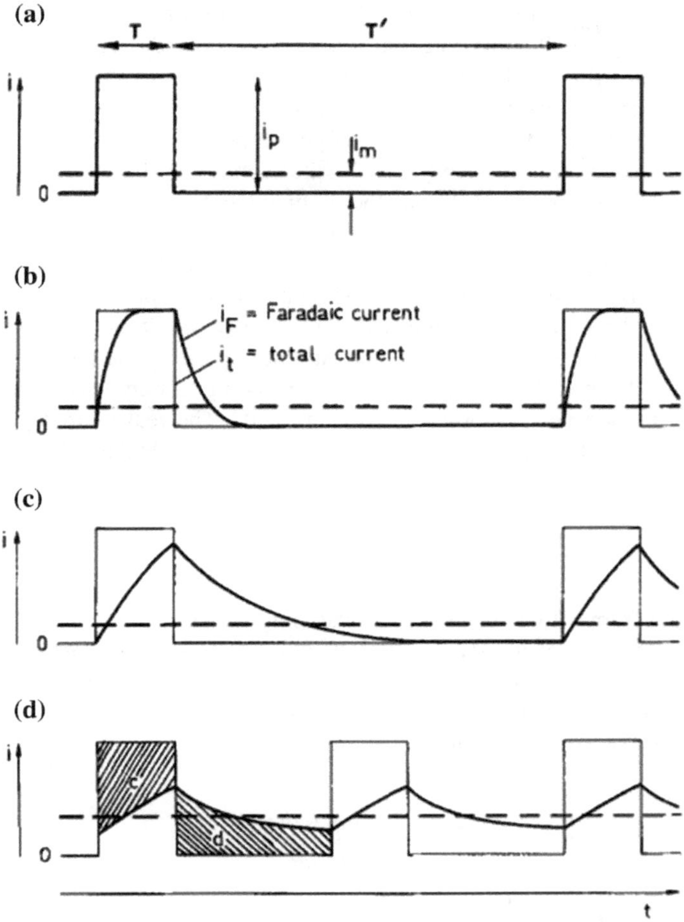

Fig. 6.4 Damping of Faradaic current. **a** $t_C < T$, no damping. **b** $t_C < T$, small damping. **c** $t_C > T$, **d** $t_C \gg T$ and $t_d \ll T'$. **c, d** strong damping. *C* and *d* denote charging and discharging, respectively. Reproduced from [10] with the permission of Springer

$$i_{total} = i_C + i_F = i_p \tag{6.4}$$

i_F is determined by the *Bulter-Volmer* equation (c.f. Chap. 2, 2.44) while i_C is expressed by:

$$i_C = \frac{dQ}{dt} = C\frac{d\eta}{dt} \tag{6.5}$$

The charging time is the time before the cathode potential reaches the value corresponding to the pulse current, i.e. until i_F becomes equal to the total current supplied by the generator (Fig. 6.4b). If the charging time (t_C) is longer than the duration of the pulse (T), this potential is never reached and i_F remains smaller than i_{total} during the whole pulse (Fig. 6.4c, d).

A similar phenomenon occurs after the end of the pulse. The double layer must be discharged and it takes some time before the potential drops to the value corresponding to zero current. Therefore, it takes some time before i_F drops to zero (Fig. 6.4b). If this time is longer than the off time the double layer is not completely discharged and i_F never decreases to zero. In general, if the charge and discharge time are negligibly small as compared to the duration of the pulse and the off time respectively, i_F is virtually equal to the total current supplied by the generator (Fig. 6.4a). Otherwise we have the cases of Fig. 6.4b, c: although the total current fed to the cell is a rectangular pulse, the current depositing the metal is not and is more or less damped. If the capacitive effect is strong i_F only oscillates around an average value and never drops to zero (Fig. 6.4d). The real current pulse curves exhibit a degree of damping or flattening (Δ) which is calculated by:

$$\Delta = \frac{\left(i_p T - \int_0^T i_p dt\right)}{i_a T'} \tag{6.5}$$

Δ is ranging from 0 to 1. If there is no damping effect then $\Delta = 0$ and the square waveform comes to reality. But if $\Delta = 1$ then the complete damping has taken pace n and pulse current becomes constant over total deposition time meaning that the pulse electrodeposition is equal to direct current electrodeposition.

A further interesting quantity is the ratio:

$$\theta = \int_0^T \frac{i_F dt}{i_p T} \tag{6.6}$$

which theoretically represents the fraction of the total amount of electricity used to deposit metal during the pulses, i.e. the current efficiency for metal deposition, assuming that metal is deposited only during the pulses. Provided that the charging and discharging times are close or greater than the pulse period, the theoretical current efficiency θ is smaller than the experimental θ [9]. This confirms the view illustrated in Fig. 6.4.

The range of variation of the capacitance of the electrical double layer at the cathode/electrolyte 10–100 μF cm^{-2} is large which makes the estimation of charging and discharging times a difficult task. Considering an average value for the capacitance about 40–50 μF cm^{-2} provides a rough estimation of charging time and discharging time for preliminary experiments as expressed by:

$$t_C = \frac{17}{i_p} \tag{6.5}$$

$$t_d = \frac{120}{i_p} \mathrm{e}$$

where t_C and t_d are measured in μs and i_p in A cm^{-2}. For precise experiments, it is essential to know capacitance and the exchange current density. For detailed calculations and mathematical aspects it is referred to [10].

Furthermore, mass transport significantly influences the pulse electrodeposition mechanism. Ilb et al. [11] have proposed a duplex diffusion layer for the case of short pulses. Figure 6.5 shows a schematic picture of the duplex diffusion layer model in which two diffusion layers exist including one layer near the electrode which pulsates and the other further outside. The diffusion layer is essentially stationary after the application of a number of pulses. The occurrence of this stationary diffusion layer is due to the fact that in the case of short pulses the pulsating diffusion layer does not penetrate into the region where there is mass transport by convection, because the hydrodynamic flow decays at the wall owing to friction forces. Through the region where convection is not effective, the species consumed at the cathode must be transported by diffusion to the boundary of the pulsating layer. The depletion in the stationary outer layer depends on the hydrodynamic conditions and on the value of the average current density i_a defined. The depletion in the stationary diffusion layer was calculated by using Nernst's model of the diffusion layer (in which a linear concentration profile is assumed) [11]:

$$\frac{C'_e}{C_0} = 1 - \frac{i_a(\delta_N - \delta_p)}{nFDC_0} \tag{6.6}$$

In which C'_e is the concentration at the outer boundary of the pulsating diffusion layer, C_0 the concentration in the bulk solution, δ_N is the thickness of Nernst's diffusion layer in DC electrodeposition under the same hydrodynamic conditions and δ_p the thickness of the pulsating layer at the end of the pulse, D is the diffusion coefficient of the metallic ions, n their charge number and F Faraday's constant.

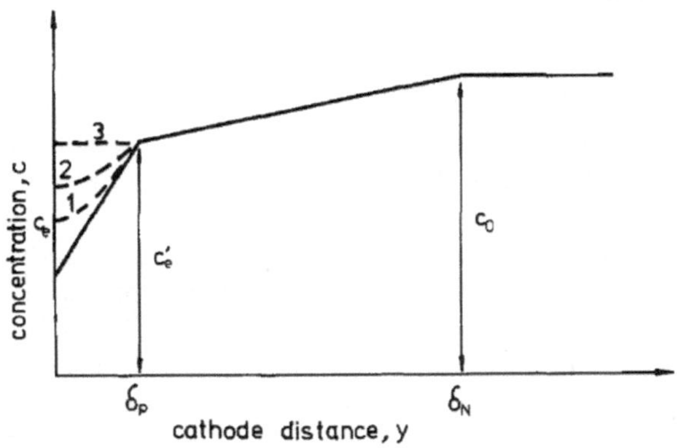

Fig. 6.5 Duplex diffusion model: Diffusion layers in pulse electrolysis (schematic): 1, 2, 3 are concentration profiles at times $t_1 < t_2 < t_3$ after current cut-off; *line* concentration profile at end of pulse; *dashed line*concentration profile during off time. Reproduced from [11] with the permission of Elsevier

Let us now turn towards the pulsating layer. Here, the concentration changes periodically with time. If the depletion in the stationary diffusion layer is small, i.e. for $C'_e \sim C_0$, the transition time τ defined as the time required to drop the interfacial concentration to zero is given by the classical equation of Sand:

$$\tau = \frac{\pi D C_0^2 (nF)^2}{4 i_p^2} \tag{6.7}$$

Whereas τ depends on the pulse current density and on the bulk concentration, the thickness δ_p which the pulsating (Nernst's) diffusion layer reaches at the end of the pulse is independent of these two quantities and simply proportional to the square root of the pulse duration T [11]:

$$\delta_p = 2\left(\frac{DT}{\pi}\right)^{\frac{1}{2}} \tag{6.8}$$

By using very short pulses one can thus realize extremely thin pulsating diffusion layers.

These all have been used to calculate the limiting current density and diffusive layer thickness. More analytical and numerical solutions of the mass transport equations for pulse electrodeposition have been proposed as explained in [12–16]. In a more recent report, [17] the calculations of the linear model of Elb was further modified to work out the pulse limiting current, i_{pL}. The pulse limiting current density is that pulse current density at which the concentration at the electrode surface of the reacting species reaches zero at the end. This is expressed by:

$$i_{pL} = i_L \left[\frac{\delta_p}{\delta}(1-\gamma) + \gamma\right]^{-1} \tag{6.9}$$

With $\delta_p = \left[\frac{4}{\pi} D t_{on} (1-\gamma)\right]^{1/2}$.

Mass transport imposes two types of limitations on the pulse electrodeposition parameters: (a) by actually limiting the rate of metal deposition and (b) by causing powder formation. If the pulse current density is higher than i_{pL} the interfacial concentration drops to zero before the end of the pulse but at least for times $t < \tau$ the current density of metal deposition is greater than i_{pL}. This is associated the concentration profiles remain constant. However, at $t = \tau$ the metal deposition switches from deposition at a constant rate to deposition with a constant interfacial concentration since for $t > \tau$ the interfacial concentration cannot drop further. Nevertheless, the metal continues to be deposited and metal ions continue to be removed from the solution so that the zone with depleted concentration in the immediate vicinity of the cathode grows. Therefore the rate of metal deposition drops and if the overall current is kept constant an increasing part of it must now be used for hydrogen evolution. The current efficiency is therefore no longer 100%.

However, in spite of the decrease in the rate of metal deposition during the interval from τ to T, the average rate of metal deposition may be somewhat greater than for electrolysis with a pulse current density equal to i_{pL}.

If the plating is carried out with limiting current pulses, a strong depletion takes place in the pulsating diffusion layer but almost no depletion in the outer stationary diffusion layer. In general, the depletion in the outer diffusion layer increases with increasing average current density. If the latter approaches the limiting current density for DC the depletion in the outer layer is strong, where the average rate of metal deposition in pulse plating cannot be larger than at the limiting current in DC electrodeposition.

Another mass transport limitation may cause the formation of very rough or powdery deposits, when the solution in the immediate vicinity of the cathode is strongly depleted with respect to the metal cations. However, in pulse electrodeposition, this is a quite general rule powder formation due to depletion effects does not lead to an upper limit on the useful range of pulse current densities: even for limiting current pulses or pulses with $i_p > i_{pL}$ and $T > \tau$ powder formation does not occur provided that the pulsating diffusion layer is sufficiently thin and the depletion in the outer stationary diffusion layer is small. In contrast, strong depletion in the outer stationary diffusion layer causes powder formation and dendrites grow in a fashion similar to that observed in DC plating. This phenomenon limits the increase in the average current density up to the limiting current density in pulse electrodeposition [9–11].

6.3.4 *Pulse Electrodeposition Conditions for Nanocrystalline Films*

In light of the concepts discussed in Chap. 3, the pulse electrodeposition may lead to the production of nanocrystalline films provided that a high nucleation rate is accompanied by slow grain growth. It must be aimed at selecting optimum parameters of pulse electrodeposition in order to produce nanograin films. In the first assumption, high cathodic pulse currents and short cathodic duty cycles are needed while limiting factors, i.e. the physical constraints and product constraints must be taken into account. Physical constraints include quality of the waveform applied, the mass transfer limitations and the capacitance effect of the double layer [9–11] as discussed in the previous section. The product constraints may relate the nucleation rate, grain size, and uniformity. Erb and co-workers [18] have carried out a sensitivity study on the effect of various pulse electrodeposition parameters on the overpotential which may be a good criterion for controlling the grain size. Table 6.2 shows the results of the sensitivity analysis.

It can be seen from Table 6.2 that significant decreases (>20%) in the overpotential occur when changes are made to the duty cycle and the cathodic charge transfer coefficient. Moderate decrease (10–20%) is observed when the limiting

Table 6.2 Results of a sensitivity analysis on the effective variables of pulse electrodeposition towards nanocrystallisation

Variable	Change	Parameter	Overpotential	Change
Decrease duty cycles	0.2–0.1	Fr = 10 Hz Fr = 50 Hz	−4.176 to −2.502 −4.998 to −3.177	[−40%] [−36%]
Increase cathodic transfer coefficient	0.25–0.50	Fr = 10 Hz Fr = 50 Hz	−3.912 to −2.502 −4.703 to −3.177	[−36%] [−32%]
Increase limiting current density (decrease diffusion layer)	0.1–0.2 A cm^{-2}	Fr = 10 Hz Fr = 50 Hz	−2.940 to −2.502 −3.671 to −3.177	[−15%] [−13%]
Decrease diffusion coefficient, D	6×10^{-6} to 3×10^{-6} $cm^2\ s^{-1}$	Fr = 10 Hz Fr = 50 Hz	−2.502 to −2.253 −3.177 to −2.878	[−10%] [−9%]
Decrease kinetic parameter	1.0–0.5	Fr = 10 Hz Fr = 50 Hz	−2.606 to −2.398 −3 305 to −3.048	[−8%] [−8%]
Increase exchange current density	0.0025–0.005 A cm^{-2}	Fr = 10 Hz Fr = 50 Hz	−2.639 to −2.502 −3.314 to −3.177	[−5%] [−4%]
Decrease frequency	20–10 Hz	$\theta_1 = 0.01$ $\theta_1 = 0.1$	−0.690 to −0.565 −2.785 to −2.502	[−18%] [−10%]

Adapted from [18] with the permission of Springer

current density is doubled while the diffusion coefficient, kinetic parameter and the exchange current density do not cause significant changes (~<10%) to the overpotential. The effect of the decrease in frequency on lowering the overpotential is more sensitive at lower duty cycle. By knowing the sensitivities of these trends, one can make the appropriate modifications to the kinetic, system or process parameters in order to affect the nucleation rates.

6.4 Electrodeposited Nanocrystalline Nickel Films

Nanocrystalline nickel is an example which has attracted much attention due to its improved surface properties such as hardness, corrosion and wear resistances, and also functional characteristics like magnetic and magneto-transport properties.

Electrodeposition of nanocrystalline nickel has been paid much attention as well using different techniques. Nasirpouri and co-workers [19, 20] have investigated the electrodeposition and properties of nanocrystalline nickel films by direct current (DC), pulse current (PC) and pulse reverse current (PRC) techniques on a copper substrate to exploit the effect of electrodeposition conditions on the structure and morphology and resulting corrosion and other properties including cathodic efficiency film, morphology, hardness, grain size, roughness and magnetic as will be discussed here.

NC nickel films were prepared by electrodeposition in a Watts bath with pure nickel (99.99%) plate (Surface Area: 10 cm^2) as the anode and copper foil (surface area: 1 cm^2) as the cathode. The Watts bath consisted of $NiSO_4{\cdot}6H_2O$ (265 g/l of H_2O), $NiCl_2{\cdot}6H_2O$ (48 g/l of H_2O), and $H_3BO_3{\cdot}$(31 g/l of H_2O). Electrodeposition was conducted in a pH of 3.7 at a temperature of 45 ± 0.5 °C. The schematic view of PRC and PC methods are presented in Fig. 6.6, representing operating parameters such as the forward (i_p) and reverse peak current density (i_n), the forward ($T_{on(p)}$) and reverse pulse on-time ($T_{on(n)}$) and the forward ($T_{off(P)}$) and reverse pulse off-time ($T_{off(n)}$). The operating parameters used in this study are summarized in Table 6.3. The variables of the pulse electrodeposition were calculated to reach the conditions discussed in above. Forward pulse on-time was less than the calculated value of

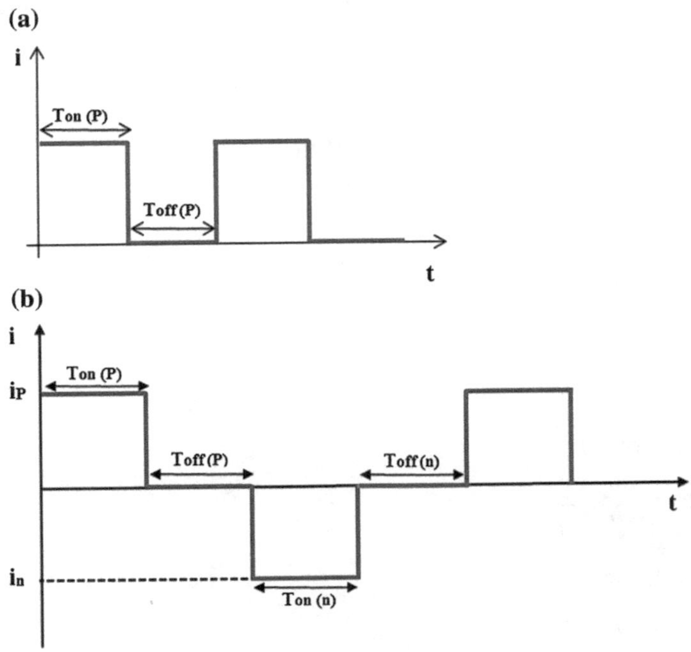

Fig. 6.6 Schematic diagram of waveforms used in electrodeposition of nanocrystalline nickel films **a** PC and **b** PRC techniques. Reproduced from [20] with the permission of Elsevier

Table 6.3 Process parameters for electrodeposition of nanocrystalline nickel films

Electrodeposition parameters	
Direct current (DC) electrodeposition	
Current density	100 mA/cm^2
Electrolyte agitation (magnetic stirring)	400 rpm
Pulse current (PC) electrodeposition	
Peak current density	100 mA/cm^2
Electrolyte agitation (magnetic stirring)	400 rpm
pulse on-time	2 ms
pulse off-time	18 ms
Pulse reverse current (PRC) electrodeposition	
Peak current density of forward pulse	100 mA/cm^2
Peak current density of reverse pulse	100 mA/cm^2
Electrolyte agitation (magnetic stirring)	400 rpm
Forward pulse on-time	2 ms
Forward pulse off-time	18 ms
Reverse pulse on-time	1 ms
Reverse pulse off-time	4 ms

Reproduced from [20] with the permission of Elsevier

migration time τ (the time needed for interfacial concentration to reach zero, but was more than the charge time of double layer as calculated.

6.4.1 Surface Morphology and Roughness of Nanocrystalline Ni Films

Surface morphology of DC, PC, and PRC electrodeposited nanocrystalline nickel films examined by SEM is illustrated in Fig. 6.6. It is seen that DC electrodeposited Ni film (Fig. 6.7a) has a non-uniform surface structure with some micropores. But,

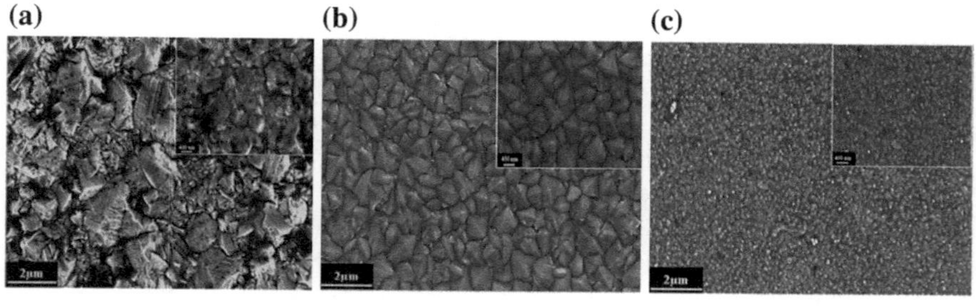

Fig. 6.7 Surface micrographs from nanocrystalline nickel films are deposited by **a** DC, **b** PC, and **c** PRC electrodeposition methods. Reproduced from [20] with the permission of Elsevier

Ni films electrodeposited by PC (Fig. 6.7b) and PRC (Fig. 6.7c) methods have a uniform surface morphology with defined repetitive patterns across the surface. Ni films electrodeposited by PC reveal a surface structure with polygonal texture having well defined facets, while those electrodeposited by PRC mention semi-spherical conical surface morphology.

The evolution of surface structural patterns is a complicated process to understand and little information exists on this subject, but it has been shown to be influenced by the deposition overvoltage, the type of metal ions to be deposited and other electrolyte conditions [21, 22]. However, for the morphology evolution in the electrodeposited nickel films, there is a narrow literature indicating that [23] the evolution of hydrogen and incorporation of hydrogen single atoms dissolved in Ni plays an important role in the change of structure, as is evident in our investigations too, which will be discussed in the next section.

Another feature observed in SEM images, is that the surface micropores or cavities are significantly avoided by both PC and PRC methods. Cavities are assumed to be generally caused by hydrogen evolution forming gas bubbles which remain attached onto the surface of the Ni electrodeposit. This observation may be due to the increase in the cathodic efficiency obtained for PC and PRC methods in respect to that obtained for DC electrodeposition. Current efficiency quantities of DC, PC and PRC methods were measured using weight measurements. PC technique exhibits the maximum cathodic current efficiency ($\sim$98%) and DC technique has the lowest cathodic efficiency ($\sim$88%). The reason for higher efficiency of PC and PRC (95%) methods is the reduction of hydrogen evolution. Indeed, during off pulse (t_{off}) the adsorbed hydrogen atoms are separated from the surface of the Ni electrodeposits and avoid the progress of evolution of H_2.

In order to study the evolution of surface morphology in more details, we compared the surface morphology and roughness of Ni electrodeposits obtained by the three different electrodeposition methods. Figure 6.8 illustrates atomic force micrographs taken from Ni films electrodeposited by PRC and PC techniques. The variation of the roughness parameters (R_a and R_z) with electrodeposition current mode measured using AFM and RT are summarized in Table 6.4. Results of both methods are consistent implying that Ni films electrodeposited using PRC method exhibited the lowest surface roughness values, whereas the maximum values recorded for surface roughness correspond to the films produced by DC method. The change in surface irregularities during electrodeposition owes to the nature of the kinetic and mass transfer processes. For the electrodeposition of nickel from Watts bath [24], deposition is a diffusion-controlled process. Diffusion in the electrolyte gives rise to a morphological instability in the sense that any surface irregularities are amplified rather than damped during growth. This instability causes significant changes in the surface metal ion concentration around a protuberance on the cathode so that the diffusion flux will be greatest near the tip of the protuberance, and as a result so does the rate of deposition. Hence diffusion can cause peaks on the cathode surface to grow more rapidly than valleys, amplifying any pre-existing roughness (c.f. Chap. 2). In contrast, when the reverse pulse is induced, it inversely affects the surface irregularities and reduces the surface

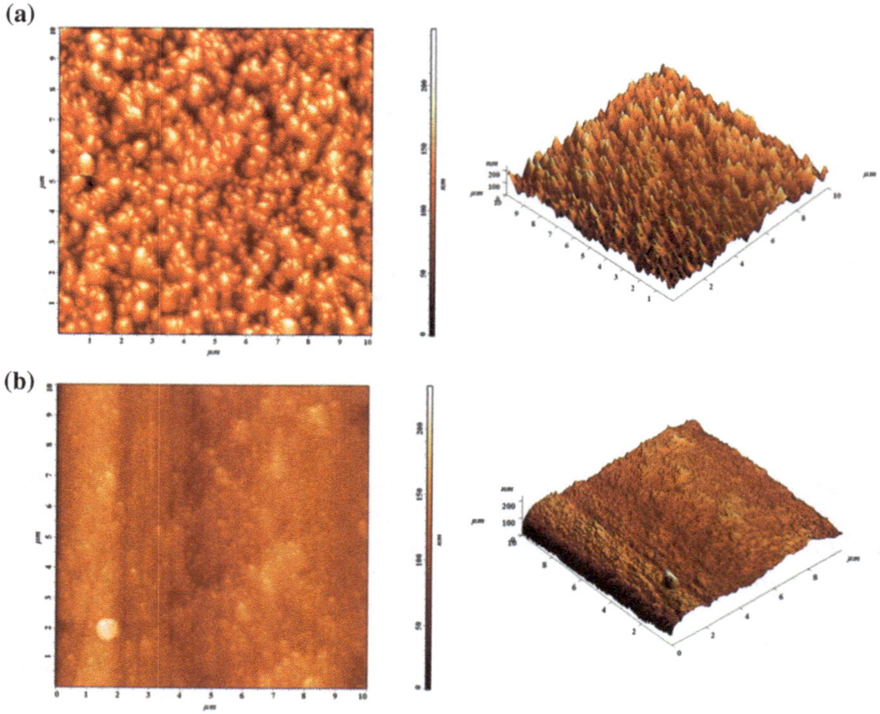

Fig. 6.8 AFM images of Ni films electrodeposited by PC, and PRC electrodeposition methods representing a measure for surface roughness. Reproduced from [20] with the permission of Elsevier

Table 6.4 Roughness values of Ni films electrodeposited by DC, PC and PRC methods measured by atomic force microscope (AFM) and roughness tester (RT)

Electrodeposition technique	RMS (nm)		R_z (nm)		R_a (nm)		R_y (nm)	
	AFM	RT	AFM	RT	AFM	RT	AFM	RT
PC	30.692	–	118.62	170.0	24.61	26.0	245.1	–
PRC	16.672	–	115.92	120.0	12.75	18.0	232.0	–
DC	41.12	–	128.23	210.0	30.12	38.0	–	–

Reproduced from [20] with the permission of Elsevier

roughness by detracting the metal ions from the sharp protuberances and smoothen the surface [9–11].

The effect of the thickness of the films on the roughness and grain size of nickel nanocrystalline film was investigated by A. Samardak et al. [25] using AFM. Figure 6.9a–c shows AFM images of nanocrystalline nickel films prepared by PC method for different deposition times (with different thicknesses). The SEM images were used to determine the average grain size. As the film thickness increases, the

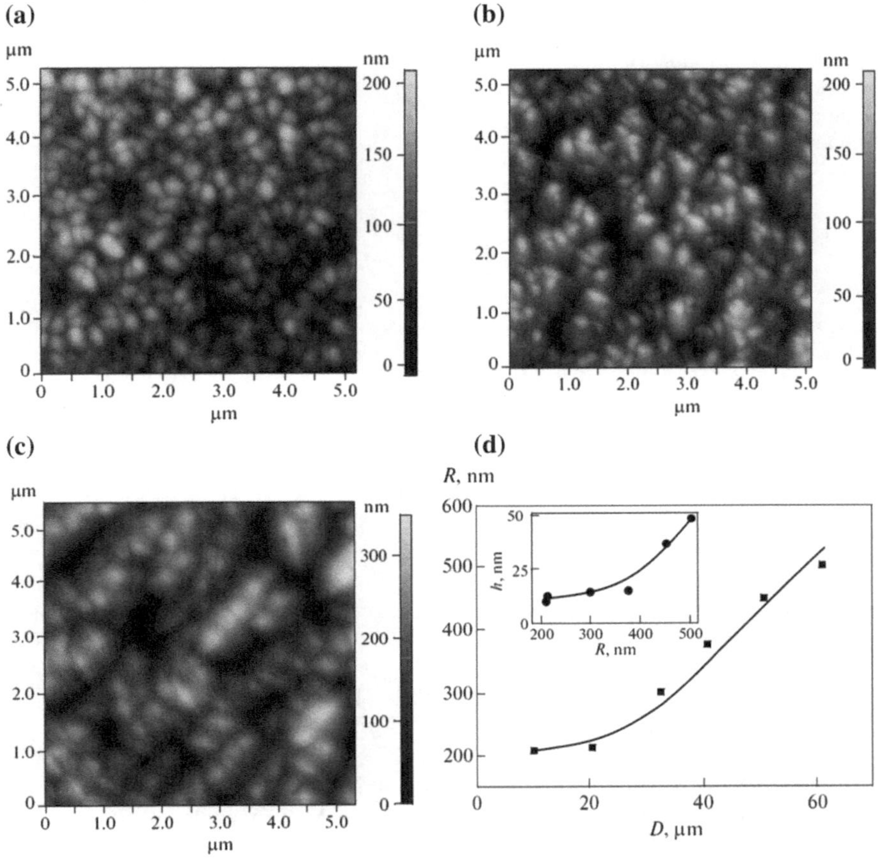

Fig. 6.9 AFM images of the surface of Ni films with thicknesses of **a** 10.2, **b** 20.4, and **c** 61.2 μm prepared by PC deposition. **d** Dependence of grain size R on thickness D of the film prepared by the PC method. The inset shows the dependence of roughness amplitude h on grain size R. Reproduced from [25] with the permission of Springer

grain size increases as shown in Fig. 6.9d. This, in turn, leads to an increase in film roughness (inset in Fig. 6.9d).

6.4.2 Crystallite Size and Crystallographic Texture

X-ray diffraction (XRD) patterns obtained for nanocrystalline nickel films electrodeposited by DC, PC, and PRC methods are illustrated in Fig. 6.10. All X-ray diffraction patterns show the main Bragg diffraction peaks of (111) and (200) of

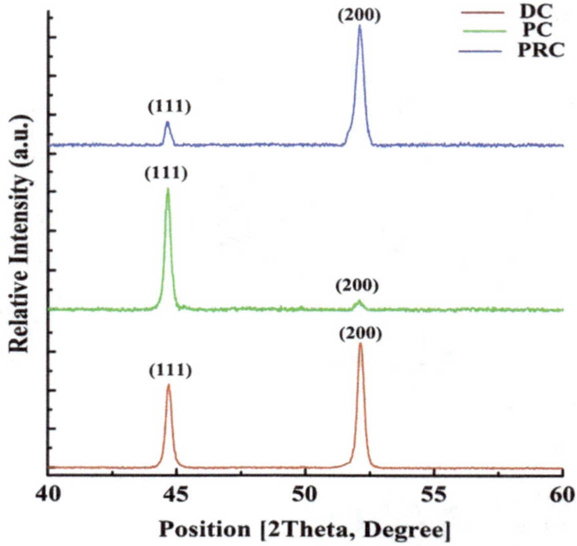

Fig. 6.10 X-ray diffraction patterns from nanocrystalline nickel films deposited by DC, PC, and PRC electrodeposition methods. Reproduced from [20] with the permission of Elsevier

nickel. On the basis of the *Scherer* equation expressed in below, the crystallite size of the Ni films was determined:

$$FWHM = \frac{K\lambda}{D\cos\theta}\frac{180^{\circ}}{\pi} \tag{6.10}$$

In which *FWHM* is full width half maxima in $2\theta^{\circ}$, *D* is the crystallite size in nm, *K* is a constant with a value of 0.94, and λ is the X-ray wavelength (0.154 nm). Average crystallite sizes obtained from (111) and (200) peaks for each Ni film are summarized in Table 6.5. This indicates that pulse electrodeposition (PC and PRC) methods reduce the crystal size.

When PC is applied, the current distribution and mass transfer are improved. Based on the Tafel equation, one can conclude that the larger the current density, the higher the overpotential. The high cathodic overpotential decreases the activation energy of nucleation, and leads to an increased nucleation rate. Consequently when the nucleation mechanism dominates the deposition process with a large number of nuclei generated on the substrate, the growth of nuclei and crystallites is strongly impeded [26]. On the other hand, it can be seen that the average crystallite size in Ni films deposited by PRC method is increased in respect to that of PC

Table 6.5 Crystallite size calculated for nanocrystalline Ni films from XRD spectrum

Electrodeposition methods	Average crystallite size (nm)
DC	41
PC	30
PRC	36

Reproduced from [25] with the permission of Elsevier

method. But still crystallite size produced by PRC is less than that of DC electrodeposited film. On the application of the reverse pulse at a constant cathodic peak current density, smaller crystals which are thermodynamically unfavorable are removed [27]. This could be the reason for the increase in crystallite for PRC method compared to PC method.

By comparing XRD patterns, there is observed a considerable change of crystallographic structure according to the intensity of (111) and (200) Bragg diffraction peaks. The intensity of the peaks in Ni electrodeposits produced by DC method is much less than that of the other deposits produced by PC and PRC methods. Therefore, the DC electrodeposited Ni film seems to assume a more random crystallographic texture. However, the greater intensity of (200) peak is revealed. In Ni films produced by PC method, the intensity of (111) peak is much higher than that of (200) indicating a preferred growth direction along ⟨111⟩ crystallographic planes. But, it changes to ⟨200⟩ direction in PRC electrodeposited Ni.

To quantify the relative crystallographic textures associated with nickel films, texture coefficients (TC) for predominant (h k l) peaks in XRD patterns were calculated. The texture coefficient (TC) for each (h k l) peak is given by (6.11) [28].

$$\mathrm{TC(hkl)} = \frac{\mathrm{I(hkl)}}{\mathrm{I_0(hkl)}}\left\{\frac{1}{\mathrm{n}}\sum\frac{\mathrm{I(hkl)}}{\mathrm{I_0(hkl)}}\right\}^{-1} \tag{6.11}$$

where *I (h k l)* is the measured intensities of (h k l) peak, *I0 (h k l)* is powder diffraction intensities of nickel, and *n* is the number of peaks used in the calculations. In this case, (111) and (200) peaks were considered for texture coefficient calculations (n = 2).

Table 6.6 summarizes the results of texture coefficient calculations for the observed (111) and (200) peaks in nanocrystalline nickel films. The texture coefficients for (111) and (200) peaks of standard randomly oriented polycrystalline nickel sample (JCPDS: 04–0850) are unity. For (111), the texture coefficient is greater than unity for PC deposition indicating strong (111) crystallographic texture in these films. However, the (200) texture coefficients for DC and PRC electrodeposited films are greater than that for PC electrodeposited films. This again indicates the existence of a strong (100) crystallographic texture in these films. With regard to the structure of the copper substrate which has no preferred orientation, thus there must not be any effect of substrate on such growth behavior.

Table 6.6 Texture coefficients (TCs) of various (hkl) planes for standard sample, and electrodeposited nanocrystalline nickel films

Electrodeposition methods	Texture coefficient (TC)	
	(111)	(200)
Standard sample	1	1
DC	0.52	1.48
PC	1.71	0.29
PRC	0.19	1.81

Reproduced from [20] with the permission of Elsevier

As we discussed earlier, we observed appreciably different surface morphology in addition to the changes seen for the crystallographic structure. In fact, these two observations are in a close relationship based on the electrodeposition mechanism. Electrodeposition of nickel from sulfate and chloride solutions proceeds simultaneously with the hydrogen discharge reaction, based on the proposed reactions:

$$\begin{aligned} &Ni + H^+ + e^- \rightarrow Ni{-}H_{\mathbf{ads}} \\ &Ni - H_{\mathbf{ads}} \rightarrow 2Ni + H_2 \\ &Ni{-}H_{\mathbf{ads}} + H^+ + e^- \rightarrow Ni + H_2 \end{aligned} \tag{6.12}$$

In Watts bath, the situation is rather complicated. There have been principle studies on the mechanism of the electrodeposition of nickel from Watts bath on vitreous carbon electrodes which in general propose the following reactions:

$$\begin{aligned} &Ni^{2+} + X^- \leftrightarrow NiX^+ \\ &Ni\,X^+ + e^- \leftrightarrow NiX_{ads} \\ &NiX_{ads} + e^- \leftrightarrow Ni + X^- \end{aligned} \tag{6.13}$$

The anion X^- has been variously assumed to be OH^- or Cl^- [29]. The changes in the texture of nanocrystalline nickel films may be attributed to the existence or formation of different interfacial inhibitors such as H_2, H_{ads} or Ni $(OH)_2$, in the nickel deposit/electrolyte interface during the electrodeposition process [C. Kollia, N. Spyrellis, Textural modification in nickel electrodeposition under pulse reversed current, Surf. Coat. Tech. 57 (1993) 71–75.]. Studies have shown that nickel film texture changes from $\langle 111 \rangle$ to $\langle 100 \rangle$ with an increase in hydrogen coverage [30]. One possible explanation is the extent of the hydrogen evolution in the electrodeposition or the H atom incorporation inside the film. This is consistent with our calculation of the hydrogen evolution which is more intensified in the PRC and DC electrodeposition method than in the PC method (Figure 6.11).

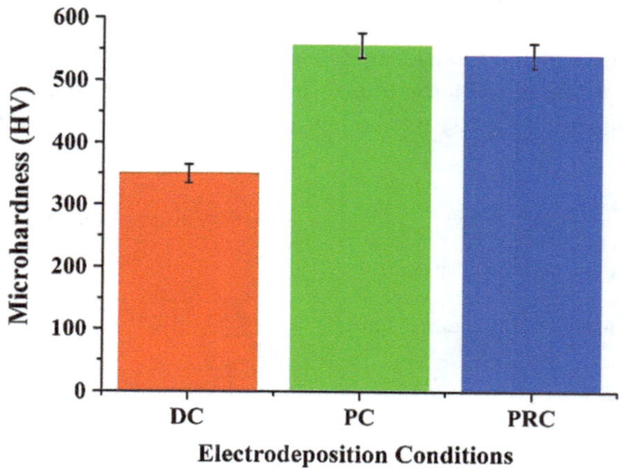

Fig. 6.11 Microhardness of nanocrystalline nickel films deposited by DC, PC and PRC electrodeposition methods. Reproduced from [20] with the permission of Elsevier

6.4.3 Microhardness of Nanocrystalline Nickel Films

The variation of microhardness of nanocrystalline nickel films with different deposition conditions (DC, PC, and PRC) are shown in Fig. 6.10. In general, hardness of the films produced by PC and PRC method are greater than that for DC films. The factors influencing the hardness of films included grain size, internal stress and the cavities. Thus the increased hardness values measured in Ni films deposited by PC and PRC methods are due to either smaller grain size (based on Hall–Petch equation [31]) or less cavities caused by hydrogen evolution. The hardness of Ni films prepared by PC is a little greater than PRC owing to the smaller grain size.

6.4.4 Magnetic Properties of Nanocrystalline Nickel Films

Magnetization curves of nanocrystalline nickel films deposited by the methods of DC, PC and PRC are shown in Fig. 6.12. The easy of magnetization lies in plane of the films which substantially arise by the shape anisotropy, but the values of coercivity and squareness change significantly for different types of Ni electrodeposits. The coercivity (H_c), squareness (M_r/M_s) and saturation field (H_s) values are extracted from the magnetic hysteresis loops are summarized in Table 6.7.

H_c of Ni films electrodeposited by PC and PRC methods increases in respect to DC method which can be attributed to a decrease in grain size. Variation of the squareness and also the saturation fields of Ni films imply that there must be another factor influencing the magnetic properties further to the shape anisotropy, for different films produced by DC, PRC and PC methods. Considering that nickel films

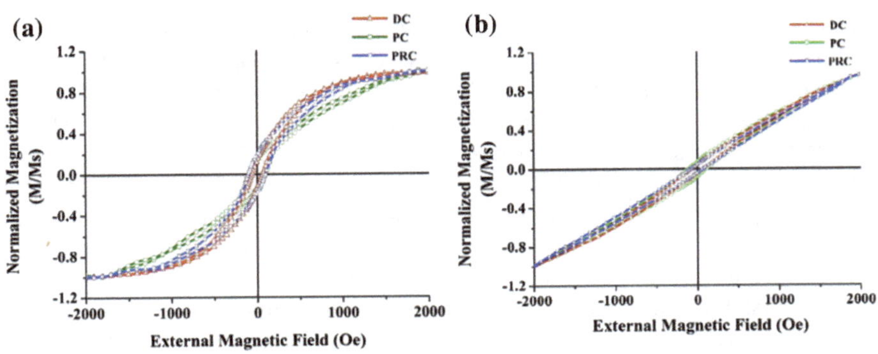

Fig. 6.12 Magnetization curves of nanocrystalline nickel films deposited by the methods of DC, PC and PRC: **a** in-plane (0°) and **b** perpendicular. Reproduced from [20] with the permission of Elsevier

Table 6.7 Coercivity (H_c), squareness (M_r/M_s) and saturation field (H_s) values of NC Ni films electrodeposited by DC, PC, and PRC electrodeposition methods

Electrodeposition methods	Coercivity (H_c, Oe)		Squareness (M_r/M_s)		Saturation field (H_s)	
	In-plane	Perpendicular	In-plane	Perpendicular	In-plane	Perpendicular
DC	64	70	0.12	0.047	1012	1740
PC	93	98	0.13	0.065	1600	1690
PRC	98	75	0.18	0.041	1200	1719

Reproduced from [20] with the permission of Elsevier

thickness is identical for all three types of films, thus the difference may only be attributed to the different crystal structures or magnetocrystalline anisotropy factors.

On the basis of our observations, different deposition methods change the crystallographic texture of the electrodeposited films. It is found that Ni films deposited by PC method has (111) crystallographic texture, while Ni films deposited by PRC mentions (100) crystallographic texture. However, Ni films obtained by DC electrodeposited exhibit an intermediate state with emphasis on greater peak intensity of (100). On the other hand, in a *fcc* nickel crystal, easy axis of magnetization lies along (111) orientation [32]. Therefore, we interpret the magnetic behavior with the crystallographic growth direction. For nickel films electrodeposited by PC, the crystals preferentially grow along $\langle 111 \rangle$ crystallographic direction which lies perpendicular to the substrate. Then, the magnetocrystalline anisotropy factor facilitates the saturation if the external magnetic field is applied perpendicular to the substrate, as is evident from the magnetization curves. But, this not the case any more for the films electrodeposited by PRC which has (100) crystallographic texture.

On the effect of film thickness or grain size on the magnetic properties of nanocrystalline nickel film, Samardak et al. [25] have performed an analytical study on the magnetism of such films. Nanocrystalline Ni films were obtained by PC electrodeposition AFM images of which were shown in Fig. 6.9. In the studied polycrystalline films, induced anisotropy is absent, while the local crystallographic anisotropy leads to the isotropy of the magnetization reversal processes. This is confirmed by the polar diagrams of the reduced residual magnetization determined from the magnetic hysteresis loops (Fig. 6.13a). For such film the grain size is in the order of 200 nm, which is comparable to the Bloch domain wall width. Hence, the grains whose sizes are comparable to the domain wall width will be single domain and, in each grain, the magnetization vector will be oriented along the crystallographic easy magnetization axis (EMA). Therefore, in films with grain sizes comparable to the domain wall width, the magnetization reversal processes occur owing to the incoherent rotation of the magnetization vector.

Nanocrystalline Ni films whose grain size is significantly larger than the domain wall width can have a two domain (multidomain) structure, which leads to a decrease in the magnetostatic energy of the grain. Therefore, in these films, the magnetization reversal process occurs due to the displacement of the domain walls (which are located in the grain) and the incoherent rotation of magnetization. The

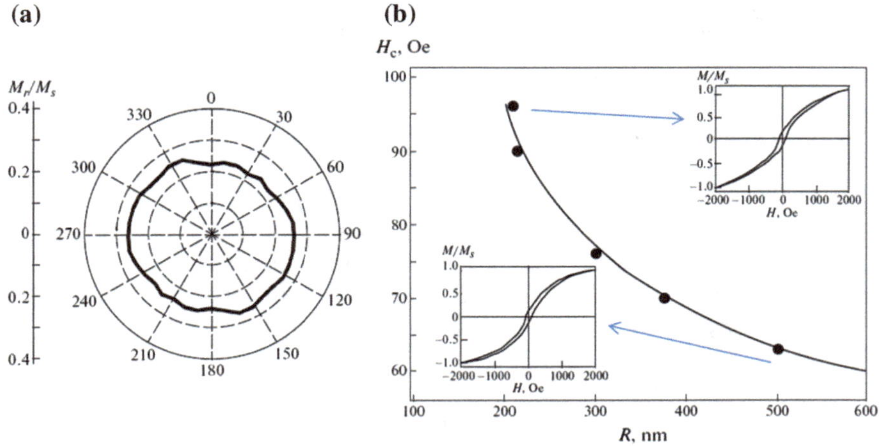

Fig. 6.13 **a** Polar diagram of the squareness ratio M_r/M_s for film with a thickness of 10.2 μm. Dependence of coercive force Hc on grain size R. The *insets* show the corresponding hysteresis loops for films with thicknesses of 10.2 (*right-top*) and 40.8 μm (*left bottom*). Reproduced from [25] with the permission of Springer

occurrence of the incoherent rotation of the magnetization vector and the displacement of the domain walls in these films is proved by the magnetic hysteresis loops (inset in Fig. 6.13b). These loops are extremely narrow and have the squareness ratio (M_r/M_s) ranging 0.11–0.24, where M_r is the residual magnetization and M_s is the saturation magnetization. Experimental results have shown that coercive force (H_c) decreases with an increase in grain size. The dependence of H_c on R is shown in Fig. 6.13b. It is known that the coercive force of polycrystalline films is determined by the pinning of the domain boundaries by structural defects. In polycrystalline films, these defects are grain boundaries, the disorientation of crystallographic EMAs, and surface roughness. More detailed analysis of the magnetic properties is available in [25].

6.4.5 *Corrosion Behaviour of Nanocrystalline Nickel Films*

Nasirpouri et al. studied the effect of DC and pulse electrodeposition techniques on the corrosion behavior of nanocrystalline Ni films in alkaline and acidic solutions as will be discussed here [20].

6.4.5.1 Electrochemical Corrosion Studies in NaOH Solution

Potentiodynamic polarization curves of the nanocrystalline (NC) Ni films electrodeposited with different methods (DC, PC and PRC) in 2 mol/l NaOH aqueous solution at room temperature are shown in Fig. 6.14a. All the Ni films show an active-passive-trans passive behavior. Based on previous studies, the formation of passive layer on the surface is one of the factors that affect the corrosion properties of nickel films. On the mechanism of passivation of Ni in alkaline environments, it has been implied that water molecules which are absorbed on the nickel surface incorporates towards the formation of $Ni(OH)_2$ passive films, according to the following different chemical and electrochemical reactions [20]:

$$\begin{aligned} &1\text{ - }Ni + H_2O \rightarrow Ni(H_2O)_{ads} \\ &2\text{ - }Ni(H_2O)_{ads} \rightarrow Ni(OH)^+ + H^+ + 2e^- \\ &3\text{ - }Ni(OH)^+ + OH \rightarrow Ni(OH)_2^- \end{aligned} \tag{6.12}$$

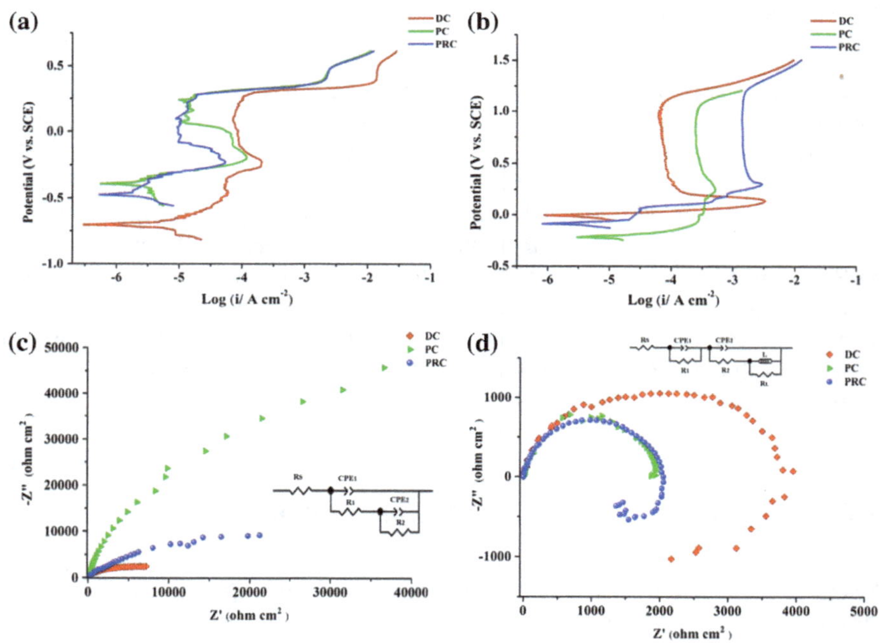

Fig. 6.14 a Potentiodynamic polarization curves and **b** Nyquist impedance diagrams and corresponding equivalent electrical circuit model at their respective open circuit potentials of NC Ni films electrodeposited by DC, PC, and PRC electrodeposition methods in 2 mol/l NaOH aqueous solution. **c** Potentiodynamic polarization curves and **d** Nyquist impedance diagrams and corresponding equivalent electrical circuit model at their respective open circuit potentials of NC Ni films electrodeposited by DC, PC, and PRC electrodeposition methods in 0.5 mol/l H2SO4 aqueous solution. Reproduced from [20] with the permission of Elsevier

The insoluble $Ni(OH)_2$ covers the surface of the corroded Ni film and creates the passivation region in polarization curves. The formation of the passive film should provide an impedance effect at the electrolyte/passive film interface.

The Nyquist impedance plots, obtained for nanocrystalline Ni films (electrodeposited by DC, PC and PRC methods) in 2 mol/l NaOH aqueous solution at room temperature, are shown in Fig. 6.14b. The Nyquist plots of all films exhibit two capacitive loops in the frequency range studied. It can be observed that by changing the deposition conditions, diameter of capacitance loop also changes. In order to fit the impedance spectra obtained for nanocrystalline Ni films, an equivalent circuit consisting of two RC circuits was applied (shown as an inset in Fig. 6.14b), that contains elements corresponding to the solution resistance (Rs), double-layer constant phase (CPE2), charge transfer resistance (R2), film constant phase (CPE1) and film resistance (R1). Circuit R_2CPE_2 are used to describes the electrode Faraday process (nickel oxidation) and corresponds to low-frequency impedance while circuit R_1CPE_1 represents the presence of a surface film and corresponds to high-frequency impedance.

It should be stated that, constant phase element may have some combination of capacitative, resistive and inductive character. This element is defined by two parameters, T and P (which are independent of frequency), in the equation for impedance [33]:

$$Z = 1/T(j\omega)^P$$

The value of the exponent P can be changed from 0 to 1. When P is 0 then the CPE is a resistor. A CPE is often used instead of a capacitor to compensate for inherent inhomogeneity in the system. Values of the equivalent circuit elements derived after fitting the data (Nyquist plot) using Z View software, for nanocrystalline Ni films are compiled in Table 6.6. The data in this Table indicate that the mean error of modulus is less than 7%, indicating a good fitting of the experimental data.

Nanocrystalline Ni film electrodeposited by PC method has the maximum diameter which decreases for nanocrystalline Ni films electrodeposited by PRC and DC techniques. This means that the corrosion resistance of nanocrystalline Ni film decreases in an order from PC to PRC and finally to DC.

The value of CPE_2-P for NC Ni film electrodeposited by the PC method is larger than those of the Ni films electrodeposited by the methods of PRC and DC, indicating the more smooth and protective nature of passive film formed on the surface.

By analyzing the data, it can be observed that the changes in the corrosion behavior of nanocrystalline nickel films depend remarkably upon the deposition technique. As a rule of thumb, we find that the deposition techniques used have significant effects on surface morphology, roughness, grain size and crystallographic texture. We interpret the change of corrosion resistance with the change of structural properties of the electrodeposited films. Corrosion resistance in nanocrystalline Ni films under our study occurs due to the interplay between surface morphology and roughness, surface crystalline texture and crystalline grain size.

The highest and lowest corrosion resistance is achieved for Ni films electrodeposited by PC and DC, respectively. Ni films prepared by DC exhibit the highest surface roughness among the three groups of Ni films. Therefore, the first reason for the increase of corrosion resistance in pulse plated films is the reduction of surface roughness. Generally, the reduction of surface roughness decreases the corrosion rate by reducing the available corrosion sites on the corroding surface. However, it should be noticed that the grain size is highly reduced by pulse plating, while nanocrystalline nickel films with smaller grain size exhibits higher corrosion resistance.

Without assuming the role of passivation on the corrosion behavior, this behavior is inconsistent with the classical theory of corrosion, which states that reducing grain size leads to increases the number of very fine electrochemical cells between grain boundaries and fields and therefore increases the corrosion rate. The more density of grain boundaries in nanocrystalline materials with smaller grain size, causing a high density of active sites for preferential corrosion (when it has been exposed to corrosive environments) leading to higher corrosion rates. Positive or negative influences of high density of grain boundaries on corrosion resistance depend on the nature of passive films or corrosion products formed on the surface. If the corrosion products are dissoluble the corrosion rate is increased; if the corrosion products are insoluble, the corrosion rate is decreased because the corrosion products act as a block layer to stalling the dissolution.

Due to the presence of a greater density of grain boundaries and intergranular dislocation in nanocrystalline nickel films with smaller grain sizes, it can be claimed that more number of sites are available for nucleation of the passive film. This causes the formation of a thicker passive film on the surface. On the other hand, it has been shown that the passive film formed on Ni in alkaline solutions composed of $Ni(OH)_2$ is insoluble. Therefore, the corrosion resistance increases with decreasing grain size in the Ni electrodeposited with PC method. Perhaps changes in the texture can also affect the corrosion behavior. According to the results of X-ray diffraction analysis, it was found that film texture changes from $\langle 111 \rangle$ to $\langle 100 \rangle$ with changing the electrodeposition conditions from PC to PRC or DC. On the other hand, research shows that low index facets have low surface energy. Plates $\{111\}$ have more energy than $\{100\}$ plates. That will be, corrosion products easier and faster on the entire surface to be generated. Because of passive film of nickel $Ni(OH)_2$ is insoluble, one may conclude that corrosion resistance increases with change in film texture from $\langle 100 \rangle$ to $\langle 111 \rangle$.

6.4.5.2 Electrochemical Corrosion Studies in H_2SO_4 Solution

Potentiodynamic polarization curves of the nanocrystalline Ni films electrodeposited with different methods (DC, PC and PRC) in 0.5 mol/l H_2SO_4 aqueous solution at room tempearture are shown in Fig. 6.14c. All the Ni film samples show an active-passive-trans passive behavior.

The Nyquist impedance plots, obtained for the Ni films (electrodeposited by DC, PC and PRC methods) in 0.5 mol/l H_2SO_4 aqueous solution at RT, are shown in Fig. 6.13. It can be seen that there are two large positive loops (capacitive loop) at higher frequency and a small loop (inductive loop) in the 4th quadrant at low frequency. An explanation for the occurrence of inductive behavior during NC Ni corrosion can be attributed to desorption intermediates and creation of sites for further adsorption and reaction.

By changing the deposition conditions, diameters of the loops also changes. The Ni film electrodeposited by DC method has the maximum diameter that this value decreases for films electrodeposited by the methods of PRC and PC, respectively. This means that the corrosion resistance of the Ni films by changing the electrodeposition method from DC to PRC and PC are reduced respectively.

The equivalent circuit for fit impedance spectra obtained for NC Ni films is shown as an inset in Fig. 6.14d. Elements of R_L and L are the resistance and inductance related to the adsorption/desorption process, respectively. The analyzing of EIS data indicates that the mean error of modulus is less than 5%, indicating a good fitting of the experimental data.

By analyzing the data obtained, it can be seen that corrosion behavior of films in the H_2SO_4 solution with the behavior observed in NaOH solution is inconsistent. This means that with decreasing grain size, corrosion resistance is reduced. This behavior can be caused due to the nature of the passive film on the surface of nanocrystalline nickel. Based on previous studies, it was found that the passive film made in H_2SO_4 solution is discontinuous and defective, the smaller the grain size, it could be worse. The imperfect passivation associated with nanocrystalline nickel could be due to intrinsic electrochemical behavior related to the disordered intercrystalline region in nickel [20, 34, 35, 36]. As a result, increasing the rate of corrosion products due to the decrease of grain size and change in texture of the film can increase the corrosion rate.

This dual behavior of observed in this study can be attributed to corrosion products. If the corrosion products are dissoluble the corrosion rate is increased, if the corrosion products are insoluble, the corrosion rate is decreased.

References

1. F. Ebrahimi, G.R. Bourne, M.S. Kelly, T.E. Matthews, Nanostruct. Mater. **11**, 343 (1999)
2. C.B. Wang, D.L. Wang, W.X. Chen, Y.Y. Wang, Wear **253**, 563 (2002)
3. H. Li, F. Ebrahimi, Mater. Sci. Eng. A **347**, 93 (2003)
4. U. Erb, Nanostruct. Mater. **6**, 533 (1995)
5. M.S. Chandrasekar, M. Pushpavanam, Electrochim. Acta **53**, 3313–3322 (2008)
6. H. Gerischer, Anal. Chem. **31**, 33 (1959)
7. D. Landolt, A. Marlot, Surf. Coat. Technol. **169–170**, 8–13 (2003)
8. G. Devaraj, S. Guruviah, S.K. Seshadri, Mater. Chem. Phys. **25**, 439–446 (1990)
9. N. Ibl, Surface Technology **10**, 81–104 (1980)
10. Puippe J.C., IBL N., 1980, J. Applied Electrochemistry, 10, 775–784

11. N. Ibl, J.C. Puippe, H. Angerer, Surface Technology **6**, 287–300 (1978)
12. H.Y. Cheh, J. Electrochem. Soc. **118**, 551 (1971)
13. H.Y. Cheh, J. Electrochem. Soc. **118**, 1132 (1971)
14. K. Viswanathan, H.Y. Cheh, J. Appl. Electrochem. **9**, 537 (1979)
15. M. Datta, D. Landolt, Surf. Technol. **25**, 97 (1985)
16. T. Chin, J. Electrochem. Soc. **130**, 1657 (1983)
17. K.M. Yin, Surf. Coat. Technol. **88**, 162 (1996)
18. R.T.C. Choo, J.M. Toguri, A.M. El-Sherik, U. Erb, J. Applied Electrochemistry **25**, 384–403 (1995)
19. M.R. Sanaeian, F. Nasirpouri, Adv. Mater. Res. **829**, 410–415 (2014)
20. F. Nasirpouri, M.R. Sanaeian, A.S. Samardak, E.V. Sukovatitsina, A.V. Ognev et al., Appl. Surf. Sci. **292**, 795–805 (2014). Reprint copyright 2016, with the permission of the Elsevier
21. F. Nasirpouri, M.A. Engbarth, S.J. Bending et al., Appl. Phys. Lett. **98**, 222506 (2011)
22. Z.L. Xiao, C.Y. Han, W.K. Kwok, H.H. Wang, U. Welp, J. Wang, G.W. Crabtree, J. Am. Chem. Soc. **126** (34), 7–16 (2004)
23. M. Motoyama, Y. Fukunaka, T. Sakka, Y.H. Ogata, J Electrochem. Soc. **153**, 502–508 (2006)
24. F. Nasirpouri, S.M. Janjan, S.M. Peighambari, M.G. Hosseini, A. Akbari, A.S. Samardak, J. Electroanal. Chem. **690**, 136–143 (2013)
25. E.V. Sukovatitsina, A.S. Samardak, A.V. Ognev, L.A. Chebotkevich, MR Sanaeian, F Nasirpouri. Nanotechnol. Russ. **9**(11–12), 723–727 (2014)
26. L. Shao, L. Du, L. Wang, Micro. Nano. Lett. **5**, 165–170 (2010)
27. K.C. Chan, D. Zhu, Microstruct. **30**, 43–53 (1982)
28. T. Borkar, S.P. Harimkar, Surf. Coat. Tech. **205**, 4124–4134 (2011)
29. A. Saraby-Reintjes, M. Fleischmann, Electrochim. Acta **29**, 557–566 (1984)
30. Y.Y. Huang, Y.C. Zhou, Physica E **41**, 1673–1978 (2009)
31. D.H. Jeong, Scripta Mater. **44**(2001), 493–499 (2001)
32. R.C. O'Handley, *Modern magnetic Materials Principles and Applications*. Massachusetts Institute of Technology (1999) pp. 287–310
33. I.D. Raistrick, D.R. Franceschetti, in: *Impedance Spectroscopy*, ed. by J.R. Macdonald (Wiley, New York, 1987), pp. 27. Gamry Instruments, Warminster, PA. Application Notes, The basics, electrochemical impedance spectroscopy: A Primer (2000)
34. R. Rofagha, S.J. Splinter, U. Erb, N.S. Mcintyre, Nanostruct. Mater. **4**, 69 (1994)
35. G. Palumbo, K.T. Aust, Mater. Sci. Eng. A **113**, 139 (1989)
36. R. Rofagha, R. Langer, A.M. El-Sherik, U. Erb, G. Palumbo, K.T. Aust, Scripta Metall. **25**, 2867 (1991)

Chapter 7
Electrodeposited Nanocomposite Films

Abstract Metal matrix-particulate nanocomposites are being successfully fabricated by co-electrodeposition technique. Matrices include usually metals, i.e. Ni, Zn, etc., and alloys, i.e. NiCo, NiFe, etc. and particulates are metal, oxide and other intermetallic compounds with a wide variety of shape, i.e. spherical, rod, tubular, etc. Co-electrodeposition of metal matrix with particulates involves the electrochemical fundamental where the mass transport and kinetics play important roles. Various theoretical models have been developed to predict the co-deposition of nanocomposites which will be explained here. The most important factors, including electrodeposition techniques, particulate characteristics, and electrolyte properties influence the mechanism of co-electrodeposition and coating properties which will be discussed in an example system; nickel-carbon nanotube nanocomposite coatings.

7.1 Introduction

A major group of bulk nanostructured coatings is nanocomposites composed of a matrix filled with nanostructures of any category. Nanoparticles, nanotubes, nanowires, nanorods, nanosheets, etc. of any desirable material are distributed into a metal or non-metal matrix. This combines the capabilities of either constituent to achieve novel and unique properties and applications. Frankly, electrodeposition is a fabrication method of these nanostructured coatings with highly desirable fabrication process and outstanding functional and mechanical characteristics of produced coatings. In this chapter, we emphasize on the electrochemical deposition of metal-matrix nanocomposites. Fundamental electrochemical aspects will be discussed and their effects will be explained by an example system; nickel-carbon nanotube nanocomposites.

F. Nasirpouri, *Electrodeposition of Nanostructured Materials*,
Springer Series in Surface Sciences 62, DOI 10.1007/978-3-319-44920-3_7

7.2 Principles of Electrodeposition of Nanocomposites

As has been mentioned in Chap. 1, nanocomposites are generally referred to multi-phase or multi-component materials in which at least one phase or component has nano-scale dimensions, on in the other word, is a nanostructure. Electrodeposition of nanocomposites is composed of co-deposition of metal ions in elemental or alloy system with suspended solid particulates in any forms of spheres, polyhedrons, tubular, rods, wires, etc. to form metal/alloy matrix nanocomposites. Many research works have been carried out in this field which have been motivated by fascinating extra-ordinary properties of the electrodeposited nanocomposites such as enhanced tribological (wear resistance and lubricating properties), corrosion resistance, higher hardness, and also device applications in for instance microelectromechanical (MEMS) systems. Several review articles and book chapters have been produced in light of the rapid progress and particular attentions for fundamental importance and potential applications [1–5].

7.2.1 Electrochemical Aspects of Co-electrodeposition of Nanocomposites

Main types of processes which may incorporate particles into metallic coatings include: adsorption, mass transport consisting of convection and diffusion, mechanical entrapment, and electrophoresis. We have introduced the fundamentals of these processes in Chap. 2 for electrically charged and neutral particles. Here we focus on the phenomena incorporating into metal-particualtes composites by electrochemical deposition and proposed models on the co-electrodeposition. Many factors and parameters influence in practice the incorporation of particles into metal deposit which may include as suspension and distribution of particles in electrolyte, particle/electrolyte interaction, e.g. the Zeta potential, mass transport mechanism, movement or convection in electrolyte, electrophoresis or electroosmosis forces, cathode/particle surface interaction, double layer characteristics, etc.

7.2.2 Models

Some studies have proposed model behaviour for the co-electrodeposition of large micrometer sized particles such as the models [6, 7] in which the Brownian motion was ignores and instead the mass transport of micron-size particles was calculated by Lagrangian approach using the Newton's second law of motion. The co-deposition mechanism of nano-size particles have been extensively studied for years. Here the Brownian motion persists and other electrokinetic effects will come into action. Under any size regime, it is crucial for particles to be residing on the cathode surface by any possible means in order to co-electrodeposit with metal ions. Hence, the electrokinetics of all possible reactions must be taken into account.

Guglielmi [8] in 1972 explained the peculiarities of the co-deposition of inert particles from electrolytic solution by a mechanism based on two successive adsorption steps. In the first step, the particles are loosely adsorbed, and they are in equilibrium with the particles in suspension. This adsorption has a physical nature in which the particles surrounded with a thin layer of electrolyte ions or solvent molecules (particle-ion complex molecules) move towards the electrode surface. They reside to the Helmholtz layer and then weakly adsorb into the cathode surface. The surrounding layer formed around the particles prevents any direct contact and interaction with the electrode surface. Thus, they remain in this loosely bonded situation until the next step. In the second step the particles are irreversibly adsorbed. This step has an electrochemical nature. At this stage under the effect of the electrical field around the cathode, a charge transfer reaction occurs between the metal ions and the cathode surface. This removes the surrounding layer around the particles making them well incorporated into the metal deposit. The strong adsorption occurring in the second step depends on the electrode potential. This model was successfully explained for co-deposition of nickel and copper with micron-size and submicron size (SiC, TiO_2 and Al_2O_3). One particular feature of this model is that the particle load in the electrodeposit increases with the particle concentration in solution as well as with the deposition current density. However, the increasing rate of particle content in the electrodeposit is impaired at higher concentration of particles and higher current densities. This limits the incorporation magnitude or the fraction of particle in the co-deposit up to a certain value likely owing to the saturation of the cathode surface by the particles. Forming a dense layer of particles on the surface of the cathode through the electrolyte prevents absorbing of further particles. Guglielmi's model did not consider the hydrodynamic effects and the influence of particle size. These made some deficiencies of this model for interpreting the co-depositing mechanisms.

In another model given by Celis et al. [9] the co-deposition mechanism was postulated by a statistical probability method including the important effect of hydrodynamics. They considered inert particles covered with an adsorbed ion-solvent layer transferred under diffusion-convection transport mechanism. Then, the particle–ion complex then adsorbs onto the electrode surface, ions are reduced, and the particle is embedded into the electrodeposited metal. The fundamental hypothesis of the model was that a particle can co-deposit with a metal provided that a certain amount of adsorbed ions is reduced. This is expressed by:

$$X_{(\%)} = \frac{Wp \cdot Np \cdot P}{\Delta Wm + Wp \cdot Np \cdot P} \cdot 100 \tag{7.1}$$

In which Wp is the weight of one particle (kg), Np amount of particles residing at the diffusion layer on the working electrode per unit of time and surface area at a certain current density i, P probability for the incorporation of one particle at the current density i, ΔWm mass of metal matrix deposited per unit of time and surface area at current density i (kg m^{-2} s^{-1}). Preliminary studies by the authors showed that this model is consistent with the experimental data the co-electrodeposition of Cu–Al_2O_3 from acidic sulfate baths, and Au–Al_2O_3 from cyanide baths and therefore it has been valid for many other composite plating systems. The probability factor P can be calculated by comparing theory and experimental data.

Valdes and Cheh's model [10] involved the mass transport of the particle and metal ion under three different kinetics. First namely "perfect sink" model, in which all the particles residing at the electrode surface are embedded. This is not the case in reality owing to the limited accommodation number of particles at a given applied potential. Second model was the surface energy term interacting with particles to co-deposit them with metal, but it still suffers from the restricted incorporation. Third, the electrode-ion-particle electron transfer model in which the effect of applied potential (or current) on the particle incorporation rate predominates. Particles surrounded with adsorbed metal ions form and the incorporation rate was shown to be directly linked with the metal deposition rate. This model predicts that at a critical current density exhibiting a maximum of the particle content in deposit the mechanism of co-electrodeposition switches from a kinetic to diffusion controlled.

Hwang and Hwang model in 1993 improved the Guglielmi's model by the development of three hypothesis at different low, intermediate, high current densities for validating the changes of the reduction of adsorbed ion on particles which included three steps of (i) forced convective of particles to surface, (2) loose adsorption on the surface and (iii) irreversible incorporation of particles by reduction of adsorbed ions [11, 12].

Bercot et al. model made a third order polynominal correcting factor to the Guglielmi's expression to consider the effects of adsorption and flow [13].

P. Searson's group have more recently developed two models for the co-electrodeposition of nano-particles into a metal matrix. In one model he particle incorporation process during the deposition of Ni/Al_2O_3 composites was simulated. They considered the effects of both diffusion and gravity forces into the incorporation of nanoparticles. The cell configuration was arranged so that the cathode surface is horizontal. Two cases were considered for incorporation of particles. One the diffusion force is acting at the same direction as the gravity and two the diffusion is acting at the opposite direction of the gravity force. The particle to metal volume ratio in the metal-particle composite strongly depends on particle size. The particle gravitational force is close to the diffusion force when the particle becomes larger than about 300 nm in radius. For particles with a radius less than 100 nm, the gravitational effects can be ignored [14]. On the other hand, the transport of nanometer size particles to the growing surface is controlled by convective diffusion as predicted by Celis et al. model. They showed that the kinetics of nanometer size particle incorporation into a growing film depends on the residence time at the surface. The model is valid if the diffusion layer thickness for particles is considerably smaller than the diffusion layer thickness for ions in solution due to the difference in diffusion coefficients and for the particle sizes smaller than the diffusion layer thickness [15].

To simplify the models presented, a five step generally admitted co-deposition mechanism has been deduced which include: (1) formation of ionic clouds on the particles, (2) convection towards the cathode, (3) diffusion through a hydrodynamic boundary layer, (4) diffusion through a concentration boundary layer and finally, (5) adsorption at the cathode where particles are entrapped within the metal deposit. Figure 7.1 schematically shows the steps mentioned in this model [16].

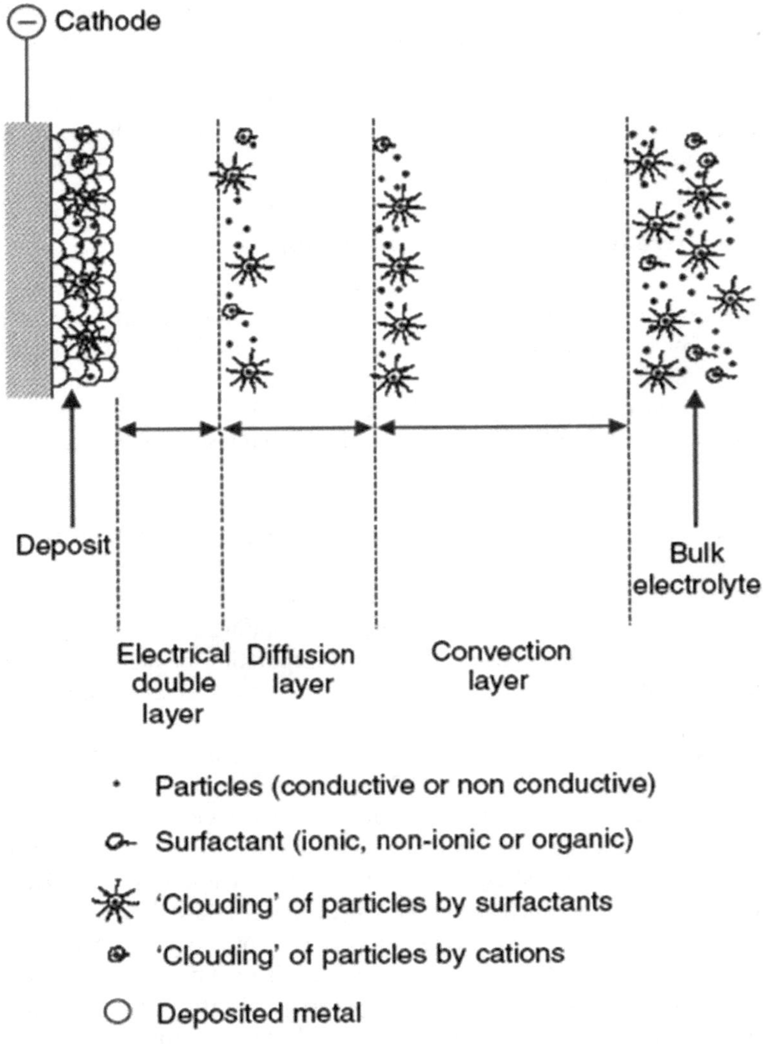

Fig. 7.1 Co-electrodeposition mechanisms of metal matrix-particulate nanocomposite into a metal deposit. The regions include: formation of ionic clouds around the particles (bulk electrolyte, typical length in cm); convective movement toward the cathode (convection layer, typical length b1 mm); diffusion through a concentration boundary layer (diffusion layer, typical dimensions of hundreds of μm); electrical double layer (typical dimensions of nm) followed by adsorption and entrapment of particles. Reproduced from [18] with the permission of Elsevier

7.3 Effective Factors in Co-electrodeposition of Nanocomposites

Currently, there is a relatively vast knowledge on the effect of several practical electrodeposition parameters on the mechanism and properties of metal matrix nanocomposite coatings. The ultimate goal in any co-electrodeposition process of the nanocomposite coatings is to achieve a homogenous well dispersion of individually separated (mono-disperse) particles in metal matrix unless the aim of the process is peculiar such as the functionally graded coatings or gradient composite coatings [17]. We can categorize the co-electrodeposition parameters into electrodeposition technique, particle properties and electroplating bath parameters.

7.3.1 Electrodeposition Control Technique

Electrodeposition techniques are basically divided into potentiostatic and galvanostatic. For co-electrodeposition techniques, the galvanostatic techniques have been mainly used. Three different techniques used for co-electrodeposition of metal matrix nanocomposite coatings are direct current (DC), pulse current (PC) and reverse pulse current (PRC). For details of the techniques see Chap. 6.

7.3.1.1 Direct Current Electrodeposition: Effect of Current Density

In many studies, DC technique is used as the most common method of co-electrodeposition of nanocomposite coatings exhibiting desirable properties. This technique is based on the concept that the incorporation of nanoparticles occurs simultaneously with the reduction reaction of an ionic species to form the metal surface [18]. The effective parameters of DC will be discussed in the next section.

Daneshvar and Nasirpouri have electrodeposited Ni-carbon nanotubes (CNT) nanocomposite coatings by DC electrodeposition. The relation between CNT content in deposit and cathodic current density is demonstrated in Fig. 7.2. This behavior was also reported for acid treated CNTs [19]. As it is observed, CNT content increases up to a certain magnitude in the current density of 4 A/dm^2 and then decreases by further increase in the current density [20, 21].

In low current densities, electrostatic forces are not adequate for proper deposition and considering that the solution is agitated there is no adequate time for the ions and complexes present in the double layer to be adsorbed onto the cathode. By increasing the current density and therefore the electrostatic forces, complexes and ions move faster towards the cathode surface and therefore the process becomes under the activation control.

Since Ni^{+2} ions move faster than the complex ions when the current density rises more than a specified value. Then, the double layer becomes rich of Ni^{+2} ions and a

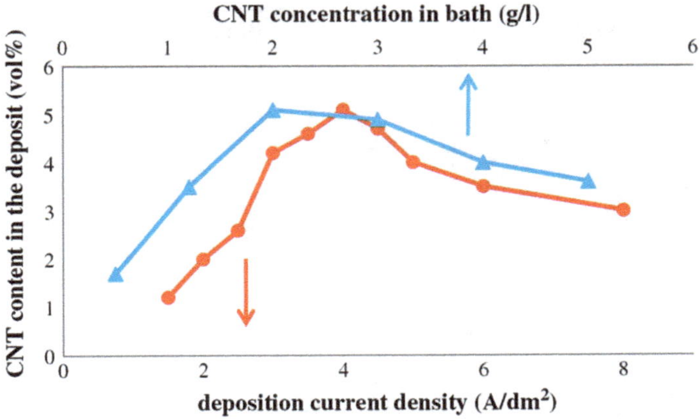

Fig. 7.2 CNT content electrodeposited in the coating as a function of *i* CNT concentration in the electrolyte at constant current density, id = 4 A/dm^2 (*blue curve—triangle symbols*), and *ii* current density at constant CNT content of 2 g/l CNT (*red curve—circle symbols*). Reproduced from [20] with the permission of Elsevier

greater number of Ni^{+2} ions reduces on the cathode surface which results in the decrease of CNT content in electrodeposits. This step is initially controlled by the diffusion of the complex. Moreover, hydrogen evolution occurs during electrodeposition of nickel as observed, even though at higher currents more hydrogen bubbles were observed to form on the surface. The hydrogen reduction during the nickel electrodeposition acts as a barrier. Consequently, it can be suggested that this hydrogen barrier can hinder the deposition of Ni^{+2} ions and especially CNTs which are physically loosely adsorbed at first steps.

Figure 7.3 clearly shows the incorporation of CNTs in electrodeposited coatings. Figure 7.3a shows that at low current densities and CNT concentrations, the electrodeposited coating has little amount of CNT. As it can be seen from Fig. 7.3b, at higher current density and higher CNT concentration, the amount of CNT incorporated into the coating increases, although the coating seems not homogenous. In Fig. 7.3c taken from a produced coating in high current density and CNT concentration in bath, agglomeration can be observed. Moreover, Fig. 7.4d demonstrates the CNTs presence within the coating which is embedded in the matrix homogenously with a unique structure. This structure can surly provide remarkable mechanical properties along the CNTs direction which is parallel to the substrate, as will be discussed later. This image demonstrates a proper connection among the CNTs and the Ni matrix.

Another effect of the current density in DC electrodeposition is on the surface roughness. By increasing the current density i_d, roughness of coatings increases according to two parameters measured including average roughness (R_a) and distance between the highest peak and lowest valley (R_z). In high current densities, CNTs induce a non-uniform distribution of current around them so that the electric fields are stronger close to CNTs, and therefore Ni reduction on these sites is

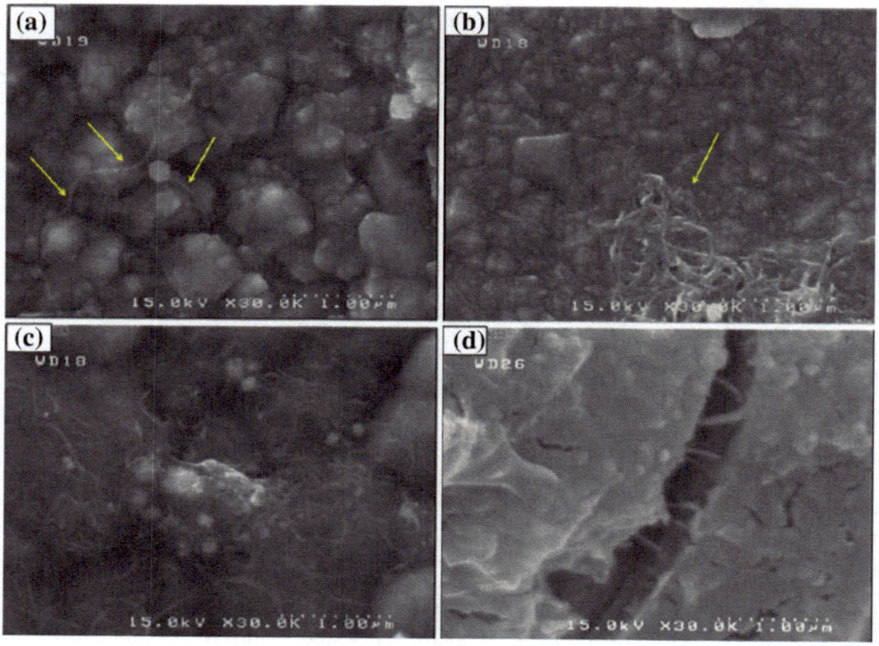

Fig. 7.3 SEM images of Ni–CNT coatings electrodeposited at **a** $i_d = 2$ A/dm^2, $C_{CNT} = 0.5$ g/l, **b** $i_d = 6$ A/dm^2, $C_{CNT} = 3.5$ g/l, **c** id = 6 A/dm^2, $C_{CNT} = 5$ g/l, and **d** $i_d = 4$ A/dm^2, $C_{CNT} = 2$ g/l. Reproduced from [20] with the permission of Elsevier

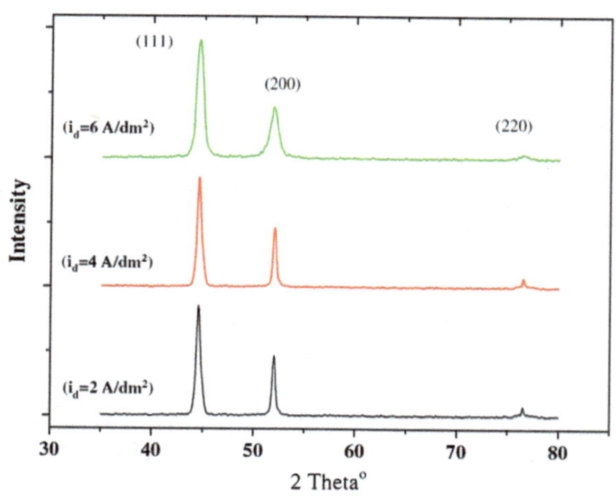

Fig. 7.4 XRD patterns of Ni-CNT nanocomposite coatings electrodeposited in a solution containing $C_{CNT} = 2$ g/l at different current densities. Reproduced from [20] with the permission of Elsevier

accelerated leading to the formation of larger grains and rougher surface. Furthermore, we have reported that hydrogen evolution which intensifies by increasing deposition current density, hinders the uniform growth of nickel, leading to a rougher surface.

The current density also influences the crystal size of metal matrix. In general, in nanocomposite coatings, although the strengthening phase in not often detected by using the XRD pattern, it would affect the matrix phase intensity peaks. The effect of presence of CNTs on nucleation, growth, size of Ni crystals was investigated by Daneshvar and Nasirpouri [20, 21]. The XRD patterns of coatings deposited from Watts bath containing 2 g/l CNT at different current densities are presented in Fig. 7.4. As it can be observed, current density has a significant effect on crystallite size as its magnitude inversely changes with the crystallite size. Furthermore, in a constant current density, inclusion of CNTs to the matrix leads to the crystal refinement. This can be attributed to two factors: (i) CNTs in the coating act as a barrier for crystals growth, and (ii) CNTs are proper sites for Ni nucleation. Mean crystallite size depends on a competition between nucleation and growth. When proper nucleation sites and/or nucleation rate increases the mean crystallite size will decrease.

7.3.1.2 Pulse Electrodeposition

PC and PRC techniques have been more recently paid much attention owing to their ability to produce more efficient incorporation of higher concentrations of nanoparticles as well as a wider range of deposit compositions and properties. The principles of PC and PRC techniques were explained in Chap. 6. The PC technique is based on alternating two or more direct cathodic currents for various deposition times whereas PRC technique has similar characteristics but combines a cathodic current and an anodic current along with off times between each pulse. The methods have been the most successful for embedding higher concentrations of nanoparticles. One reason would be the elimination a fraction of the electrodeposited metal during the off time. In PRC, the anodic pulse even dissolves more metal deposited during the cathodic on pulse. PC electrodeposition of Ni–Al_2O_3 [22], Ni–SiC [23], Zn–TiO_2 [24] and other systems was mentioned to obtain more uniform particle distributions and improved physical properties. Another effect of PC technique is to produce fine grain nanocrystalline metal matrix by inhibiting the growth of nuclei.

PRC technique provides other advantages: (1) enhanced nanoparticles fraction in a metal matrix deposit, (2) a lower required concentration of nanoparticles in the electrolyte solution and (3) selective entrapment of similar sizes of nanoparticles can be achieved. This is because of the selective entrapment of nanoparticles where the larger ones particles are being preferentially removed during the reverse pulse period, producing a finer deposit with less agglomeration of nanoparticles. The pulsed reverse current technique has been used to produce Cu–Al_2O_3 [25–27], Ni–(Al–Cu–Fe) quasi-crystals [28], Ni–CNTs [29]. Figure 7.5 illustrates TEM images of Cu–Al_2O_3 nanocomposite films electrodeposited by DC and PRC techniques.

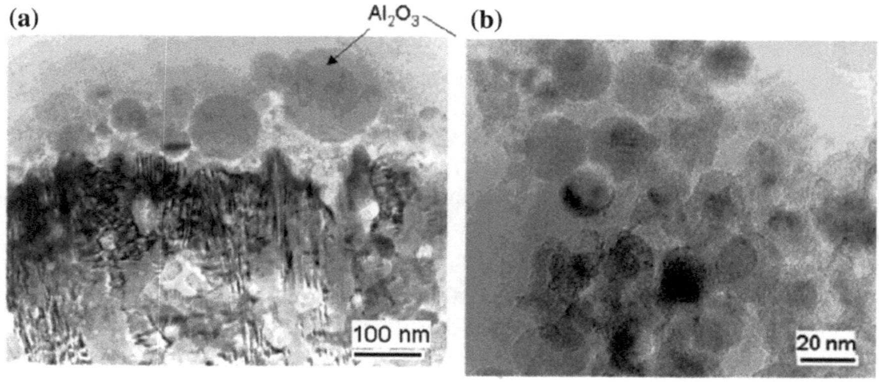

Fig. 7.5 Bright-field, cross-sectional TEM image of **a** a DC-plated Cu–Al_2O_3 composite on a rotating disk electrode with a wide particle size distribution. **b** Of a PRC-plated Cu–Al_2O_3 nanocomposite onto a rotating disk electrode showing the absence of particles greater than 100 nm. Adapted from [25] with the permission of American Chemical Society, Copyright 2016

7.3.2 Particle Properties

The most important properties of particles in co-electrodeposition of metal-matrix nanocomposites may be but not limited to particle load or concentration in the electrolyte solution, surface charge and surface physics and chemistry, type and material nature, shape and morphology and size [18]. These of characteristics interplay with other to define the amount of nanoparticles in electrolyte solution. This in turn would significantly affect the volume fraction of particles in the metal matrix.

The size of particles will affect the co-electrodeposition mechanism. In one hand, it changes the rate of incorporation into the metal deposit. Findings have suggested that as the size of nanoparticles becomes smaller, more can be incorporated into a metal deposit per unit volume. As discussed in above, the size of particles changes the mechanism of mass transport of particles from diffusion-convection for finer particles to diffusion-gravitational for larger particles [14, 15]. In the other hand, the presence of nanosized particles in a metal deposit may change the nucleation rate of metal deposit towards finer grain metal matrices.

The type of particles including oxides, or other metals, and their shape (e.g. spherical, tubular, rod-shape or sheet), and their crystalline phase structure (e.g. anatase or rutile in TiO_2 particles) may have different effects during the incorporation process [18]. For instance, a study by Li et al. [30] shows a significant effect of the crystalline structure of TiO_2 nanoparticles on the weight fraction, hardness and corrosion properties of co-deposited with nickel film.

The physical properties of particles play important role on the co-electrodeposition mechanism. Recent progresses in carbon nanostructures such as carbon nanotubes, graphene [31], etc. have made their potential use in the metal matrix composites for various applications. One important issue in co-electrodeposition of metal-CNT process is that as-produced CNTs tend to entangle and form bundles due to their high surface energy, and do not disperse properly in the electrodeposition baths. In order to tackle this problem, many methods have been employed and developed that can be categorized into two general routes: (a) covalent functionalization, and (b) non-covalent functionalization. The former is the most common process to modify CNT's surface and is performed by immersing them in harsh solutions such as H_2SO_4, HNO_3, or H_2O_2. These solutions form hydroxyl, carboxyl or carbonyl groups on the surface which causes CNTs to act as a negative ion in the electrodeposition baths. However, such harsh oxidative processes are environmentally unfriendly, time consuming and usually introduce large amount of defects [32–34] damaging the structure and reduces electrical and mechanical properties, and corrosion resistance of CNT-containing composites [35–39]. In most reports acid-treated CNT has been used in electrodeposition baths for nanocomposites. Daneshvar and Nasirpouri have recently used the non-covalent functionalization of CNTs to co-electrodeposit with nickel metal matrix [20, 21] in which magnetic ultrasound was used in several steps.

7.3.3 Electrolyte Parameters

Electrolyte bath parameters include electrolyte concentration, additives, temperature, pH, surfactant type and concentration, hydrodynamics (laminar, mixed, turbulent regimes) and electrode geometry and configuration (rotating disk electrode, rotating cylinder electrode, plate-in-tanks, parallel plate electrodes, sediment co-electrodeposition and many variations).

7.3.3.1 Agglomeration Control

One major phenomenon is the monodispersion of particles in an electrolyte. Nanoparticles with a very high surface energy tend to easily agglomerate in highly conductive metal electrolytes. This lowers the uniform distribution of the particles. The particles must be well dispersed in the electrolyte by utilizing effective mixing methods. Erler et al. [40] have compared the effect of two mixing methods on the dispersion and agglomeration of Al_2O3 and TiO_2 nanoparticles in nickel ion electrolytes with different pH and concentrations. It was shown that the mechanical rotating rod had only a very slight effect on agglomerate size, whereas treatment of the suspension with an ultrasonic sonotrode at low amplitudes broke the agglomerates down to a size of approximately 150 nm for Al_2O_3 and 300 nm for TiO_2. At higher amplitudes of ultrasound, the agglomerate size was a little increased. The effect of ultrasonic waves on the reduction of agglomeration has been also reported elsewhere [41], see also Chap. 2.

The average agglomerate size and magnitude of agglomeration of alumina and titania particles was measured by considering effect of concentration of nickel sulphate, the zeta potential, the pH value and the concentration of nickel chloride.

Even a low concentration of $NiSO_4$ results in enhanced growth of agglomerates up to 4 mm for Al_2O_3 and 1.5 mm for TiO_2. It also results in a sharp reduction of zeta-potential from positive values down below zero. The pH value decreases from approximately 6.6–5.2 for Al_2O_3 and from 6 to 4.6 for TiO_2 particles. Concentrations of $NiCl_2$ above 0.1 mol/l leads to less growth of Al_2O_3 and TiO_2 agglomerates than with $NiSO_4$. Likewise, the zeta-potential drops only slightly with increased concentration of $NiCl_2$. With TiO_2 particles, the pH-value of the electrolyte is approximately constant with increasing content of $NiCl_2$, whereas it decreases if Al_2O_3-particles are used. The proportion of incorporated particles in deposited nickel layers increases with increasing particle content of the electrolyte and decreasing deposition current density. The deviation between measurements under same conditions was found to be 10–15%. Above and beyond this, the agglomerate size in the electrolyte prior to deposition has a marked influence on the incorporation rate and thus also any mechanical treatment (ultrasound, dispersing rod) and the hydrodynamic conditions in front of the cathode. All these factors have to be kept constant during plating experiments. Larger and "denser" agglomerates facilitate a higher incorporation rate. It can be deduced from SEM examinations that the relatively high particle content at 0.5 g/l Al_2O_3 and 50 g/l TiO_2 is caused by such "dense" agglomerates [40].

Two common techniques have been used to keep particles suspended in the electrolyte, namely (1) 'physical' dispersion of particles by bath agitation and (2) 'chemical' dispersion of particles by surfactants.

7.3.3.2 Surfactants

A combination of cationic, anionic and non-ionic surfactants can be used to achieve control of surface charge on the particles and the degree of suspension in the electrolyte. The influence of various surfactants on the dispersion and properties of Ni–SiC nanocomposite films have been recently reported [42].

Daneshvar and Nasirpouri [20, 21] have used Sodium dodecyl sulfate ($CH_3(CH_2)11SO_4Na$) or SDS as surfactant for dispersion of CNTs under magnetic sonication. In order to optimize the required amount of SDS surfactant for suspending the CNTs, different SDS/CNT ratio of 1:10, 1:5, 1:3, and 1:2 were employed. Figure 7.6 shows that after 24 h, SDS had no effect in 1:10 ratio. However, in 1:5 ratio, the dispersion was improved significantly but not yet stable enough for the co-electrodeposition. In higher ratios the solution was completely homogeneous; therefore, 3:1 ratio was selected for suspending CNTs for electrodeposition. The solution was stable for more than 2 weeks.

7.3.3.3 Particle Concentration

Another parameter related with the electrolyte is the particle load or concentration in the electrolyte. The effect of nanoparticle concentration in the electrolyte on their

Fig. 7.6 Stability and dispersion of CNT-acetone suspension in different SDS/CNT fractions obtained after 24 h: **a** 1:10, **b** 1:5, **c** 1:3, and **d** 1:2. Reproduced from [20] with the permission of Elsevier

concentration in the metal deposit has been predicted to follow a Langmuir isotherm adsorption phenomenon. Two general observations can be made: (1) the volume fraction of nanoparticles in the metal deposit increases substantially when the amount of nanoparticles concentration in the solution increases and (2) a saturation state of particles in the deposit is reached at high nanoparticle concentration in the solution.

For co-electrodeposition of Ni-CNT nanocomposite, the effect of CNT concentration was studied in details. Figure 7.3 illustrates the variation of CNT content in the coating as a function of CNT load in the electrolyte. As is evident, CNT content increases as CNT concentration increases up to 2 g/l and then decreases. For lower CNT concentrations, as the concentration increases, the amount of CNT available in the Helmholtz layer, and as a result, possibility of trapping and deposition of CNTs in the coating increases. We interpret the subsequent decrease of CNT content with the formation of CNT-Ni complex. On the basis of the Celis-Roos-Buelens model deposition of particles on the cathode depends on the formation of an ionic cloud around them. When the amount of ions and particles is far from the equilibrium, co-deposition process is hindered. In our results, Ni^{+2} ions are attracted to the negatively charged CNTs due to the adsorbed anionic SDS surfactant and consequently CNT–Ni^{+2} complex forms. This complex is absorbed to the cathode surface by electrostatic forces. However, if CNT concentration increases extensively in the electrolyte where a constant amount of Ni^{+2} ions exists, CNTs content in the Helmholtz layer significantly increases and as a result complex formation and therefore CNT deposition is hindered. In addition, in higher concentrations, agglomeration likely occurs, making it difficult to precipitate in the deposition process. Therefore, amount of proper CNTs for electrodeposition in Helmholtz layer decreases. Variation of CNT content in coating as a function of CNT concentration in the electrolyte can be divided into three stages: (I) by increasing the concentration in the bath, more CNTs would be near cathode surface and co-deposition increases; (II) CNT concentration in the double layer is saturated and there is not enough Ni^{+2} to form CNT-Ni^{+2} complex, and consequently

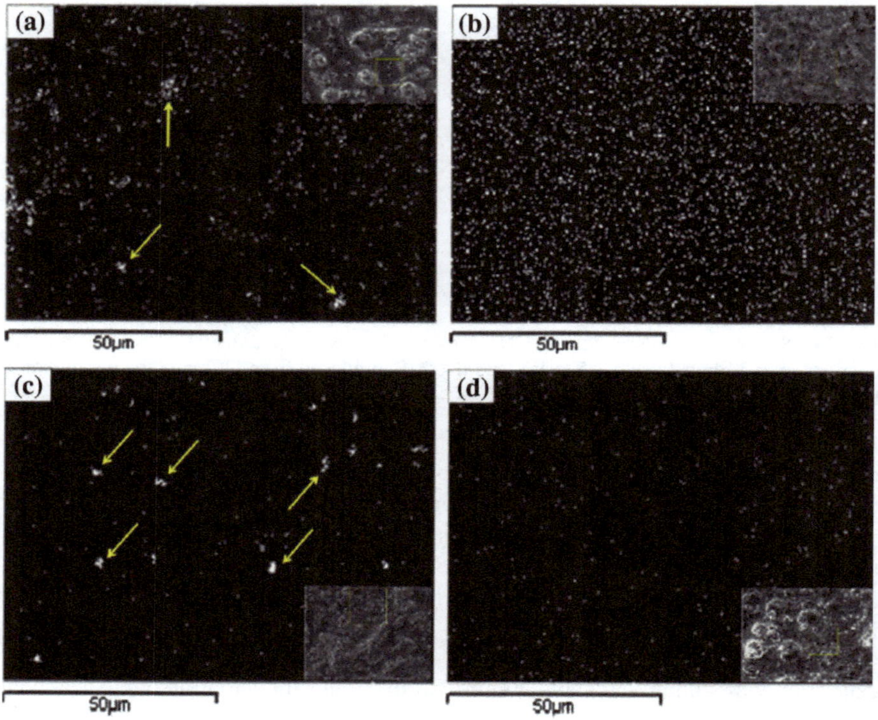

Fig. 7.7 EDX mapping and the corresponding SEM images of Ni–CNT coatings electrodeposited at **a** $i_d = 8$ A/dm^2, $C_{CNT} = 2$ g/l, **b** $i_d = 4$ A/dm^2, $C_{CNT} = 2$ g/l, **c** $i_d = 4$ A/dm^2, $C_{CNT} = 5$ g/l, and **d** $i_d = 4$ A/dm^2, $C_{CNT} = 0.5$ g/l. Reproduced from [20] with the permission of Elsevier

co-deposition decrease slightly; (III) agglomeration occurs, heavy particles do not take part in co-deposition, and CNT content in the deposit decrease more considerably.

Figure 7.7 represents map analysis of coatings electrodeposited at different conditions. As it can be observed for the Ni-MWCNT coating electrodeposited under optimum conditions (Fig. 7.7b), CNTs are uniformly distributed in the matrix which indicates an appropriate dispersion process obtain under the optimum deposition conditions. However, in higher CNT concentration in the bath and low current densities, CNTs tend to agglomerate resulting in a weak co-deposition.

CNT concentration in the bath also significantly affects the surface roughness. In low CNT concentrations, there is no sufficient CNT to uniformly cover the whole surface of the cathode. Since Ni prefers to reduce on CNTs, the coating grows non-uniformly on surface. By increasing CNT content, they are deposited more uniformly and the surface becomes smoother. In higher concentrations, however, agglomeration occurs and the agglomerates cause local increase in the current density; therefore, Ni is preferably deposits on these sites, and causes a rougher surface.

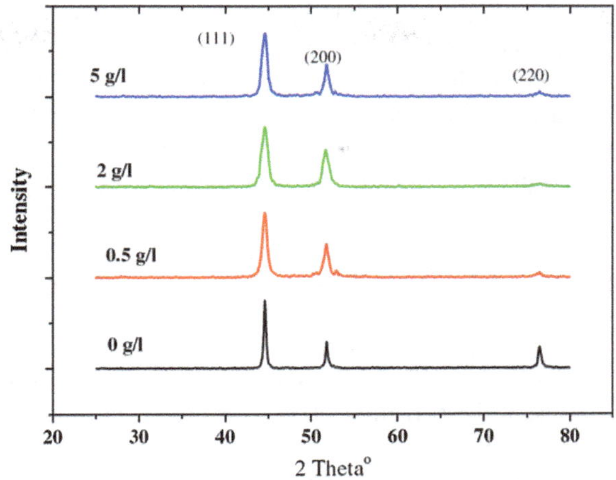

Fig. 7.8 XRD patterns of Ni-CNT nanocomposite films electrodeposited at 4 A/dm^2 in solutions containing different CNT loads. Reproduced from [20] with the permission of Elsevier

The effect of particle load in the electrolyte on the crystal structure of metal matrix is also quite remarkable. Figure 7.8 shows XRD patterns of coatings electrodeposited at a current density of 4 A/dm^2 and different CNT load in the electrolyte. The preferred growth orientation for pure nickel is observed to exhibit (111) and (200) crystalline orientations. By increasing CNT content in the coating, (200) orientation intensifies slightly so that its proportional intensity to the maximum peak increases from 40% in pure sample to 65% in the nanocomposite containing 5.1 vol.% CNT. It was clearly observed that addition of MWCNT to Ni matrix leads to an intensified (200) peak. On the other hand, it should be noticed that by increasing CNT concentration in the electrolyte, the SDS content in the bath needs to be increased. Guo et al. [29] have studied the surfactant effect on preferred growth orientation and reported that SDS addition in the Ni electroplating bath changes the preferred orientation from single (111) to a combination of (111) and (200). Therefore, it can be concluded that the change of preferred orientation can also be attributed to the presence of SDS. Moreover, it has been claimed that CNT defects interfere with preferable nucleation sites; considering the use of acid treated CNTs in previous works, which causes significant amount of defects, it can be expected that nucleation and growth process would be different comparing to this work.

7.3.3.4 Electrolyte Agitation

Influence of bath agitation and cathode movement have been vigorously investigated as it is normally carried out for two purposes, namely to keep the particles suspended in the electrolyte and to transport the particles to the cathode surface. Many investigations have found that increased agitation generally enhances the amount of particles in the metal deposit. However, excessive agitation may lead to a lower quantity of particles in the metal deposit. This may be explained by vigorous hydrodynamic forces in the electrolyte removing the particles from the cathode surface before they can be entrapped in the metal deposit.

The rotating disk electrode, rotating cylinder electrode and parallel plates channel flow provide controllable hydrodynamics for quantitative characterisation of electrolytes containing particle suspensions. Three different flow regimes can be distinguished: laminar, transition (laminar to turbulent) and turbulent.

7.3.3.5 Electrolyte Temperature

The effect of temperature of electrolyte on the co-electrodeposition of nanocomposites is also quite important. In an study on NiCo-Sic nanoparticles, it was shown that the weight fraction of SiC increases with the temperature up to 40 °C. It was argued that the thermal energy activates the particles movement towards the electrode until a certain temperature (40 °C in the study) above which the thermodynamic movement of the ions decays their transport and incorporation into the deposit. A similar result is observed by Kim and Yoo [43]. At higher temperatures it is in accordance with the Langmuir adsorption theory that the adsorption of the particulates is degraded. This decreases the overpotential of the cathode and the electric field and as a result SiC particles are hardly embedded in the Ni metal matrix.

7.4 Properties of Nanocomposite Coatings

Properties of many electrodeposited metal matrix nanocomposite coatings have been studied and reported in review articles. Here we focus on one of our group's recent works on the properties of Ni-CNT nanocomposite films.

7.4.1 Mechanical Properties of Ni-CNT Nanocomposite Films

Hardness and elastic modules of coatings were measured using microhardness tester and nanoindentation instrument. The results for Vickers microhardness and nano-indentation test in various electrodeposition conditions are given in Table 7.1.

It is clear that the addition of CNT in the coating remarkably increases the hardness of coatings as observed to be increased by 70% for Ni-5.1 vol.% CNT coating in respect to pure Ni. However, we find that CNT content of coating is not the only effective parameter on the coating hardness. For instance, although the coating deposited in $i_d = 6$ A/dm^{2}has less CNT than coating deposited in $i_d =$ 4 A/dm^2, the hardness values are almost the same. This can be attributed to the structure refinement (lower grain size) developed by increasing the current density

Table 7.1 Micro-hardness and nano-hardness test results for the electrodeposited Ni-MWCNT coatings

Deposition conditions		CNT content (vol.%)	Hardness (HV)	Nanohardness (GPa)	Elongation modules (GPa)
i_d = 4 A/dm^2	C_{CNT} = 0 g/l	0	320 ± 5	3.10 ± 0.05	235
	C_{CNT} = 0.5 g/l	1.7	374 ± 10	–	–
	C_{CNT} = 2 g/l	5.1	547 ± 10	5.20 ± 0.1	278.3
	C_{CNT} = 3.5 g/l	4.2	452 ± 10	4.39 = 0.15	264.7
	C_{CNT} = 5 g/l	3.6	409 ± 10	4.10 = 0.15	256.5
i_d = 2 A/dm^2	C_{CNT} = 2 g/l	2	368 ± 10	3.56 ± 0.15	250.6
i_d = 4 A/dm^2		5.1	527 ± 10	5.20 ± 0.1	278.3
i_d = 6 A/dm^2		4.1	463 ± 10	4.43 ± 0.15	268.2
i_d = 8 A/dm^2		3	402 ± 10	3.97 ± 0.15	254.1

Reproduced from [20] with the permission of Elsevier

(i_d). On the other hand, for coating deposited in i_d = 8 A/dm^2, severe agglomeration results in weakening of mechanical properties. In composites, strengthening is calculated by Orawan equation [44]:

$$\Delta\sigma_p = \frac{\beta_{or} b \mu M (f_v)^{1/2}}{d_p} \tag{7.2}$$

in which μ is shear modules of matrix, b is Berger's vector, f_v and d_p are volume fraction and size of dispersed particulate phase, respectively. M is Taylor factor which is about 1.5–2 for FCC metals and β_{or} is calculated by (7.3) as follows:

$$\beta_{or} = 1/2\pi \ln\left[\frac{d_p(\pi/6f_v)^{1/2}}{b}\right] \tag{7.3}$$

Based on (7.2) and (7.3) strengthening is caused by grain refinement and increases with relative volume fraction. Nevertheless, agglomeration which means not only heterogeneous structure but also an increase in size of dispersed particles (d_p), decreases the mechanical properties. By any increase in current density, crystallite and grain size refinement occur which results in increase in grain boundaries fraction. Grain boundaries hinder dislocation motion and this is one of the main causes of strengthening. Moreover, heterogeneous introduction of CNT particles into the nickel matrix can also act as barriers and impede dislocation motion. It is worth noting that an important aspect of strengthening by grain refinement is that it leads to improvement in toughness simultaneous with that in strength which was confirmed later by nanoindentation. On the other hand, as is evident from XRD results, the addition of CNT to Ni imposes crystallographic evolution where CNT addition decreases the maximum peak intensities and intensify the (200) Bragg diffraction. Previous studies have reported the effect of crystallographic texture on the hardness in the electrodeposited coatings. In pure Ni electrodeposited from Watts Bath, a polycrystalline texture is found. Changing the texture to one particular orientation achieved by different electrodeposition regime or by the addition of CNT as is the case in the current study changes the hardness of the coating. This could be considered as a further reason for increase of hardness of CNT-Ni nanocomposite films.

Nanoindentation test confirmed Vickers microhardness test results even for coatings deposited in high CNT bath load and i_d, the results were relatively scattered likely due to the agglomeration of CNTs and non-uniformity in the coating structure.

The elastic modules of the coatings were improved slightly by increasing CNT content of coatings. In the utmost condition, the elastic modulus increases by 18% in respect to pure Ni electrodeposit. With respect to significant hardness increase for nanocomposite coating (optimum coating: 70%), it can be concluded that CNT incorporation improves the coating toughness. Since the EDX-mapping images confirm good uniformity of the composite structure, we interpret the improved mechanical properties by relatively more uniform deposited CNT across the coating. In addition, one may assume that the orientation of CNTs in respect to the substrate in the nanocomposite perhaps affect the overall final properties and performance of coating. For instance, we observed non-covalently functionalized CNT aligned with the substrate (Fig. 7.3d) which response to load transversely and might be a possible assumption for enhanced mechanical properties. Thus, these coatings are expected to show a remarkable tensile strength in comparison with coatings containing acid treated CNTs [20].

7.4.2 *Corrosion Behavior of Ni-MWCNT Coatings in 3.5% NaCl*

There are several works that have studied the corrosion properties of electrodeposited metal matrix composites [20, 45, 46]. However, the subject needs more careful discussion and analysis. We investigated the corrosion behavior of non-covalently functionalized CNT (in comparison with covalently functionalized reported with others)-Ni nanocomposite coating using potentiodymanic polarization and EIS curves to understand the effect of CNT into Ni films. Table 7.2 shows potentiodymanic polarization data extracted from the curves measured on Ni-MWCNT coating electrodeposited at $i_d = 4$ A/dm^2 from Watts bath with different CNT concentrations and electrodeposited from a solution containing $C_{CNT} = 2$ g/l at different current densities in NaCl solution. It is clear that Ni-CNT nanocomposite coatings exhibit less negative corrosion potentials compared to pure

Table 7.2 Corrosion data on Ni-CNT nanocomposite coatings

Deposition conditions		R_p (kΩ)	E_{corr} (mV)	i_{corr} (A/cm^2)
$i_d = 4$ A/dm^2	$C_{CNT} = 0$ g/l	5.6	−352	4.47×10^{-6}
	$C_{CNT} = 0.5$ g/l	13.1	−205	2.75×10^{-6}
	$C_{CNT} = 2$ g/l	67.5	−193	0.45×10^{-6}
	$C_{CNT} = 3.5$ g/l	38	−244	1.45×10^{-6}
	$C_{CNT} = 5$ g/l	36.5	−297	1.96×10^{-6}
$i_d = 2$ A/dm^2	$C_{CNT} = 2$ g/l	21.7	−200	0.99×10^{-6}
$i_d = 4$ A/dm^2		67.5	−193	0.45×10^{-6}
$i_d = 6$ A/dm^2		60.2	−249	0.87×10^{-6}
$i_d = 8$ A/dm^2		55.3	−265	1.13×10^{-6}

Reproduced from [20] with the permission of Elsevier

Ni coating. The potential also shifts towards more positive values by increasing the CNT content of the coating. This means that the presence of CNT hinders the matrix dissolution; consequently, significantly improve the coating corrosion resistance.

It is clear that beside corrosion potential improvement, the corrosion rate also decreases by CNT addition to the coating. As it can be observed, generally, by CNT addition to Ni matrix, corrosion resistance increases as is seen for our samples. The best corrosion property is obtained for the coating electrodeposited under $i_d = 4$ A/dm^2 and $C_{CNT} = 2$ g/l which contains the maximum CNT content. The variation of CNT content of coatings and corrosion resistance suggest that corrosion resistance is proportional to CNT content in the coating. Surface of electrodeposited Ni matrix contains microholes and gaps which act as proper sites for initiation and development of corrosion due to the hydrogen evolution as discussed in above. In nanocomposite coatings, CNTs can fill the surface defects, acting as a barrier and hindering the corrosion attack. On the other hand, it is worth noting that grain refinement also affects the corrosion resistance.

In Chap. 6, previously, it was shown that Ni coatings with different grain size so different corrosion behaviour. They claimed that in nanostructured coatings grain boundaries act as proper sites for nucleation and growth of passive layers. Therefore, by structure refining more stable and uniform passive layers are formed and corrosion resistance increases. XRD results showed that the crystallite size decreases by increasing of CNT content of the coating. Thus, it can be deduced that CNT incorporation can also indirectly improve the corrosion resistance by means of structure refining. Therefore, at $i_d = 6$ A/dm^2 higher degrees of agglomeration and consequently, marked increase of corrosion current is expected. However, microstructure refinement compensates the loss of corrosion resistance caused.

Furthermore, as it was observed in the EDS and FE-SEM results, CNTs tend to deposit non-uniformly due to bundles or agglomerates at higher current densities and bath CNT concentrations. These particles may not be able to fill the submicron defects on the matrix surface. Besides, CNT exhibits a more positive standard

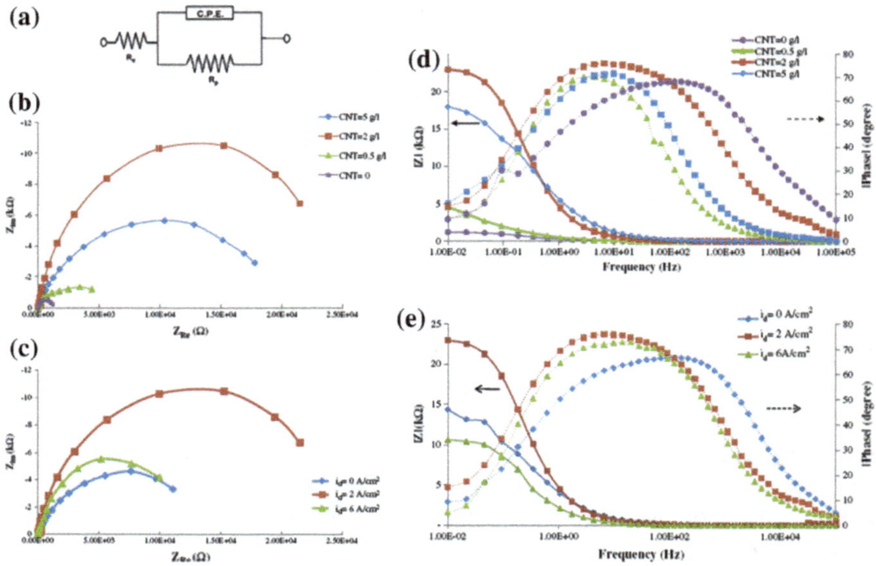

Fig. 7.9 **a** The equivalent circuit model used for fitting the experimental results; and Nyquist plots of coatings electrodeposited at **b** different deposition current densities, and **c** different bath CNT concentrations in 3.5 wt% NaCl solution. Bode plots of coatings electrodeposited at **d** different deposition current densities, and **e** different bath CNT concentrations in 3.5 wt% NaCl solution. Reproduced from [20] with the permission of Elsevier

potential than that of nickel. Therefore, the agglomerates formed on the nickel surface may operate as cathode and nickel does as anode, leading to rapid localized corrosion occurrence. On the other hand, by uniform electrodeposition of CNTs, homogenous corrosion takes place and corrosion resistance is improved.

The EIS results further approved that the resistance of nanocomposite coatings are generally higher than pure nickel coating. The equivalent circuit and Nyquist plot of pure nickel and nanocomposite coatings are presented in Fig. 7.9. Due to non-ideal behavior of double layer capacitor, instead, a constant phase element was used. The Nyquist plots have a similar shape including depressed semicircles with a circle not lying on the x-axis with different sizes. Curves with larger radius encompass higher polarization resistance (R_p) which is proportional to the corrosion resistance. As it can be observed, the coating electrodeposited $\text{ini}_d = 4\ \text{A/dm}^2$ from a solution containing $C_{CNT} = 2$ g/l has the largest radius, hence the highest corrosion resistance. Moreover, the identical shapes of the Nyquist plots indicate that the corrosion mechanism of the coating is constant and the solution was not able to enter through coating and reach the substrate.

The corrosion properties of Ni-MWCNT coatings were further discussed by Bode plots. The advantages of these plots are that by studying inflection point and the slopes of Bode-impedance and Bode-phase diagrams, the rate-determining step and corrosion mechanism can be easily distinguished. Moreover, the initial and final points of frequency-impedance curve indicate $R_p + R_s$ and R_s values, respectively. R_s is the electrolyte and device components resistance and mostly depends on the distance between the samples and reference electrode. It is clear that there is just one inflection point and one maximum point in both Bode-impedance and Bode-phase plots. Thus, it can be deduced that the corrosion reaction mechanism of the coating is the same. Furthermore, among nanocomposite coatings, the coating electrodeposited at $i_d = 4$ A/dm^2 from a solution containing $C_{CNT} = 2$ g/l which contains the highest amount of CNT, exhibits the highest maximum in Bode-phase diagram, close to maximum of the pure nickel coating. It means that this coating (the optimum coating) has a denser and more uniform structure.

R_s is similar for all experiments and the small difference can be attributed to changes in distance of the working and references electrodes. As it can be seen, the optimum coating has the highest R_p, about 30 times greater than that of pure nickel coating, showing that the presence of CNT in nickel coating significantly improves its corrosion resistance in 3.5 wt% solution. Furthermore, it is worthy to mention that the double layer capacitance is an indicator of coating capability in protecting the substrate. This parameter has the highest value for pure nickel coating and the lowest for optimum coating. Capacitance increase is a result of Cl^- ion diffusion into the coating diminishing coating's protective capability. In other words, the thickness of the coating that has the capability to protect the substrate decreases as the capacitance increases. This parameter is inversely proportional to CNT content in the coatings.

Comparing with the literature results, non-covalently functionalized CNTs remarkably improve the corrosion resistance of the nickel coating. Comparing pure nickel and the optimum coating, corrosion rate and R_p of the optimum coating have been improved 10 and 30 times, respectively. This remarkable performance in comparison to what mentioned in the literature can be attributed to three issues: (a) the size of CNTs which decreases in acid treating process, (b) structure weakening of CNTs by acid treatments, and (c) the embedded CNT direction, which is parallel to substrate and provide more effective barrier against the corrosive environment.

References

1. A. Hovestad, L. Janssen, J. Appl. Electrochem. **25**, 519–527 (1995)
2. M. Musiani, Electrochim. Acta **45**, 3397–3402 (2000)
3. J.P. Celis, J.R. Roos, C. Buelens, J. Fransaer, Trans. Inst. Met. Finish. **69**, 133–139 (1991)
4. C. Kerr, D. Barker, F. Walsh, J. Archer, Trans. Inst. Met. Finish. **78**, 171–178 (2000)
5. F.C. Walsh, C.P.de Leon, Trans. IMF, **92**, 83–98 (2014)
6. J.P. Celis, J.R. Roos, J. Electrochem. Soc. **139**, 413–425 (1992)
7. G. Maurin, A. Lavanant, J. Appl. Electrochem. **25**, 1113 (1995)
8. N. Guglielmi, J. Electrochem. Soc. **119**, 1009–1012 (1972)
9. J.J.P. Celis, J.R. Roos, C. Buelens, J. Electrochem. Soc. **134**, 1402–1408 (1987)
10. J.L. Valdes, H.Y. Chen, J. Electrochem. Soc. **133**, C322 (1986)

11. B.J. Hwang, C.S. Hwang, J. Electrochem. Soc. **1986**(133), 259–264 (1986)
12. B.J. Hwang, C.S. Hwang, J. Electrochem. Soc. **1993**(140), 979–984 (1993)
13. P. Bercot, E. Pena-Munoz, J. Pagetti, Surf. Coat. Technol. **157**, 282–289 (2002)
14. I. Shao, P.M. Vereecken, R.C. Cammarata, P.C. Searson, J. Electrochem. Soc. C610–C614 (2002)
15. P.M. Vereecken, I. Shao, P.C. Searson, J. Electrochem. Soc. **147**, 2572–2575 (2000)
16. J.R. Roos, J.P. Celis, J. Fransaer, C. Buelens, J. Met. **42**, 60 (1990)
17. Y. Miyamoto, W.A. Kaysser, B.H. (Eds.), *Functionally Graded Materials: Design, Processing and Applications* (Springer, 1999)
18. C.T.J. Low, R.G.A. Wills, F.C. Walsh, Surface Coatings Technology **201**, 371–383 (2006)
19. C. Guo, Y. Zuo, X. Zhao, J. Zhao, J. Xiong, Surf. Coat, Tech. **202**, 3246–3250 (2008)
20. F. Daneshvar-Fatah, F. Nasirpouri, Surf. Coat. Technol. **248**, 63–73 (2014)
21. F. Nasirpouri, F. Daneshvar-Fattah, A. Samardak, E. Sukovatitsina, A. Ognev, L. Chebotkevich, IEEE Trans. Magn. **51**(1–4), 2006804
22. Ferkel H. Steinbach, Scr. Mater. **44**, 1813 (2001)
23. B. Gyftou, M. Stroumbouli, E.A. Pavlatou, N. Spyrellis, Trans. IMF **80**, 88–91 (2002)
24. A. Gomes, M.I. da Silva Pereira, M.H. Mendonc, F.M. Costa, J. Solid State Electrochem. **9**, 190–196 (2005)
25. E.J. Podlaha, Nano Lett. **1**, 413 (2001)
26. A. Panda, E.J. Podlaha, Electrochem. Solid-State Lett. **6**, C149 (2003)
27. E.J. Podlaha, D. Landolt, J. Electrochem. Soc. **144**, L200 (1997)
28. N.S. Qu, K.C. Chan, D. Zhu, Trans. IMF. **90**, 103–105 (2003)
29. C. Guo, Y. Zuo, X. Zhao et al., Surf. Coat. Technol. **201**, 9491–9496 (2007)
30. J. Li, Y. Sun, X. Sun, J. Qiao, Surf. Coat. Technol. **192**, 331–335 (2005)
31. F. Nasirpouri, H. Pourmahmoudi, F. Abbasi, S. Littlejohn, A.S. Chauhan, A. Nogaret, J. Electron. Mater. **44**(10), 3512–3522 (2015)
32. C.R. Carpenter, P.H. Shipway, Y. Zhu, D.P. Weston, Surf. Coat. Tech. **205**, 4832–4837 (2011)
33. C. Kahattha, P. Woointranont, T. Chodjarusawad, W. Pecharapa, J. Microsc. Soc. Thail. **24**, 133–135 (2010)
34. V. Datsyuka, M. Kalyvaa, K. Papagelisb, J. Partheniosa, D. Tasisb, A. Siokoua, I. Kallitsisa, C. Galiotisa, Carbon **46**, 833–840 (2008)
35. N.I. Kovtyukhova, T.E. Mallouk, L. Pan, E.C. Dickey, J. Am. Chem. Soc. **125**, 9761–9767 (2003)
36. I. Capek, Adv. Coll. Interface Sci. **150**, 63–89 (2009)
37. Z. Špitalskýa, C.A. Krontirasb, S.N. Georgab, C. Galiotis, Composites A **40**, 778–883 (2009)
38. P.C. Ma, N.A. Siddiqui, G. Marom, J.K. Kim, Composites A **41**, 1345–1367 (2010)
39. D. Vairavapandian, P. Vichchulada, M.D. Lay, Anal. Chim. Acta **626**, 121 (2008)
40. F. Erler, C. Jakob, H. Romanus, L. Spiess, B. Wielage, T. Lampke, S. Steinhauser, Electrochim. Acta **48**, 3063–3070 (2003)
41. S. Kuo, Y. Chen, M. Ger, W. Hwu, Mater. Chem. Phys. **86**, 5 (2004)
42. F.C. Walsh, C.T.J. Low, J.O. Bello, Trans. IMF **93**(3), 147–156 (2015)
43. S.K. Kim, H.J. Yoo, Surf. Coat. Technol. **108**(109), 564–568 (1998)
44. C.R. Carpenter, P.H. Shipway, Y. Zhu, Surf. Coat. Tech. **205**, 5059–5063 (2011)
45. B. Bakhit, A. Akbari, F. Nasirpouri, M.-G. Hosseini, Appl. Surf. Sci. **307**, 351–359 (2014)
46. Q. Li-yuan, L. Jian-she, J. Qing, Trans. Nonferrous Met. Soc. China **20**, 82–89 (2010)

Chapter 8
Miscellaneous Electrodeposited Nanostructures

Abstract Various nanostructures have been synthesized by electrodeposition through particular methods and techniques. In this chapter a few of them are explained including STM-assisted electrodeposition, lithographically-patterned nanowire electrodeposition, electrodeposition of mesoporous in liquid lypothorpic crystals and nanostructures by galvanic displacement.

8.1 Scanning-Tunneling Assisted Electrodeposition of Nanostructures

The application of scanning probe microscopy (SPM) techniques in electrochemistry has been extensively studied. One interesting application is to use them as a localised method of materials deposition or dissolution. Most of them have employed scanning tunnelling microscopy (STM) to nanostructuring on electrode surfaces. The various routes for electrochemical nanostructuring by an STM are classified as *mechanical impact, removal of a tarnishing film, double-layer cross talk, pulse double layer charging, burst like dissolution, scanning electrochemical microscopy, and jump to contact* [1]. Figure 8.1 shows schematically different approaches based on the STM-assisted electrochemical deposition.

Figure 8.2a shows mechanical impact technique in which surface defects are generated by the high voltage pulses applied by a STM tip at preferential sites. These may be obtained by a mechanical contact between tip and substrate (tip crash) or by some sort of sputtering process. These artificially created defects then acted as nucleation centres for metal deposition, which allowed the decoration of electrode surfaces by metal clusters on a nanometer scale (Fig. 8.2a).

IN another technique, removal of a tarnishing film, the substrate is not damaged whereas a semiconductor or organic overlayer which causes a high overpotential for metal deposition is locally removed by the tip of an STM or AFM sliding across the surface. This method has been mostly used for decoration of metal on metal yet, but its potential applications remain on decoration of semiconductors and organic materials with metal nanostructures, See Fig. 8.2b.

F. Nasirpouri, *Electrodeposition of Nanostructured Materials*,
Springer Series in Surface Sciences 62, DOI 10.1007/978-3-319-44920-3_8

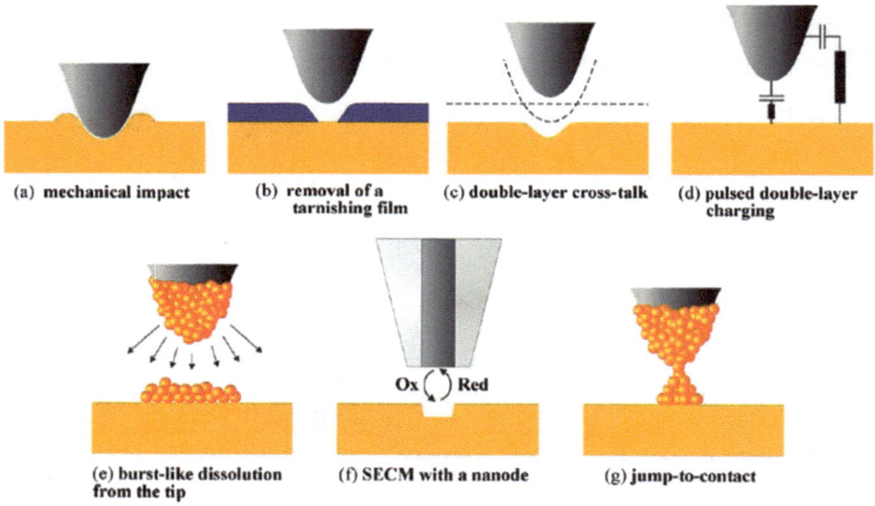

Fig. 8.1 Schematic representation of the various approaches to nanostructuring electrode surfaces via an STM. Reproduced from [1] with the permission of Elsevier, copyright 2016

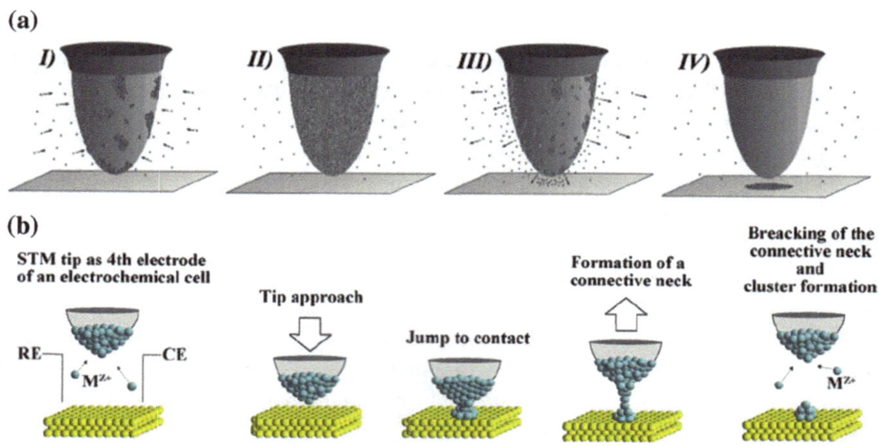

Fig. 8.2 Schemes of growth stages for **a** burst-like dissolution (from *left* to *right* it shows the nanofabrication procedure, Reproduced from [6] with the permission of the AIP publishing **b** jump to contact reproduced from [1] with the permission of Elsevier

Figure 8.2c shows the double layer cross talk technique. The distance between the STM tip and substrate electrode is so reduced (less than 500 μm) that the double-layers of tip and substrate begin to merge, and the assumption of a non-interacting tip is no longer valid. This proximity between the tip and surface may lead to different processes such as selective dissolution from and deposition on the surface.

Pulse double layer charging by the use of STM tips has been shown as a means of for electrochemical machining of electrode surfaces on a micrometer length scale. This is achieved by controlling the nanosecond voltage pulses between the tip and surface. Pulse voltages and time duration must be controlled to have significant effect of tip/surface interaction. This has been explained in Chap. 6 how pulse time affect the double layer structure. This is shown in Fig. 8.2d.

Scanning electrochemical microscopy (SECM) has been demonstrated to ideally create electrochemical redox reactions via a STM tip on the substrate on a micrometer length scale. Heinze et al. [2] have reported the possibility of application of SECM for nanostructuring on nanoelectrodes. Thus SECM can be used for confined surface areas on electrodes for nanostructing.

The burst-like dissolution of metal from the tip, (Fig. 8.2e) onto which metal been deposited from solution, the re-deposition occurs onto the substrate within a narrow region directly underneath the tip. The key lies in the momentarily high metal ion concentration after the sudden metal dissolution at the tip, which causes a more positive Nernst potential for the surface region underneath the tip [3–5]. This method has been successfully used to deposit cobalt nanomagnets [6].

The electrochemical reduction and oxidation potentials of desired metal on the STM tip can be defined first by cyclic voltammetry. Prior to each experiment the STM tip is held at a constant potential more positive than the onset potential of deposition of the metal. The tip-substrate distance must be high enough to avoid any unexpected touch during the process. The first step in the process is to initially coat the tip by the metal by holding the tip potential negative than the onset potential of deposition for a definite time according to the required thickness. The thickness can be estimated via the Farady's law. Then the tip approaches the electrode surface until a close proximity is achieved. The second step of the process to anodically dissolve the metal from the tip. This Co dissolution results in a local oversaturation of the electrolyte by metal ions at the substrate surface underneath the STM tip. This is readily obtained by holding the tip potential at a value more positive than the dissolution onset. The Nernst's equation predicts the change of substrate electrode potential under the local change of metal ions. This local potential is changed to those required to deposit the metal. Therefore, the metal ions will be locally deposited onto the substrate surface, while other areas of the substarte remain unchanged. Last step would be to retract the tip from the surface and put the tip in the tunneling mode. The process steps are shown in Fig. 8.2a.

The "jump-to-contact" [3–7] [between tip and substrate (Fig. 8.2g) allows the formation of metal clusters in quick succession and without destroying the single crystallinity of the substrate. By applying an electrode potential to the STM tip negative to the bulk deposition potential for the metal ions in solution, metal is deposited from the electrolyte onto the tip. Then, the metal-loaded tip is made to approach the surface close enough, so that the jump-to-contact occurs. This leads to the formation of a connective neck, a metal bridge between tip and substrate, which will break upon the subsequent retreat of the tip, leaving a small metal cluster on the substrate surface. The tip is automatically "reloaded" because of the ongoing metal deposition and hence, is ready for the next cluster formation. This sequence is shown in Fig. 8.2b [1].

8.2 Lithographically Patterned Nanowire Electrodeposition

Lithographically patterned nanowire electrodeposition (LPNE) was discovered in Penenr's group in 2006, and is a fascinating method for the synthesis of nanowires on substrate surfaces. LPNE combines the advantages of photolithography such as pattern fabrication capabilities, easy to perform with the diversity and flexibility of bottom-up electrodeposition to fabricate a wide variety of patterns and geometries of metal nanowires. Figure 8.3 shows a schematic presentation of several stages employed in LPNE. The process involves the preparation of a sacrificial metal nanoband electrode like nickel using photolithography technique (or mask aligner) and the subsequent electrodeposition of a noble-metal (Au, Pt or Pd) nanowire at this electrode.

The first and second steps involve the deposition of a continuous sacrificial layer, e.g. nickel, on an insulating substrate using evaporation or sputtering technique (Fig. 8.3a, steps 1). The next two steps (Fig. 8.3a, steps 2, 3) are dedicated to lithographically pattern the nickel layer with a desirable feature shape and size. In step 4, an undercut section is produced in the nickel film underneath the photoresist by an anodic dissolution method. The electrochemical approach to the anodic dissolution is shown in Fig. 8.3c, d. For instance, the potential for oxidizing nickel ranges from −0.085 to −0.10 V versus a saturated calomel reference. The anodic dissolution leads to produce undercut surface by 100–300 nm (Fig. 8.3c) which can be tuned by anodic current following the Farady's law. This step produces a horizontal trench or recessed regions under the photoresist which subsequently act as templates for further electrodeposition (Fig. 8.3a, step 5). A noble-metal nanowire can be electrodeposited into the recessed region. The growth of noble metal nanowires follows a three-step mechanism. The width and thickness of the metal nanowire can be controlled by the electrodeposition parameters and the sacrificial film thickness, respectively. For the width of nanowires($W(t)$), it has been expressed by:

$$W(t) = \frac{i_D t_D V_m}{nF} \tag{8.1}$$

where i_D is the galvanostatic electrodeposition current density, t_D is the deposition time, V_m is the molar volume of the deposited metal, n is the number of electrons required to reduce each metal complex ion and F is the Faraday constant (96,485 C mole−1).

Several patterns and wire geometries have been synthesized, for instance height modulated gold nanowires [9] It is a competitive to other currently available

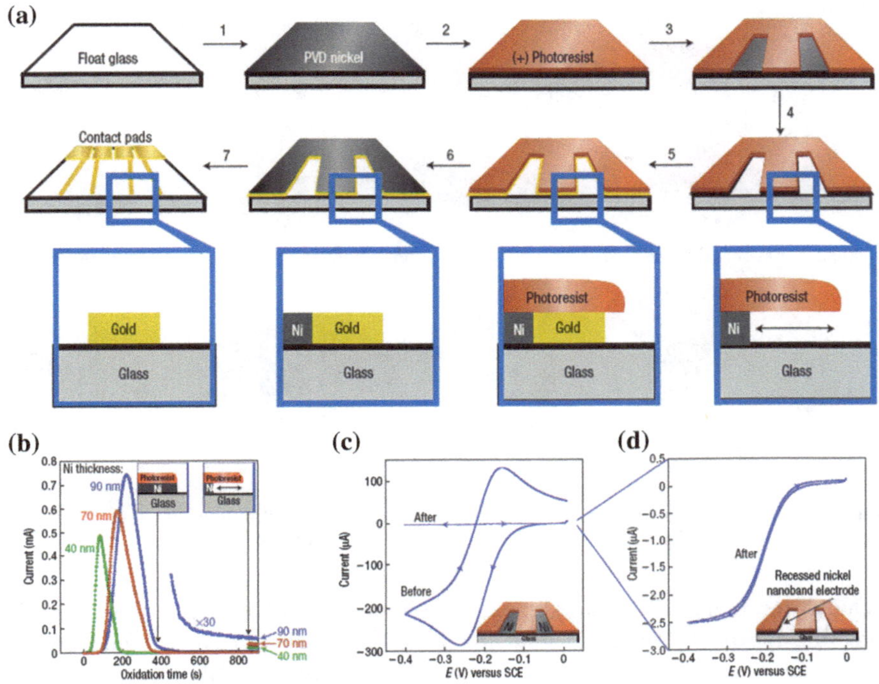

Fig. 8.3 **a** Seven-step process flow for metal-nanowire fabrication using LPNE. Anodic dissolution of the sacrificial nickel layer producing a horizontal trench originating at the edge of the photoresist. **b** Oxidation current versus time during the removal of nickel from lithographically patterned surfaces by potentiostatic electro-oxidation (reaction: $Ni \rightarrow Ni^{2+}+2e^{-}$) at −0.10 V versus SCE in aqueous 0.1 M KCl at pH = 1.0. Three transients are shown for nickel layers with approximate thicknesses of 40, 70 and 90 nm, as shown. **c**, **d**, Cyclic voltammogram for $Ru(NH_3)_6^{3+}$ at 5 mV s^{-1} in aqueous 0.1 M NaCl before the removal of nickel (**c**) and after the removal of nickel (**d**) by electro-oxidation, resulting in the formation of a *horizontal trench* ≈200 nm in depth, terminated by a nickel nanoband 40 nm in thickness. Reprinted from [8] with the permission of Nature publishing, copyright 2016

methods and open the door to the fabricate nanowires of materials accessible using electrodeposition ranging from metal oxides, to polymers, to semiconducting compounds, Further details are given in [10].

8.3 Electrodeposited Mesoporous Nanostructures by Lyotropic Liquid Crystals

Over the last decade the need to produce mesoporous materials and films has been increased due to need for high performance devices and instruments for gas generation and sensing, catalysis, supercapacitors, bioelectronics, etc. [11]. It is essential to

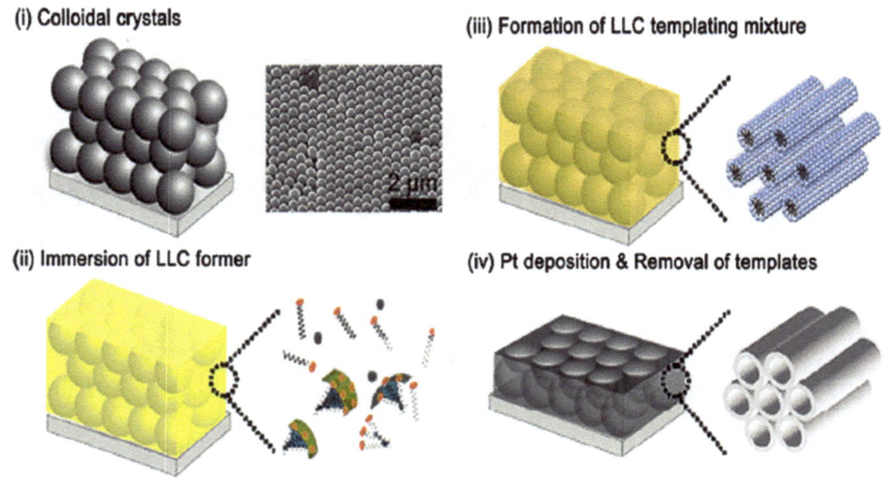

Fig. 8.4 Fabrication process for preparing a hierarchically porous Pt film using a typical lyotropic liquid crystal (LLC) template. From [13] with the permission of Elsevier

achieve the highest possible surface to volume ratio which can be achieved by ordering the pore structure within the whole material. Lyptropic liquid crystals (LLC) have been recently developed to act as a soft template. Electrodeposition has been used to fill the template pores to generate meso and nanostructured porous materials. LLCs are made of various surfactants in aqueous and organic solvents containing the electrodeposition electrolyte. A phase diagram of ternary compositions (surfactant/water/metal species) usually is used to find the appropriate phase required for the electrodeposition. Many metal and alloy systems have been electrodeposited by LLC templates such as nanoporous Pt, Ni, Co, NiCo and nickel oxide. For example, cetyltrimethylammonium bromide (CTAB) in an aqueous solution [12].

Mesoscopic periodicity, shape and size can be simply controlled by both the choice of surfactants and the addition of lipophilic substances. Figure 8.4 shows a schematic presentation of the formation of LLC hexagonal mesophase [13].

8.4 Nanostructures by Galvanic Displacement

Electrodeposited nanostructure obtained by galvanic displacement has been also realised. The galvanic displacement process is a well know electrodeposition process. The thermodynamic driving force of this process can be well understood by the relationship between free Gibbs energy and electrode potential difference, as a starting material acting as electron-donating species scarifies at the expense of reduction of electroactive species. In other words, in general, redox couple with more noble standard electrode potential are reduced by other redox couple with less noble standard electrode potentials. This is called the galvanic displacement which lasts until the whole surface of the starting material is covered with the reduced one, as a self-terminating process. The kinetics of the galvanic displacement process is also simple. It has been shown [14] that the initial stages of the process is charge-transfer controlled while for longer times where the starting material is partially covered the transport of reducing electroactive species is limited by diffusion and therefore, the mechanism becomes diffusion controlled.

The galvanic displacement has found many applications in nanostructuring of materials for functional, catalytic, electronic and bio applications. It is crucial to have a less noble starting metal as nano-scale template to be displaced by a more noble metal. A procedure has been demonstrated for synthesizing porous Cu/Ni/AuNi nanostructures via the galvanic displacement [15]. The starting material was NiZn coating electrodeposited on Ni film in which Zn/Zn^{2+} with the standard potential of −0.762 V versus SHE acts as the less noble scarified material. An alkaline aqueous solution containing NaOH and $KAu(CN)_2$ was as the precursor of Au/Au^+ reducing redox couple with the standard reduction potential of 1.691 V versus SHE. It can be deduced that the displacement reaction is:

$$Zn + 2Au(CN)^{2-} \rightarrow 2Au + Zn^{+2} + 4CN^- \tag{8.2}$$

Tellurium and bismuth telluride nanowires on nickel or copper coated λ-DNA templates have been achieved by the galvanic displacement [16]. First, continuous Ni and Cu nanowires were deposited on silver-seeded DNA templates by electroless deposition. Second, the metallic nanowires were then galvanically displaced Te or Bi_2Te_3 is deposited from an acidic solution containing $HTeO_2^+$ ions or a combination of $HTeO_2^+$ and Bi^{3+} ions. For instance for nickel template the reactions cn be written as:

$$\begin{aligned} &Ni \rightarrow Ni^{2+} + 2e^-, E^0 = -0.250\,\text{V}\ vs.\ \text{SHE} \\ &HTeO_2^+ + 3H^+ + e^- \rightarrow Te + H_2O, E^0 = +0.551\,\text{V}\ vs.\ \text{SHE} \end{aligned} \tag{8.3}$$

Metal nanowires have been fabricated by the galvanic displacement method. Self-standing Pd nanowires were also fabricated into gold-coated anodic alumina membranes (AAO templates) by the galvanic displacement deposition. Under the same argument as mentioned above, polycrystalline Pd nanowires were displaced from a solution containing $Pd(NH_3)_4(NO_3)_2$ as precursor. Morphological analysis showed the formation of perfectly aligned nanowires with a uniform diameter throughout the entire length. This last parameter was controlled by both the deposition time and the ratio between the anodic area (active metal) and the cathodic area (pore bottom) [17]. Tapered BiTe nanowires with controlled periodicity and composition were synthesized by galvanic displacement of compositionally modulated NiFe nanowires. The Fe content in NiFe nanowires was controlled by sweeping the applied deposition potentials where more negative cathodic potential resulted in Fe-rich NiFe nanowires. Depending upon Fe content of NiFe starting template, the Bi/Te ratio was shown to change [18].

References

1. D.M. Kolb, F.C. Simeone, Electrochim. Acta **50**, 2989 (2005)
2. N. Baltes, L. Thouin, C. Amatore, J. Heinze, Angew. Chem. Int. Ed. **43**, 1431 (2004)
3. R.T. Potzschke et al., J. Electrochem. Soc. **146**, 141 (1999)
4. W. Schindler, J. Appl. Phys. **87**, 7007 (2000)
5. W. Schindler et al., J. Electrochem. Soc. **148**, C124 (2001)
6. D. Hofmann et al., Appl. Phys. Lett. **73**, 3279 (1998)
7. D.M. Kolb et al., Science **275**, 1097 (1997)
8. E.J. Menkke, M.A. Thompson, C. Xiang, L.C. Yang, R.M. Penner, Nat. Mater. **5**, 914 (2006)
9. H. Kim, D.K. Taggart, C.X. Xiang, R.M. Penner, E.O. Potma, Nano Lett. **2008**(8), 2373–2377 (2008)
10. C. Xiang, Y. Yangw, R.M. Penner, Chem. Commun. **2009**, 859–873 (2009)
11. P. Innocenzi, I. Malfatti, Chem. Soc. Rev. **42**, 4198–4216 (2013)
12. P.N. Bartlett, D. Pletcher, T.F. Esterle, C.T. John, J. Electroanal. Chem. **688**, 232–236 (2013)
13. Y. Yamauchi, K. Kuroda, Electrochem. Commun. **8**, 1677–1682 (2006)
14. N. Massee, D.L. Piron, J. Electrochem. Soc. **140**, 2818 (1993)
15. M.-G. Hosseini, M. Abdolmaleki, F. Nasirpouri, Electrochim. Acta **114**, 215–222 (2013)
16. J. Liu, B. Uprety, S. Gyawali, A.T. Woolley, N.V. Myung, J.N. Harb, Langmuir **29**, 11176–11184 (2013)
17. R. Inguanta, S. Piazza, C. Sunseri, Electrochem. Commun. **11**, 1385–1388 (2009)
18. T. Suh, K.-H. Nam, H. Jung, C.-H. Kim, J.-G. Kim et al., Electrochim. Acta **90**, 582–588 (2013)

Index

F. Nasirpouri, *Electrodeposition of Nanostructured Materials*,
Springer Series in Surface Sciences 62, DOI 10.1007/978-3-319-44920-3

FSC
www.fsc.org
MIX
Papier | Fördert
gute Waldnutzung
FSC® C083411

Kolibri 360 GmbH
Ferdinand-Jühlke-Straße 7
99095 Erfurt, Deutschland
produktsicherheit@kolibri360.de